Mathematics Study Resources

Volume 1

This series comprises direct translations of successful foreign language titles, especially from the German language.

Powered by advances in automated translation, these books draw on global teaching excellence to provide students and lecturers with diverse materials for teaching and study.

Ludger Rüschendorf

Stochastic Processes and Financial Mathematics

Ludger Rüschendorf
Institute of Mathematical Stochastics
University of Freiburg
Freiburg, Germany

ISSN 2731-3824 ISSN 2731-3832 (electronic)
Mathematics Study Resources
ISBN 978-3-662-64710-3 ISBN 978-3-662-64711-0 (eBook)
https://doi.org/10.1007/978-3-662-64711-0

This book is a translation of the original German edition „Stochastische Prozesse und Finanzmathematik" by Ludger Rüschendorf, published by Springer-Verlag GmbH Germany in 2020.

The translation was done with the help of artificial intelligence (machine translation by the service DeepL.com) and in a subsequent editing, improved by the author. Springer Nature works continuously to further the development of tools for the production of books and on the related technologies to support the authors.

This Springer imprint is published by the registered company Springer-Verlag GmbH, DE part of Springer Nature.
The registered company address is: Heidelberger Platz 3, 14197 Berlin, Germany

Contents

Introduction

This book provides an introduction to recent topics in stochastic processes and stochastic analysis and combines them with an introduction to the basics of financial mathematics. Through a selective choice of topics and a motivating (not technically elaborate) presentation, it allows one to gain an insight into the fundamental developments, ideas and models of stochastic processes and stochastic analysis and their importance for an understanding of the issues and concepts of financial mathematics. Unlike other textbooks, it does not resort to a highly reductive presentation of the subject (e.g., in the context of discrete time models).

The book is particularly suitable as a foundation or companion text to a more advanced 4-hour lecture on these topics in graduate mathematics. The aim is to present the particularly interesting developments and ideas in an extent suitable for a lecture or a seminar. With its motivating form of presentation and linguistic design, it is aimed in particular at students and offers a wealth of illustrative examples and applications. It is therefore not only particularly well suited for self-study but also provides many interesting additions for interested researchers and also lecturers.

A prerequisite of the book is a good knowledge of a more advanced probability theory course, in particular of discrete-time processes (martingales, Markov chains) as well as introductory continuous-time processes (Brownian motion, Poisson process, Lévy processes, processes with independent increments, and Markov processes) as taught in many existing good representations of probability theory; for their scope, see Rüschendorf (2016).

In Chap. 1, an introduction is given in a non-technical way to the basic principles of the theory of arbitrage-free prices, the hedging principle and the risk-neutral price measure, leading from the binomial price formula by approximation to the Black–Scholes formula. The necessary link from processes in discrete time (binomial model) to those in continuous time (Black–Scholes model) is given by approximation theorems for stochastic processes such as the fundamental Donsker theorem and the associated Skorohod embedding theorem in Chap. 2. Theorems of this type allow interpretation of continuous financial market models using simple discrete models such as the Cox–Ross–Rubinstein model.

The first main part of the book in Chap. 3 is devoted to the introduction of the stochastic integral (Itô integral) and the related theory of martingales and semimartingales. K. Itô had introduced this integral for Brownian motion in several papers in 1944–1951. It is thus possible—despite the infinite variation of the paths of Brownian motion—to give integral expressions of the form $\int_0^t \varphi_s \mathrm{d}B_s$ for a stochastic process (φ_s) a significant meaning. Also, Itô had recognized that, through the corresponding stochastic differential equations of the form $\mathrm{d}X_t = a(t, X_t)\mathrm{d}t + b(t, X_t)\,\mathrm{d}B_t$ by analogy with the ordinary differential equations, a fundamental tool is given to construct, by local properties, in the above example the drift a and the volatility b, an associated stochastic model; in the above case a diffusion process; compare the works of Itô and McKean (1974) and Itô (1984).

Semimartingales are a generalization of the class of diffusion models described by a (local) drift, a continuous (local) diffusion part (martingale part) and a jump part. They form a fundamental class of models in particular of importance also for the relevant models of financial mathematics.

The second main part of the textbook in Chap. 4 is devoted to stochastic analysis, its importance for model building and its analysis. Essential building blocks are the partial integration formula and the Itô formula with numerous applications, e.g., to Lévy's characterization of Brownian motion. The stochastic exponential is a solution of a fundamental stochastic differential equation and shows its importance in the characterization of equivalent measure changes (Girsanov's theorem). Together with the martingale representation theorem (completeness theorem) and the Kunita–Watanabe decomposition theorem, these form the foundation for arbitrage-free pricing theory in financial mathematics in the third part of the textbook.

A detailed section in this part is devoted to stochastic differential equations. In particular, a comparison of the different approaches to diffusion processes is presented, namely 1. the Markov process (semigroup) approach, 2. the PDE approach (Kolmogorov equations) and 3. the approach by stochastic differential equations (Itô theory). Moreover, the particularly fruitful connection between stochastic and partial differential equations (Dirichlet problem, Cauchy problem, Feynman–Kac representation) is also elaborated.

Chapter 5 introduces the foundations of general arbitrage-free pricing theory. The foundations are the first and second fundamental theorems of asset pricing and the associated risk-neutral valuation formula, which is based on valuation by means of equivalent martingale measures. For their construction, Girsanov's theorem proves to be very useful. For the standard options, it can be used to determine the corresponding pricing formulas (Black–Scholes formulas) in a simple way. The determination of the corresponding hedging strategies leads in the Black–Scholes model (geometric Brownian motion) to a class of partial differential equations, the Black–Scholes differential equations, going back to the fundamental contributions of Black and Scholes. The connection with stochastic calculus goes back to a derivation by Merton in 1969, which led to the development of a general theory of arbitrage-free pricing for continuous-time price processes and the important role of equivalent martingale measures for this purpose by Harrison, Kreps and Pliska in the period 1979–1983.

Central topics of this general theory are the completeness and incompleteness of market models, the determination of associated arbitrage-free price intervals via the equivalent martingale measures and the corresponding sub- or super-hedging strategies (optional decomposition theorem).

In incomplete models, the no-arbitrage principle does not uniquely identify an arbitrage-free price. Therefore, additional criteria must be applied to select an arbitrage-free price. Chapter 6 gives an introduction to the determination (resp. selection) of option prices via minimum distance martingales as well as to pricing and hedging via utility functions. In addition, the problem of portfolio optimization is treated. For exponential Lévy models, these methods are characterized in detail for a number of standard utility functions.

Chapter 7 is devoted to the determination of optimal hedging strategies by the variance-minimal strategy criterion. In incomplete market models, not every claim H is hedgeable. Therefore, a natural question is: How good is H hedgeable? The answer to this question is based on the Föllmer–Schweizer decomposition, a generalization of the Kunita–Watanabe decomposition, and the associated minimal martingale measure.

This textbook is based on lectures by the author on stochastic processes and financial mathematics, given repeatedly over many years since about the mid-1990s, and on related transcripts and notes by Sascha Frank and Georg Hoerder (2007, unpublished), Anna Barz (2007, unpublished) and Janine Kühn (2013, unpublished). Chapter 6 is based in large part on the elaboration of Sandrine Gümbel (2015). We would like to express our sincere thanks to all of them. Special thanks also go to Monika Hattenbach for preparing some parts of the text as well as for the habitual exquisite final text corrections and text layout.

1 Option Pricing in Discrete Time Models

This chapter gives an introduction to basic concepts and ideas of option pricing in the context of models in discrete time. In this frame the basic no-arbitrage principle and the determination of the fair (arbitrage-free) price resulting from it by means of a suitable hedging argument can be explained in a simple way. This approach makes it possible to introduce essential concepts and methods with little technical effort and,allows for example, to derive the Black–Scholes formula by means of an approximation argument.

Let $(S_t)_{t\geq 0}$ be the price process of a security (stock) that is traded in a market. Further, let there be in the market a risk-free, fixed-interest investment $(B_t)_{t\geq 0}$ (bond), with interest factor $r \geq 0$. The value of the bond at time $t = 0$ is therefore $\frac{1}{(1+r)^t}$ of the value at time t:

$$B_0 = \frac{1}{(1+r)^t} \cdot B_t .$$

Securities traded in the market are those on underlying commodities (e.g. wheat, oil, gold) as well as derivatives, i.e. contracts (functionals) on underlying commodities. Derivatives include, for example, forwards, futures and options such as puts and calls.

Forwards and Futures

Forwards are contracts that give the market participant the right to buy or sell an underlying or financial asset at a time T in the future or in a future time period $[T, T']$ at a fixed price K. A distinction is made between

- **long position:** Entering into a contract to buy
- **short position:** Entering into a sales contract

L. Rüschendorf, *Stochastic Processes and Financial Mathematics*, Mathematics Study Resources 1, https://doi.org/10.1007/978-3-662-64711-0_1

Forwards and futures involve hedging for their settlement, i.e. the market participant has both the right and the obligation to buy or sell. While futures are traded on financial markets, forwards are based on an individual agreement between the participants without market intervention.

Options, Call, Put

A **call option** (call for short) gives the buyer of the option the right to buy a financial asset at a future date T at an agreed price. However, the buyer is not obliged to exercise the contract. In contrast, a **put option** (put for short) gives the buyer the right to sell a financial asset at a fixed price at a time point T (Fig. 1.1).

A **European Call** is the right (but not the obligation) to buy a security at time T at price K. The value of a European Call at time T is

$$Y = \left(S_T - K\right)_+.$$

As an alternative there is the **American Call**. This American Call gives the buyer the right to buy a security at any time in the time interval τ in the time interval $[0, T]$ at the price K. The time τ is a stopping time from a mathematical point of view. An American Call therefore has the value at the time of execution

$$Y = \left(S_\tau - K\right)_+.$$

A **put** denotes the right to sell a security at time T at a specified price K. The value of the put at time T is therefore

$$Y = \left(K - S_T\right)_+.$$

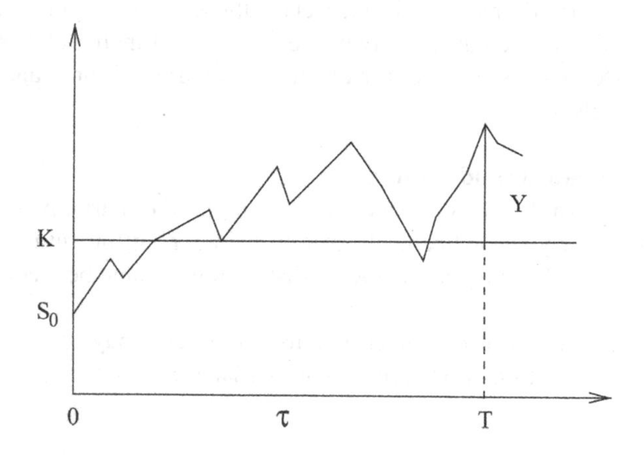

Fig. 1.1 European Call, $Y = (S_T - K)_+$

The question arises: What is a correct premium for a call Y? The classic answer to this is the expected value

$$EY = E(S_T - K)_+$$

Black and Scholes, in 1973, introduced a compelling new method to calculate the premium. In the following this premium is called the *fair price* or *Black–Scholes price*.

The basic idea of the Black–Scholes pricing principle can be explained by means of a simple pricing model.

Simple Pricing Model
Be $S_0 = 100$ the price of the security at the time $t = 0$. We consider a one-period discrete model. Up to time $T = 1$ the price of the security can with probability $p = 0.4$ rise to $S_1^a = 130$ and with probability $p = 0.6$ fall to $S_1^b = 80$.

$p=0.4$ 130

$S_0 = 100$

$1-p=0.6$ 80

$T = 1,\ r = 0.05,\ K = 110,$
$Y = (S_1 - K)_+$

A call secures the buyer the right to purchase the security at time $T = 1$ at the price (strike) $K = 110$. Accordingly, the call has at $T = 1$ the value $Y = (S_1 - K)_+$. As an alternative, there is a risk-free investment at the interest rate $r = 0.05$.

With the classic valuation, the price of the call $t = 1$ would be determined as follows:

$$E_p Y = 20 \cdot 0.4 = 8$$

According to this, the value of the call at $t = 0$ is

$$C_0 = \frac{1}{1+r} \cdot E_p Y = \frac{8}{1.05} = 7.62.$$

The Black–Scholes formula gives a different valuation: $C_0^{BS} = 9.52$. This valuation method is based on the following consideration: Determine a portfolio of shares of the stock and of the bond, which generates at $T = 1$ the same payoff as the call. Such a portfolio is called a **duplicating portfolio**. Once one has determined such a duplicating portfolio, one has already found the fair price for the option. Both instruments, the call and the duplicating portfolio, have at $T = 1$ the same payoff. Therefore, they should also have the same price.

$$\text{price of the call } Y \overset{!}{=} \text{price of the duplicating portfolio.}$$

Calculating a duplicating portfolio in the above example:

Let π be a portfolio

$$\pi = (\Delta, B),$$

where Δ denote the stock shares and B the bond shares. Choose specifically

$$\Delta = 0.4 \text{ and } B = -30.48$$

Here, the quantity B is negative, denoting a loan. The price of the portfolio at time $t = 0$ is easy to calculate in this example:

$$V_0(\pi) = \Delta \cdot 100 + B = 40 - 30.48 = 9.52.$$

The value of the portfolio at the time $T = 1$ is:

$$\begin{aligned} V_1(\pi) &= \Delta \cdot S_1 + B \cdot (1+r) \\ &= \begin{cases} 0.4 \cdot 130 - 30.48 \cdot 1.05, & S_1 = 130 \\ 0.4 \cdot \ \ 80 - 30.48 \cdot 1.05; & S_1 = \ \ 80 \end{cases} \\ &= \begin{cases} 20, & S_1 = 130 \\ 0, & S_1 = \ \ 80 \end{cases} \\ &= Y \end{aligned}$$

The portfolio generates the same payoff as the call Y at time $T = 1$, i.e., the portfolio duplicates the call Y. Thus, the *fair price* of the call at time is $t = 0$ is equal to the price of the portfolio: $C_0^{BS} = 9.52$.

If one were to price the call differently, then a risk-free profit (arbitrage) would be possible. This fair pricing is therefore based on the *no-arbitrage principle*:

In a real market model there is no risk free gain possible;

NFLVR = No free lunch with vanishing risk

This NFLVR principle implies that two market instruments with the same payoff also have the same price.

General Discrete Models for Securities

Let $S_n = (S_n^0, S_n^1, \ldots, S_n^d)$ describe the d development of d equity securities; let S_n^i denote the price of the ith stock at time n, $S_n^0 = (1+r)^n$.

The dimension d of a large bank's securities holdings can be quite large, e.g., in the case of Deutsche Bank $d \approx 500{,}000$.

Some basic terms used to describe securities trading are summarized below.
A **portfolio** describes the proportions of different securities in the portfolio.

A **trading strategy** $\Phi = (\Phi_n) = \big((\Phi_n^0, \dots, \Phi_n^d)\big)$ describes the development of the portfolio: $\Phi_n^i \sim \sharp$ shares of security no. i at time n.

The **value** of the portfolio at time n is

$$V_n(\Phi) = \Phi_n \cdot S_n = \sum_{i=0}^{d} \Phi_n^i S_n^i. \tag{1.1}$$

Φ is called **self-financing,** if

$$\Phi_n \cdot S_n = \Phi_{n+1} \cdot S_n. \tag{1.2}$$

Changes to the portfolio result only from regrouping. No additional capital is required for the change. No withdrawal of value takes place.
The increase in value is $\Delta V_n(\Phi) = V_{n+1}(\Phi) - V_n(\Phi) = \Phi_{n+1} \cdot \Delta S_n$.
The no-arbitrage principle implies: There is **no-arbitrage strategy** Φ, i.e. there is no strategy Φ such that

$$V_0(\Phi) = 0,\ V_n(\Phi) \geq 0 \text{ and } P\big(V_n(\Phi) > 0\big) > 0, \tag{1.3}$$

or equivalent to it:

$$V_0(\Phi) \leq 0,\ V_n(\Phi) \geq 0 \text{ and } P\big(V_n(\Phi) > 0\big) > 0.$$

Cox–Ross–Rubinstein Model
A simple and basic model for price evolution in discrete time is the Cox–Ross–Rubinstein model. This describes the time evolution of the price of a security by subdividing the time interval $[0, T]$ into n sub-intervals of length h, $T = nh$.

The CRR model is based on the following assumption for price movements in the subintervals (Fig. 1.2).

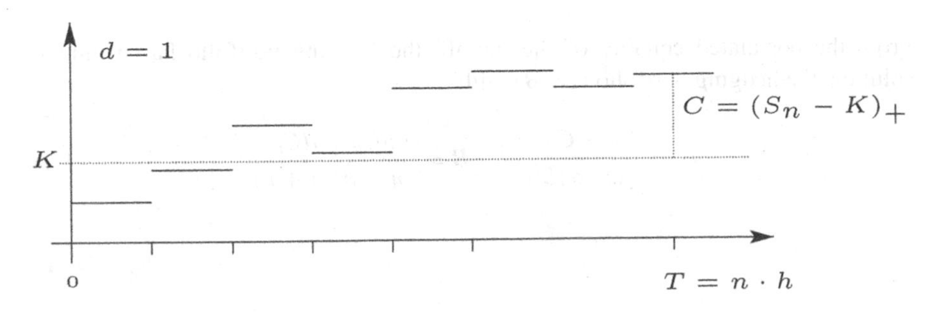

Fig. 1.2 Cox–Ross–Rubinstein model for $d = 1$

Assumption

A The price trend in a step is constant.

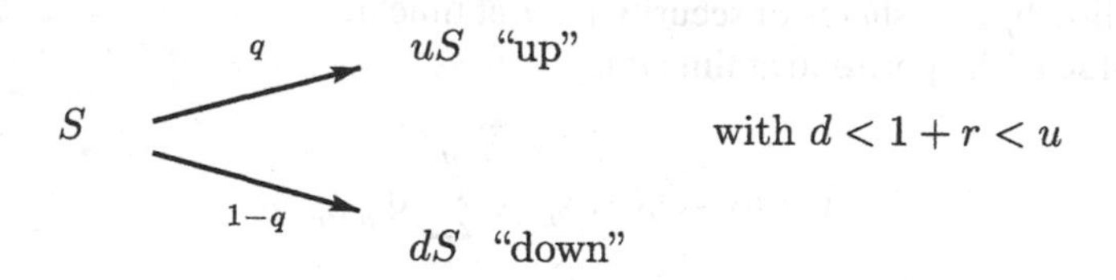

The price S_k after k substeps is therefore of the form $u^\ell d^{k-\ell} S_0$. The parameters of this model are u, d, q, n. The basic question is: "What is the value of a call $Y := (S_n - K)_+$?"

To answer this question, we first consider the development of the value C in one period.

Step 1: Development of value in one period

$C \xrightarrow{q} C_u = (uS - K)_+$

$C \xrightarrow{1-q} C_d = (dS - K)_+$

To determine C, the hedging portfolio $\pi = (\Delta, B)$ (with Δ: stocks, B: bonds) is chosen so that the payoff at the end of the period is identical to that of the call. The development of the portfolio $\pi = (\Delta, B)$ in one period is as follows:

$\Delta S + B \xrightarrow{q} \Delta uS + (1+r)B \overset{!}{=} C_u$

$\Delta S + B \xrightarrow{1-q} \Delta dS + (1+r)B \overset{!}{=} C_d$

value of hedging portfolio π, $t = 0$ — value at the end of the period

From the postulated equality of the payoff, the hedging portfolio has a unique solution, the hedging portfolio (Δ, B) with

$$\Delta = \frac{C_u - C_d}{(u-d)S}, \qquad B = \frac{uC_d - dC_u}{(u-d)(1+r)} \tag{1.4}$$

From the no-arbitrage principle, the *fair price* C of the call thus follows:

$$\begin{aligned}
C &\overset{!}{=} \Delta S + B \\
&= \frac{C_u - C_d}{(u-d)} + \frac{uC_d - dC_u}{(u-d)(1+r)} \\
&= \Bigg(\underbrace{\left(\frac{1+r-d}{u-d}\right)}_{=:p^*} C_u + \underbrace{\left(\frac{u-(1+r)}{u-d}\right)}_{=:1-p^*} C_d \Bigg) \Big/ (1+r) \qquad (1.5) \\
&= \big(p^* C_u + (1-p^*) C_d\big)/(1+r) \\
&= E_{p^*} \frac{1}{1+r} Y, \qquad Y = Y_1, d = 1.
\end{aligned}$$

The value development of the price process in a period with respect to the risk-neutral measure P^* is, with $S_0 = S$

$$\begin{aligned}
E_{p^*} S_1 &= p^* uS + (1-p^*)\, \mathrm{d}S = \frac{1+r-d}{u-d} uS + \frac{u-(1+r)}{u-d} \mathrm{d}S \\
&= (1+r)S.
\end{aligned}$$

Equivalent to this is:

$$E_{p^*} \frac{1}{1+r} S_1 = S_0 = S. \tag{1.6}$$

With respect to the risk neutral measure P^* the expected discounted value of S_1 is equal to that of S_0 (equilibrium). Equation (1.6) states that the discounted price process is $\left(S_0, \frac{S_1}{1+r}\right)$ is a martingale with respect to P^* e.

We note two features of this process:

(a) Evolution of the pricing process:

With $p^* = \frac{(1+r)-d}{u-d}$ it holds: $p^* uS + (1-p^*) dS = (1+r)S$

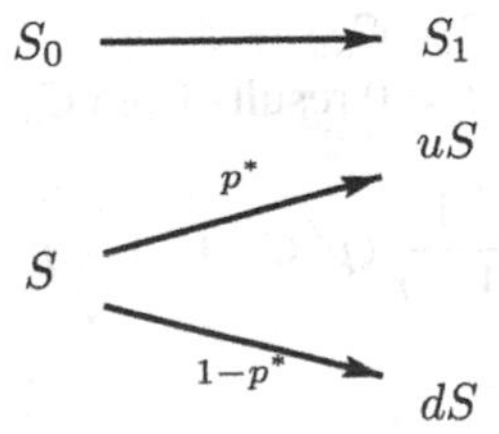

$S_0 \longrightarrow S_1$

$S \xrightarrow{p^*} uS$

$S \xrightarrow{1-p^*} dS$

(b) The *fair price* $C = E_{p^*} \frac{1}{1+r} S_1$ is independent of the underlying *statistical* price model!

The value of the call at time $t = 0$ is equal to the expected value of the discounted call with respect to the risk neutral measure.

Step 2: Two-period model

In the two-period-model the evolution of the price process $S_0 \to S_1 \to S_2$ and the corresponding evolution of the value process $C = C_0 \to C_1 \to C_2$ are as follows:

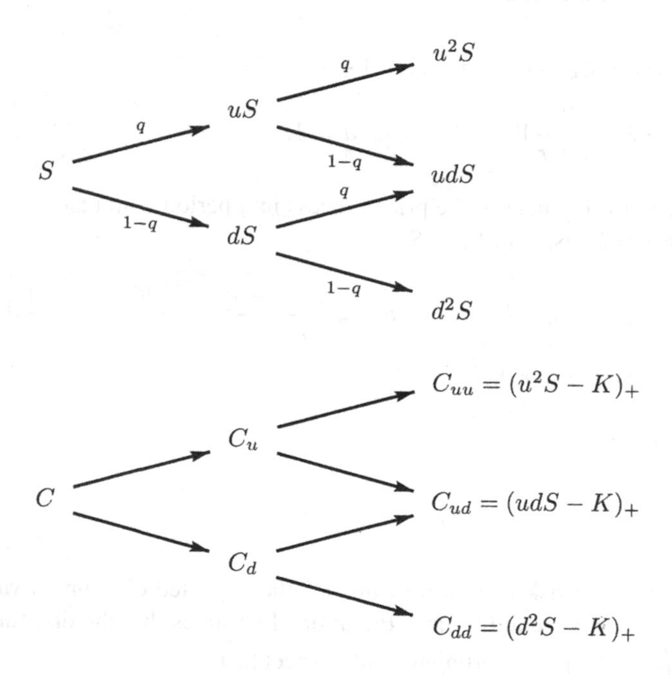

The value C is obtained by backward induction through iterated application of the first substep. C_u results from C_{uu}, C_{ud}

$$C_u = \frac{1}{1+r}(p^* C_{uu} + (1 - p^*)C_{ud}).$$

C_d is obtained similarly from C_{ud}, C_{dd},

Then the value of C at time $t = 0$ results from C_u, C_d as follows:

$$C = \frac{1}{1+r}(p^* C_u + (1 - p^*)C_d).$$

More generally, by backward induction,one obtains the value $C = C_n$ of the call for n periods. The result is the **binomial pricing formula**

$$C_n = \frac{1}{(1+r)^n} \sum_{j=0}^{n} \binom{n}{j} (p^*)^j (1-p^*)^{n-j} \big(\underbrace{u^j d^{n-j} S}_{=S_n} - K\big)_+ \tag{1.7}$$

$$= E_{P^*} \frac{1}{(1+r)^n} (S_n - K)_+$$

Here $P^* = (p^*, 1-p^*)$ is the risk-neutral measure:

$$p^* u + (1-p^*) d = 1 + r.$$

The value of the call $Y = (S_n - K)_+$ is given as the expected discounted value of the call with respect to the risk neutral measure P^*. The probabilities for $P(S_n = u^j d^{n-j} S)$ are just the binomial probabilities. The development of the price is multiplicative with the factors u, d.

The binomial pricing formula can also be rewritten by logarithmic transformation into the form

$$C_n = S\,\Phi_n(a, p') - K(1+r)^{-n} \Phi_n(a, p^*),$$

where $p^* = \frac{1+r-d}{u-d}$, $p' = \frac{u}{1+r} p^*$, $a = \Big[\Big(\log \frac{K}{S d^n}\Big) / \big(\log \frac{u}{d}\big)\Big]_+$, and $\Phi_n(a, p) := P(X_{n,p} \geq a)$, $X_{n,p} \sim \mathfrak{B}(n, p)$.

The transition to the additive binomial model can be described as follows. It holds that

$$\log \frac{S_n}{S} = \sum_{k=1}^{n} X_{n,k}$$

with $X_{n,k} = \xi_{n,k} U + (1 - \xi_{n,k}) D$, where $U = \log \frac{u}{1+r} > 0 > D = \log \frac{d}{1+r}$, $(\xi_{n,k})_k$ is an i.i.d. $\mathcal{B}(1, p^*)$ distributed sequence. (S_n) is therefore an *exponential pricing model*.

$$S_n = S e^{\sum_{k=1}^{n} X_{n,k}} \tag{1.8}$$

with a sum of i.i.d. terms as exponents.

As a result, we have an explicit value formula using the binomial distribution. By choosing the parameters u, d appropriately, we thus obtain an approximation of the value formula by the normal distribution from the central limit theorem.

Let $u = u_n = e^{\sigma\sqrt{\frac{t}{n}}}$, $d = d_n = e^{-\sigma\sqrt{\frac{t}{n}}}$, $h = \frac{t}{n}$. Then it follows: $p^* \sim \frac{1}{2} + \frac{1}{2}\frac{\mu}{\sigma}\sqrt{\frac{t}{n}}$ and it holds:

$$C_n \to C \tag{1.9}$$

with

$$\begin{aligned} C &= S\,\Phi(x) - K(1+r)^{-t}\,\Phi\left(x - \sigma\sqrt{t}\right) \\ x &= \frac{\log(S/K(1+r)^t)}{\sigma\sqrt{t}} + \frac{1}{2}\sigma\sqrt{t}. \end{aligned} \tag{1.10}$$

This is the famous **Black–Scholes formula**. It thus describes approximately the arbitrage-free valuation of a call in a binomial model (Cox–Ross–Rubinstein model). The additive binomial model (as a stochastic process) can be approximated by a Brownian motion (see the Donsker theorem in Chap. 3). Therefore, it is natural to assume that the above pricing formula (Black–Scholes formula) will also result from a suitable derivation of the price of a call in the analogous model in continuous time (Black–Scholes model). This model in continuous time is an exponential Brownian motion (geometric Brownian motion).

Analogous pricing formulas in general models in continuous time (such as exponential stable distributions, Weibull distributions, or hyperbolic distributions) by approximation with Cox–Ross–Rubinstein models are given in Rachev and Rüschendorf (1994).

2 Skorohod's Embedding Theorem and Donsker's Theorem

Subject of this chapter is a description of the interaction of discrete-time and continuous-time models. Functionals of discrete-time sum processes can be approximatively described by functionals of a continuous-time limit process. Conversely, functionals of the limit process can be simulated by those of discrete-time approximating processes. By this connection, by means of simple laws in the discrete time model, suitable continuous-time models and their regularities can be justified by means of approximation. A particularly important example of this connection is the Donsker theorem and the closely related Skorohod embedding theorem. Hereby the simple binomial model (or more generally the Cox–Ross–Rubinstein model) can be used to approximate a geometric Brownian motion (or more generally an exponential Lévy process) as a suitable continuous-time limit model.

2.1 Skorohod's Embedding Theorem

We consider an i.i.d. sequence of real random variables (X_i) with $EX_i = 0$ and $EX_i^2 = 1$ and with partial sums $S_n := \sum_{i=1}^n X_i$.

Linear interpolation produces a continuous function from the partial sum sequence (see Fig. 2.1). One can also set the function constant from a fixed time n on constant. Then the curve is rescaled by linear interpolation in the time dimension and by the factor $\frac{1}{n}$ to the interval $[0, 1]$ and rescaled in the spatial dimension with the factor $\frac{1}{\sqrt{n}}$. We obtain the random function $\left(S_t^{(n)}\right)$, with $t \in [0, 1]$. The statement of Donsker's theorem is that the process $\left(S_t^{(n)}\right)$ converges on the unit interval to the Brownian motion:

$$\left(S_t^{(n)}\right)_{0 \le t \le 1} \xrightarrow{\mathcal{D}} \left(B_t\right)_{0 \le t \le 1}.$$

L. Rüschendorf, *Stochastic Processes and Financial Mathematics*, Mathematics Study Resources 1, https://doi.org/10.1007/978-3-662-64711-0_2

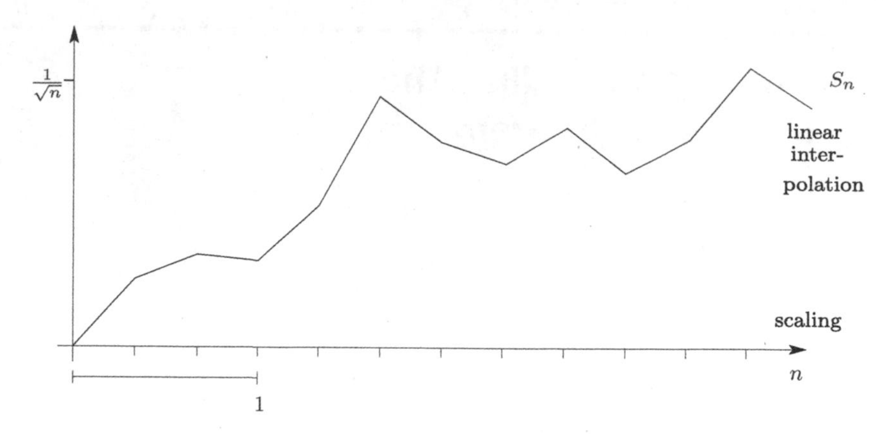

Fig. 2.1 Simulation of the Brownian motion

Convergence in distribution of a stochastic process is explained in more detail below. The process $(S_t^{(n)})$ is introduced in the following two steps:

1. continuous interpolation
2. scaling in the temporal and spatial dimension

$(S_t^{(n)})$ reflects the whole partial sum sequence $\left(\frac{S_k}{\sqrt{n}}\right)$. Using Donsker's theorem, one can approximate the behavior of functionals of partial sums by the behavior of functionals of the Brownian motion. The Brownian motion model has the advantage of being analytically tractable. Because of the above connection, the Brownian motion is a universal model that can be used to approximate the behavior of functionals of the partial sum process. In particular, as a consequence of this approximation one obtains the strong law of large numbers the central limit theorem and the law of the iterated logarithm.

In the opposite direction, Donsker's theorem provides a standard method for simulating a Brownian motion through the scaled partial sum sequences. This method is used in many simulation programs.

Continuous processes induce measures on the space of continuous functions. One can treat these based on a general theory of convergence in distribution in function spaces. However, there is another way to obtain such approximation theorems in a direct way. This is based on the Skorohod representation (or embedding).

Theorem 2.1 (Skorohod Representation) *Let X be a real random variable with $EX = 0$ and finite second moment $EX^2 < \infty$ and be $B = \left(B_t\right)$ a Brownian motion starting in zero, $P = P_0$. Then there exists a stopping time τ such that*

$$B_\tau \stackrel{d}{=} X \quad \textit{and} \quad E\tau = EX^2. \tag{2.1}$$

The first point is trivial to achieve because $\limsup B_t = \infty$, resp. $\liminf B_t = -\infty$. Therefore every value is reached by the stopping time τ: stop B_t when $X(\omega)$ is reached. Then $B_\tau = X$. Thus the random variable X is reproduced. However, in general $E\tau$ for the above stopping time is not finite.

We construct in the following proof an *extended stopping time* with property (2.1).

Proof

(1) **first case**

For two numbers a and b with $a < 0 < b$ we define the two-point measure

$$\mu_{a,b}(\{a\}) := \frac{b}{b-a} \text{ and } \mu_{a,b}(\{b\}) := \frac{-a}{b-a}.$$

Further let $\mu_{0,0}(\{0\}) = 1$. We assume, in the first case, that the distribution of X is the two-point measure above:

$$Q := P^X = \mu_{a,b}.$$

The probabilities are such that

$$EX = 0 \text{ and } EX^2 = a^2 \frac{b}{b-a} - b^2 \frac{a}{b-a} = -ab.$$

We define

$$\tau := \inf\{t \geq 0 : B_t \notin (a,b)\} = \tau_{a,b},$$

i.e. we stop when one of the limits a or b is reached. Using Wald's equation, it follows:

$$B_\tau \stackrel{d}{=} X, \quad E\tau = -ab = EB_\tau^2 = EX^2,$$

i.e., using the properties of Brownian motion, we get a solution for the first case by a suitable two-sided stopping time.

(2) **general case**

In the general case, the distribution of X is a general probability measure:

$$Q = P^X \in M^1(\mathbb{R}^1, \mathcal{B}^1) \quad \text{with} \quad EX = 0 = \int_{-\infty}^{\infty} x \, dQ(x).$$

We trace this case back to the first case as follows. Let

$$c := \int_{-\infty}^{0} (-u)\,\mathrm{d}Q(u) = \int_{0}^{\infty} v\,\mathrm{d}Q(v).$$

Q can then be written as a mixture of two-point measures $\mu_{a,b}$. For the proof let $\varphi \in \mathcal{L}_b$ be a bounded measurable function with $\varphi(0) = 0$, then it holds:

$$\begin{aligned} c\int \varphi\,\mathrm{d}Q &= \int_0^\infty \varphi(v)\,\mathrm{d}Q(v) \int_{-\infty}^{0} (-u)\,\mathrm{d}Q(u) + \int_{-\infty}^{0} \varphi(u)\,\mathrm{d}Q(u) \int_0^\infty v\,\mathrm{d}Q(v) \\ &= \int_0^\infty \mathrm{d}Q(v) \int_{-\infty}^{0} \mathrm{d}Q(u)\big(v\varphi(u) - u\varphi(v)\big). \end{aligned}$$

Thus, the mixture representation of Q follows

$$\begin{aligned} E\varphi(X) &= \int \varphi\,\mathrm{d}Q \\ &= \frac{1}{c}\int_0^\infty \mathrm{d}Q(v) \int_{-\infty}^{0} \mathrm{d}Q(u)(v-u)\underbrace{\Big(\frac{v}{v-u}\varphi(u) + \frac{-u}{v-u}\varphi(v)\Big)}_{=\int \varphi\,\mathrm{d}\mu_{u,v}} \end{aligned} \tag{2.2}$$

as a mixture of the two-point measures $\mu_{u,v}$

$$\mu_{u,v}\big(\{u\}\big) := \frac{v}{v-u} \text{ and } \mu_{u,v}\big(\{v\}\big) := \frac{-u}{v-u}.$$

For this purpose ν be the measure on $\mathbb{R}_+ \times \mathbb{R}_-$ defined by

$$\nu\big(\{0,0\}\big) := Q\big(\{0\}\big) \text{ and for a subset } A \text{ of the plane with } (0,0) \notin A$$

by $\nu(A) := \frac{1}{c}\int\int_A (v-u)\,\mathrm{d}Q(u)\,\mathrm{d}Q(v)$. Then (with $\varphi \equiv 1$) $1 = \int 1\,\mathrm{d}Q \overset{(2.2)}{=} \nu(\mathbb{R}_+ \times \mathbb{R}_-)$; thus ν is a probability measure.

Formula (2.2) can also be read as follows: Let (U, V) be two random variables on a suitable probability space $(\Omega, \mathcal{A}, P)$ independent of the Brownian motion $B = (B_t)$ such that $P^{(U,V)} = \nu$. Because of (2.2) it then holds with $\varphi = (\varphi - \varphi(0)) + \varphi(0)$

$$\int \varphi\,\mathrm{d}Q = E\int \varphi(x)\,\mathrm{d}\mu_{U,V}(x), \quad \varphi \in \mathcal{L}_b. \tag{2.3}$$

Thus, any distribution with zero expectation can be represented as a mixture of two-point measures $\mu_{a,b}$.

$\tau_{U,V}$ is in general not a stopping time with respect to the σ-algebra $\mathcal{A}^B$. But it is a stopping time with respect to the enlarged σ-algebra $\mathcal{A}_t := \sigma\big(B_s, s \le t, U, V\big)$. Therefore, $\tau_{U,V}$ is called an extended stopping time. With $\tau_{U,V}$ one obtains the assertion. It holds:

$$B_{\tau_{u,v}} \stackrel{d}{=} \mu_{u,v}. \tag{2.4}$$

Because of the independence of (U, V) from the Brownian motion B it follows because of (2.3):

$$B_{\tau_{U,V}} \stackrel{d}{=} \mu_{U,V} \stackrel{d}{=} X.$$

Because of $P^X = Q$ it follows: The Brownian motion, stopped at $\tau_{U,V}$ yields the distribution Q. Further, after the first step of the proof, it follows that

$$E\tau_{U,V} = EE\big(\tau_{U,V} \,|\, U = u,\ V = v\big) = EE\tau_{u,v} \stackrel{(2.4)}{=} EB^2_{\tau_{U,V}} = EX^2.$$

□

Remark 2.2 The construction of a (non-extended) stopping time goes back to Azema and Yor (1979). The more elaborate proof can be found, for instance, in Rogers and Williams (2000, pp. 426–430) or in Klenke (2006, pp. 482–485).

Random variables can thus be reproduced by stopping a Brownian motion at appropriate times. This procedure can also be undertaken for entire sequences of random variables.

Theorem 2.3 (Skorohod's Embedding Theorem) *Let be (X_i) be an i.i.d. sequence with $P^{X_i} = \mu$, $EX_i = 0$, and finite second moment, $EX_1^2 < \infty$. Then there exists a sequence of independent, identically distributed random variables $(\tau_n)_{n\ge 0}$ with $E\tau_n = EX_1^2$ such that the partial sum sequence (T_n) with $T_0 := 0$ and $T_n := \sum_{i=1}^n \tau_i$ is an increasing sequence of stopping times with respect to a Brownian motion B with*

(1) $$P^{B_{T_n} - B_{T_{n-1}}} = \mu, \quad \forall n,$$

(2) $$\big(B_{T_n} - B_{T_{n-1}}\big) \stackrel{d}{=} (X_n),$$

(3) $$\big(S_n\big) \stackrel{d}{=} \big(B_{T_n}\big), \tag{2.5}$$

i.e. the sequence (S_n) is 'embedded' in (B_t).

Consequence Because of the embedding $(S_n) \stackrel{d}{=} (B_{T_n})$ a wealth of asymptotic properties of the partial sum sequence can be obtained using a Brownian motion.

Proof According to Skorohod's representation theorem, there exists a stopping time τ_1 with $B_{\tau_1} \stackrel{d}{=} X_1$ and $E\tau_1 = EX_1^2$. The increments of a Brownian motion after the time point τ_1, $(B_{t+\tau_1} - B_{\tau_1})$, are a Brownian motion again according to the strong Markov property. One can apply the previous result to this process:

$$\exists \tau_2 : B_{\tau_2+\tau_1} - B_{\tau_1} \stackrel{d}{=} X_2 \text{ and } E\tau_2 = EX_2^2.$$

Because of the independent increment property of Brownian motion τ_2 is independent of $\mathcal{A}_{\tau_1}$ and is in particular independent of τ_1.

Inductively, we obtain a sequence of stopping times T_n with

$$T_n = T_{n-1} + \tau_n, \quad B_{T_n} - B_{T_{n-1}} \stackrel{d}{=} X_n \quad \text{and} \quad E\tau_n = EX_n^2,$$

and so that τ_n is independent of $(\tau_i)_{i \le n-1}$. Accordingly, the increments $B_{T_i} - B_{T_{i-1}}$ are independent of $\mathcal{A}_{T_{i-1}}$. Thus it follows that $\left(B_{T_n} - B_{T_{n-1}}\right) \stackrel{d}{=} (X_i)$. From this it follows that

$$\left(S_n\right) \stackrel{d}{=} \left(B_{T_n}\right) \quad \text{and} \quad ET_n = \sum_{k=1}^{n} EX_k^2.$$

□

Remark 2.4 The embedding holds in an analogous way for non-identically distributed sum sequences.

$$(S_n) = \left(\sum_{i=1}^{n} X_i\right) \stackrel{d}{=} \left(B_{T_n}\right) \quad with \quad ET_n = \sum_{k=1}^{n} EX_k^2. \tag{2.6}$$

A direct corollary of Skorohod's embedding theorem is the central limit theorem:

Theorem 2.5 (Central Limit Theorem) *Let be (X_i) be an i.i.d. sequence with $EX_1 = 0$, $EX_1^2 = 1$, then it holds:*

$$\frac{S_n}{\sqrt{n}} \xrightarrow{\mathcal{D}} N(0, 1). \tag{2.7}$$

Proof By Theorem 2.3 is $S_n \stackrel{d}{=} B_{T_n}$ with a stopping time $T_n = \sum_{i=1}^{n} \tau_i$ for an i.i.d. sequence (τ_i) with $E\tau_1 = EX_1^2 = 1 < \infty$. Using the standard scaling property of Brownian motion, it follows

$$\frac{S_n}{\sqrt{n}} \stackrel{d}{=} \frac{B_{T_n}}{\sqrt{n}} \stackrel{d}{=} B_{\frac{T_n}{n}}$$

By the strong law of large numbers, it follows: $\frac{T_n}{n} \to 1\,[P]$. Since a Brownian motion has continuous paths, it follows that

$$B_{\frac{T_n}{n}} \longrightarrow B_1 \stackrel{d}{=} N(0,1).$$

□

In the following theorem, we show that Skorohod's embedding theorem can be used to transfer the (relatively easy to prove) law of the iterated logarithm for Brownian motion to the proof of Hartmann–Wintner's theorem for partial sum sequences.

Theorem 2.6 (Hartmann–Wintner's Law of the Iterated Logarithm) *Let (X_i) be an i.i.d. sequence of random variables with $EX_i = 0$ and $\operatorname{Var} X_i = 1$, then it follows*

$$\limsup \frac{S_n}{\sqrt{2n \log\log n}} = 1\ [P]. \tag{2.8}$$

Proof By Theorem 2.3 it holds that $(S_n) \stackrel{d}{=} (\widetilde{S}_n)$ with $\widetilde{S}_n := B_{T_n} = \sum_{\nu=1}^{n}(B_{T_\nu} - B_{T_{\nu-1}})$, $B_{T_0} := 0$.

According to the law of the iterated logarithm for the Brownian motion it holds that (Fig. 2.2)

$$\limsup_{t\to\infty} \frac{B_t}{\sqrt{2t \log\log t}} = 1\ [P].$$

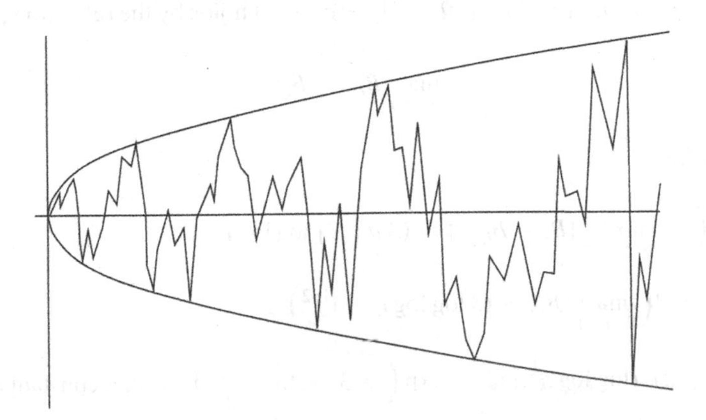

Fig. 2.2 Hartmann–Wintner's law of the iterated logarithm

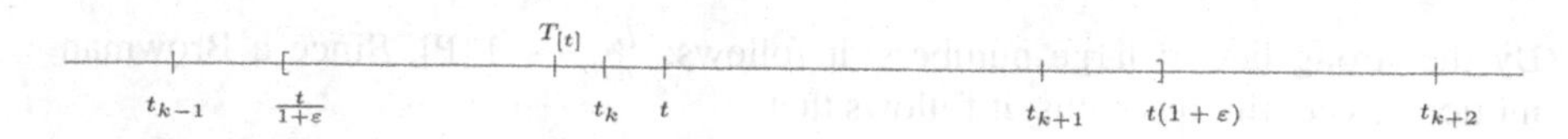

Fig. 2.3 Proof of Hartmann–Wintner's law

Assertion:

$$\lim_{t\to\infty} = \frac{B_t - \widetilde{S}_{[t]}}{\sqrt{2t \log\log t}} = 0\,[P].$$

Hartmann–Wintner's theorem then follows from this assertion. The proof uses the strong law of large numbers:

$$\frac{T_n}{n} \longrightarrow 1\,[P]. \tag{2.9}$$

Thus for $\varepsilon > 0$ and $t \geq t_0(\omega)$:

$$T_{[t]} \in \left[\frac{t}{1+\varepsilon}, t(1+\varepsilon)\right].$$

Let $M_t := \sup\left\{|B_s - B_t|,\ \frac{t}{1+\varepsilon} \leq s \leq t(1+\varepsilon)\right\}$ (Fig. 2.3) and consider the subsequence $t_k := (1+\varepsilon)^k \uparrow \infty$.

For $t \in [t_k, t_{k+1}]$ it holds:

$$\begin{aligned} M_t &\leq \sup\{|B_s - B_t|;\ t_{k-1} \leq s \leq t_{k+2}\} \\ &\leq 2\sup\{|B_s - B_{t_{k-1}}|;\ t_{k-1} \leq s \leq t_{k+2}\}. \end{aligned}$$

Because of $t_{k+2} - t_{k-1} = \vartheta t_{k-1}$, $\vartheta = (1+\varepsilon)^3 - 1$, it holds by the reflection principle

$$\max_{0\leq r\leq t} B_r \stackrel{d}{=} |B_t|, \tag{2.10}$$

and, therefore,

$$\begin{aligned} &P\Big(\max_{t_{k-1}\leq s\leq t_{k+2}} |B_s - B_{t_{k-1}}| > (3\,\vartheta\, t_{k-1} \log\log t_{k-1})^{1/2}\Big) \\ &= P\Big(\max_{0\leq r\leq 1} |B_r| > (3\log\log t_{k-1})^{1/2}\Big) \\ &\leq 2\,\kappa\,(\log\log t_{k-1})^{-1/2} \exp\Big(-3\log\log\frac{t_{k-1}}{2}\Big), \text{ with a constant } \kappa. \end{aligned}$$

It follows: $\sum_k P(\ldots) < \infty$. Therefore, the Borel–Cantelli lemma implies

$$\overline{\lim_{t\longrightarrow\infty}} \frac{|\tilde{S}_{[t]} - B_t|}{\sqrt{t \log\log t}} \leq (3\,\vartheta)^{1/2}\ [P].$$

With $\vartheta \longrightarrow 0$ the assertion follows. □

2.2 Functional Limit Theorem

For applications Donsker's invariance principle is of special importance. It states that continuous functionals of the partial sum sequence converge in distribution to the corresponding functional of a Brownian motion.

Idea:

$$\text{functional of } \left(\frac{S_k}{\sqrt{n}}\right)_{0\leq k\leq n} \xrightarrow{\mathcal{D}} \text{functional of } (B_t).$$

The space of continuous functions on $[0, 1]$, $C = C[0, 1]$ equipped with the supremum metric is a complete separable metric space. We equip this space with the σ-algebra generated by the projections. This is the same σ-algebra as the one generated by the topology of uniform convergence and also generated by the topology of pointwise convergence on C, the Borel σ-algebra on C.

$$\mathcal{E} = \mathfrak{B}_u(C) = \mathfrak{B}_p(C) =: \mathfrak{B}(C). \tag{2.11}$$

Definition 2.7 (Convergence in Distribution) A sequence of probability measures $\mu_n \in M^1(C, \mathcal{B}(C))$ converges in distribution to a measure $\mu \in M^1(C, \mathcal{B}(C))$,

$$\mu_n \xrightarrow{\mathcal{D}} \mu$$

if and only if for all real, continuous, bounded functions on C, $\varphi : C \to \mathbb{R}$, it holds:

$$\int \varphi\, d\mu_n \longrightarrow \int \varphi\, d\mu. \tag{2.12}$$

Also for measures in general metric spaces one introduces the notion of convergence in distribution by the convergence of integrals of continuous bounded functions. Correspondingly, one defines convergence in distribution for stochastic processes with continuous paths

$$X^{(n)} = \left(X_t^{(n)}\right)_{0\leq t\leq 1} \xrightarrow{\mathcal{D}} X,$$

if the associated distributions converge, equivalently, if the expectations of continuous bounded functions Ψ converge, $E\Psi(X^{(n)}) \longrightarrow E\Psi(X)$.

Remark 2.8

(1) If $X^{(n)} \xrightarrow{\mathcal{D}} X$ and if $\Psi : C \longrightarrow \mathbb{R}$ is P^X-almost surely continuous, then it follows

$$\Psi\big(X^{(n)}\big) \xrightarrow{\mathcal{D}} \Psi(X),$$

because for $\varphi \in C_b$, $\varphi \circ \Psi$ is a bounded P^X-almost surely, continuous function.

(2)

$$\mu_n \xrightarrow{\mathcal{D}} \mu \iff \int \psi d\mu_n \longrightarrow \int \psi d\mu, \quad \forall \psi \in C_b, \psi \text{ uniformly continuous.} \tag{2.13}$$

Now let (X_i) be an i.i.d. sequence with $EX_i = 0$ and $EX_i^2 = 1$. We introduce a suitable scaling of the partial sum process in two steps.

(a) **Linear interpolation:** Let

$$\widetilde{S}^{(n)}(u) := \begin{cases} \frac{S_n}{\sqrt{n}}, & u \geq n \\ \frac{1}{\sqrt{n}}\big(S_k + (u-k)(S_{k+1} - S_k)\big), & u \in [k, k+1), 0 \leq k \leq n-1, \end{cases}$$

i.e. $\widetilde{S}^{(n)}$ is the normalized partial sum process defined on $[0, n]$ by linear interpolation.

(b) **Temporal scaling**: The partial sum process $S^{(n)} = (S_t^{(n)})_{0 \leq t \leq 1}$ is obtained from $\widetilde{S}^{(n)}$ by:

$$S_t^{(n)} = \widetilde{S}^{(n)}(nt), \quad 0 \leq t \leq 1.$$

By linearly interpolation of the scaled partial sum sequence and by rescaling it to the time interval $[0, 1]$, we have obtained a stochastic process with continuous paths in $C[0, 1]$. An important corollary of Skorohod's embedding theorem is Donsker's invariance principle.

Theorem 2.9 (Donsker's Invariance Principle) *Let (X_i) be an i.i.d. sequence with $EX_i = 0$ and $EX_i^2 = 1$, then the partial sum process $S^{(n)}$ converges to a Brownian motion in $C[0, 1]$,*

$$S^{(n)} \xrightarrow{\mathcal{D}} B. \tag{2.14}$$

Meaning Thus, one can simulate a Brownian motion by a partial sum sequence (for example, by a random walk). The theorem is called the invariance principle because the limit of $S^{(n)}$ is independent of the distribution of X_i. The fundamental consequence of the theorem is the possibility of describing the distribution of a functional of a partial sum process approximately by the distribution of the functional of a Brownian motion. This is a generalization of the Central Limit Theorem, which has many fundamental applications in probability theory and mathematical statistics.

Proof of Theorem 2.9 $(S_t^{(n)})$, the scaled and interpolated sum process encodes the sequence $(\frac{S_m}{\sqrt{n}})$. It holds:

$$\left(\frac{S_m}{\sqrt{n}}\right) \stackrel{d}{=} \left(\frac{B_{T_m}}{\sqrt{n}}\right) \stackrel{d}{=} \left(B_{\frac{T_m}{n}}\right).$$

The normalized stopping time transformation converges a.s.,

$$\frac{T_{[ns]}}{n} \longrightarrow s \text{ a.s.}$$

From this follows pointwise convergence of $\frac{S_{[ns]}}{\sqrt{n}}$ in distribution

$$\frac{S_{[ns]}}{\sqrt{n}} \stackrel{d}{=} B_{\frac{T_{[ns]}}{n}} \longrightarrow B_s \text{ a.s.}, \quad 0 \le s \le 1.$$

However, even uniform stochastic convergence of $B_{\frac{T_{[ns]}}{n}}$ holds. □

Lemma 2.10 *The process* $\left(B_{\frac{T_{[ns]}}{n}}\right)$ *converges stochastically to B uniformly in* $[0, 1]$, *i.e.*

$$\sup_{t\in[0,1]} \left| B_{\frac{T_{[nt]}}{n}} - B_t \right| \xrightarrow[P]{} 0.$$

Proof By a simple monotonicity argument, the following holds

$$P\left(\sup_{0\le s\le 1} \left|\frac{T_{[ns]}}{n} - s\right| < 2\delta\right) > 1 - \varepsilon$$

for $n \ge N_{\delta,\varepsilon}$. As in the proof of Theorem 2.6, using the maximal inequality for Brownian motion, the assertion follows. □

Proof of Theorem 2.9 (Continued) It is to show: Let $\varphi : C[0,1] \longrightarrow \mathbb{R}$ be uniformly continuous and bounded; then it holds

$$E\varphi(S^{(n)}) \longrightarrow E\varphi(B).$$

Because of $\|S_t^{(n)} - \frac{S_{[nt]}}{\sqrt{n}}\|_\infty \underset{P}{\longrightarrow} 0$ and $\left(\frac{S_{[nt]}}{\sqrt{n}}\right) \overset{d}{=} \left(\frac{B_{T_{[nt]}}}{\sqrt{n}}\right) \overset{d}{=} \left(B_{\frac{T_{[nt]}}{n}}\right)$ it suffices with $B_t^{(n)} := B_{\frac{T_{[nt]}}{n}}$ to show:

$$E\varphi(B^{(n)}) \longrightarrow E\varphi(B).$$

But it holds with $\Delta_n := \|B^{(n)} - B\|$ because of the uniform continuity of φ:

$$\begin{aligned}
&\left|E\left(\varphi(B^{(n)}) - \varphi(B)\right)\right| \\
&\le E(\varphi(B^{(n)} - \varphi(B))\mathbb{1}_{\{\Delta_n > \delta\}} + |E(\varphi(B^{(n)} - \varphi(B))\mathbb{1}_{\{\Delta_n \le \delta\}}| \\
&\le 2\sup|\varphi|\, P(\Delta_n > \delta) + \varepsilon \qquad \text{for } \delta \le \delta_0.
\end{aligned}$$

By Lemma 2.10 the first term converges to 0. This implies that

$$B^{(n)} \xrightarrow{\mathcal{D}} B$$

and thus the assertion. □

Remark and Examples

(a) $\Psi : C \longrightarrow \mathbb{R}^1$, $\Psi(\omega) = \omega(1)$ is a continuous function. Thus it follows from Donsker's invariance principle:

$$\Psi\big(S^{(n)}\big) = \frac{S_n}{\sqrt{n}} \xrightarrow{\mathcal{D}} \Psi(B) = B_1 \overset{d}{=} \mathcal{N}(0,1).$$

This is precisely the central limit theorem.

(b) $\Psi(\omega) := \sup\{\omega(t), 0 \le t \le 1\}$ is a continuous function. Thus it follows

$$\Psi\big(S^{(n)}\big) = \max_{0\le m\le n} \frac{S_m}{\sqrt{n}} \xrightarrow{\mathcal{D}} M_1 = \sup_{0\le t\le 1} B_t. \tag{2.15}$$

The maximum of the partial sum process is found at the discrete time points because of the linear interpolation. According to André's reflection principle

$$P_0(M_1 \ge a) = P_0(\tau_a \le 1) = 2\,P_0(B_1 \ge a) = P_0\big(|B_1| \ge a\big).$$

Thus $M_1 \stackrel{d}{=} |B_1|$. M_1 is distributed like the absolute value of a standard normally distributed random variable.

(c) Let (S_n) be a symmetric random walk, and let R_n = be range of S_n, i.e. the number of points that are visited by the time n of S_n.

$$R_n = 1 + \max_{m \leq n} S_m - \min_{m \leq n} S_m. \tag{2.16}$$

The question is: Which interval will be visited by the time n ? To answer the question, we need to normalize suitably: $\frac{R_n}{\sqrt{n}} = \Psi(S^{(n)})$ is a functional of the partial sum process with Ψ as in (2.16). Therefore, it follows

$$\frac{R_n}{\sqrt{n}} \xrightarrow{\mathcal{D}} \Psi(B), \quad \Psi(B) = 1 + \sup_{t \leq 1} B_t - \inf_{t \leq 1} B_t. \tag{2.17}$$

(d) An example where one needs the extension to a.s. continuous functionals. We consider (Fig. 2.4)

$$\Psi(\omega) := \sup\{t \leq 1; \omega(t) = 0\}.$$

$\Psi(\omega)$ denotes the last time t before 1 at which $\omega(t) = 0$. This functional is not continuous, because $\Psi(\omega_\varepsilon) = 0, \forall \varepsilon > 0$. Further it holds that $\|\omega_\varepsilon - \omega_0\| \longrightarrow 0$ but $\Psi(\omega_0) = \frac{2}{3}$.

But for $\omega \in C$ with $\Psi(\omega) < 1$ and such that ω has in each neighborhood $U_\delta(\Psi(\omega))$ positive and negative values for all $\delta > 0$, it holds that Ψ is continuous in ω. These ω have measure 1 with respect to Brownian motion. So Ψ is P_0 almost surely continuous.

Let

$$L_n := \sup\{m \leq n; S_{m-1}S_m \leq 0\}$$

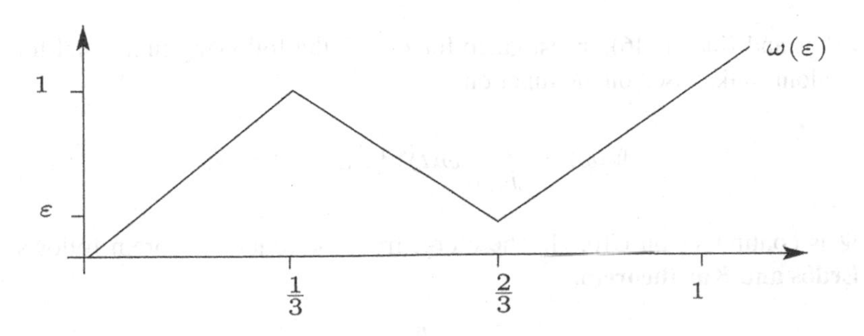

Fig. 2.4 Example not continuous

be the index of the last change of sign before n of the random walk. Analogously, let

$$L := \sup\{0 \le t \le 1;\ B_t = 0\}$$

be the time of the last zero of the Brownian motion before $t = 1$. The distribution of L is the arc sine distribution. We now obtain as a corollary.

Corollary 2.11 (Arc-Sine Law) *For the random walk (S_n) the normalized time of the last sign change before n, $\frac{L_n}{n}$ converges in distribution to the arc-sine distribution*

$$\frac{L_n}{n} \xrightarrow{\mathcal{D}} L. \tag{2.18}$$

(e) **Positivity range** Let $\Psi(\omega := \lambda\!\!\lambda^1([t \in [0, 1]; \omega(t) > 0\})$. Ψ is not continuous on $C[0, 1]$, but Ψ is continuous in ω if

$$\lambda\!\!\lambda^1(]t \in [0, 1]; \omega(t) = 0\}) = 0.$$

The set of these exceptional points has measure 0 with respect to P_0, the distribution of the Brownian motion, because according to Fubini

$$E_0 \lambda\!\!\lambda^1(\{t \in [0, 1]; B_t = 0\}) = \int_0^1 P_0(B_t = 0) \mathrm{d}\lambda\!\!\lambda^1(t) = 0.$$

Thus Ψ is P_0-a.s. continuous and it follows from the Skorohod theorem

Corollary 2.12 (Convergence of Positivity Ranges)

$$\frac{|\{m \le n; S_m > 0\}|}{n} \xrightarrow{\mathcal{D}} \lambda\!\!\lambda^1(\{t \in [0, 1]; B_t = 0\}) \tag{2.19}$$

(f) Erdös and Kac (1946), investigated for $k = 2$ the following functional for the random walk, based on the function

$$\Psi(\omega) := \int_{[0,1]} \omega(t)^k \mathrm{d}\lambda\!\!\lambda^1(t), \quad k \in \mathbb{N}.$$

Ψ is continuous on $C[0, 1]$. Therefore, from Skorohod's theorem follows the **Erdős and Kac theorem:**

$$\Psi(\widetilde{S}^{(n)}) = n^{-1-\frac{k}{2}} \sum_{m=1}^{n} S_m^k \xrightarrow{\mathcal{D}} \int_0^1 B_t^k \,\mathrm{d}t. \tag{2.20}$$

It is worth noting that also for $k \geq 2$ the above approximation in (2.20) only needs the assumption $EX_i^2 = 1$, $EX_i = 0$, but not the stronger assumption $E|X_i|^k < \infty$. In the case $k = 1$ it follows from (2.20) that

$$n^{-\frac{3}{2}} \sum_{m=1}^{n} (n+1-m)X_m \xrightarrow{\mathcal{D}} \int_0^1 B_t \, \mathrm{d}t \stackrel{d}{=} N(0, 1).$$

3 Stochastic Integration

The aim of this chapter is to introduce the stochastic integral $\int_0^{\cdot} \varphi_s \, \mathrm{d}X_s$ for semimartingales X and stochastic integrands φ and to describe its properties. If one interprets φ as a trading strategy and X as the price process of a security, then the stochastic integral can be understood to be the accumulated gain. The stochastic integral is constructed in a series of steps and extended to general function classes and process classes. In the first part of the chapter, after an introduction to martingales, the stochastic integral for a Brownian motion is introduced as a model case. The construction is then extended in several steps to the integration of general semimartingales.

The stochastic integral is a particularly important tool for stochastic model building. In analogy to the ordinary differential equations it allows by means of the corresponding stochastic differential equation e.g. of the form $\mathrm{d}X_t = a(t, X_t)\,\mathrm{d}t + b(t, X_t)\,\mathrm{d}B_t$ to construct from local properties, here the drift a and the volatility b, a stochastic model; in the above case, a diffusion process. Semimartingales generalize this principle and can be described by local characteristics, the (local) drift, the continuous diffusion part (martingale part) and a (local) jump part. They form a fundamental class of models and are of particular importance for models in financial mathematics.

3.1 Martingales and Predictable Processes

Let $(\Omega, \mathfrak{A}, P)$ be a complete martingale probability space and $(\mathfrak{A}_t)_{t\geq 0} \subset \mathfrak{A}$ a filtration in Ω. We postulate in general a regularity property of filtrations:

L. Rüschendorf, *Stochastic Processes and Financial Mathematics*, Mathematics Study Resources 1, https://doi.org/10.1007/978-3-662-64711-0_3

General Condition (Conditions Habituelles)

$$(\mathfrak{A}_t) \text{ is right-continuous, i.e. } \mathfrak{A}_{t+} = \bigcap_{s>t} \mathfrak{A}_s = \mathfrak{A}_t$$
$$\mathfrak{A}_0 \text{ contains all (subsets of) } P \text{ zero sets.} \tag{3.1}$$

Let X be a d-dimensional stochastic process, $X : [0, \infty) \times \Omega \longrightarrow \mathbb{R}^d$.
X is called **adapted** on the filtration $(\mathfrak{A}_t)$, if $X_t \in \mathfrak{L}(\mathfrak{A}_t)$, for all t, $X_t = X(t, \cdot)$.

In the standard case based on natural filtration $\mathfrak{A}_t := \bigcap_{\varepsilon>0} \sigma(\sigma(X_s, s \le t+\varepsilon) \cup \mathcal{N}_P) =: \mathfrak{A}_t^X$ the general condition is fulfilled.

The central concept of this chapter and of stochastic integration theory is the notion of a martingale.

Definition 3.1 $M = (M_t, \mathfrak{A}_t)$ is a **martingale** if

(1) M is $(\mathfrak{A}_t)$-adapted
(2) $E|M_t| < \infty$, for all t.
(3) $E(M_t \mid \mathfrak{A}_s) = M_s \ [P]$, for all $s \le t$

M is called **submartingale** if (1), (2), and
(3') $E(M_t \mid \mathfrak{A}_s) \ge M_s \ [P]$

If in (3') "$\le$" holds, then M is called a supermartingale.
X is called a continuous process if the paths of X are continuous
X is called **càdlàg process** if the paths are right-continuous and have limits on the left-hand side, i.e. $X_t = \lim_{s\downarrow t, s>t} X_s$, $X_{t-} = \lim_{s\uparrow t, s<t} X_s$ exists.

càdlàg is an abbreviation for *continue à droite, limite à gauche* (continuous from the right, with left limits).

Remark 3.2 Let X, Y be càdlàg processes with $P(X_t = Y_t) = 1 \ \forall t$, i.e. Y is a **version** or **modification** of X, then it follows $P(X_t = Y_t, \forall t) = 1$, i.e. X and Y are **'indistinguishable'**.

Under weak regularity assumptions, there exists a càdlàg version of submartingales.

Theorem 3.3 (Regularity Theorem) *Let X be a submartingale and $t \longrightarrow EX_t$ is right-continuous.*
Then X has a càdlàg version which is an $(\mathfrak{A}_t)$-submartingale.

Proof The proof is based on the following convergence theorems for submartingales in discrete time.

(a) (X_n) submartingale, $\sup EX_n^+ < \infty \Longrightarrow X_n \longrightarrow X_\infty$ almost surely
(b) (X_n) inverse submartingale $\Longrightarrow X_n \longrightarrow X_\infty$ almost surely
(c) (X_n) inverse martingale $\Longrightarrow X_n \longrightarrow X_\infty$ in L^1 and almost surely
(d) X_n inverse submartingale, $\sup E|X_n| < \infty$
$\iff EX_n \geq K, \forall n$, i.e. EX_n is bounded from below
$\Longrightarrow X_n \longrightarrow X_\infty$ in L^1 and almost surely (Revuz and Yor (2005, p. 58)).

It follows: $\forall t \in \mathbb{R}_+$: $\underbrace{\lim_{r\uparrow t, r\in\mathbb{Q}} X_r(\omega)}_{\text{submartingale}}$ exists almost surely, because $EX_r^+ \leq EX_t^+$,
and $\underbrace{\lim_{r\downarrow t, r\in\mathbb{Q}} X_r(\omega)}_{\text{inv. submartingale}}$ exists almost surely.

Define $X_{t+} = \overline{\lim_{r\downarrow t, r\in\mathbb{Q}}} X_r$. X_{t+} is right-hand continuous and has left-hand limits, so is càdlàg.
Assertion: $(X_{t+}, \mathfrak{A}_t)$ is a submartingale, and it holds that $X_t = E(X_{t+} \mid \mathfrak{A}_t) = X_{t+}$ $[P]$.
To prove this assertion, note that for an antitone sequence $t_n \searrow t$ it holds:

$$(X_{t_n}) \text{ is inverse submartingale and } EX_{t_n} \geq EX_t.$$

By the convergence theorem for submartingales, part (d) follows:
$X_{t+} \in L^1$, $X_{t_n} \longrightarrow X_{t+}$ in L^1 and almost surely.
Because of the submartingale property $X_t \leq E(X_{t_n} \mid \mathfrak{A}_t)$ it follows $X_t \leq E(X_{t+} \mid \mathfrak{A}_t)$.
Because of L^1-convergence and the condition of right-continuity of $t \longrightarrow EX_t$ we obtain as a result

$$EX_{t+} = \lim EX_{t_n} = EX_t.$$

Thus one obtains

$$X_t = E(X_{t+} \mid \mathfrak{A}_t) = X_{t+} \; [P].$$

In particular (X_{t+}) is a càdlàg-version of (X_t) and is adapted. □

General Assumption In the following we always assume càdlàg versions of the submartingale.

Definition 3.4 A martingale M is called ***L^2*-martingale** if $EM_t^2 < \infty, \forall t < \infty$. Define

$\mathcal{M}^2 := \{M; MisL^2martingale\}$,
$\mathcal{M}_0^2 =: \{M \in \mathcal{M}^2; M_0 = 0\}$
$\mathcal{M}_c^2 = \{M \in \mathcal{M}^2; Mhascontinuouspaths\}$.

Stopping times are a fundamental tool of martingale theory.

Definition 3.5 (Stopping Times)

(a) $\tau : \Omega \longrightarrow [0, \infty]$ is a **stopping time** with respect to $(\mathfrak{A}_t)$,

$$\textit{if } \{\tau \leq t\} \in \mathfrak{A}_t, \quad \forall t < \infty.$$

τ is a finite stopping time if $\tau < \infty$ $[P]$.

(b) $\mathfrak{A}_\tau := \{A \in \mathfrak{A}_\infty; A \cap \{\tau \leq t\} \in \mathfrak{A}_t, \forall t\}$ is the **σ-algebra of the τ-past.**

Non-finite stopping times are also called **Markov times**. A weaker definition of the notion of stopping time requires that $\{\tau < t\} \in \mathfrak{A}_t$ for right-continuous filtrations, this is equivalent to the strong definition of stopping time.

Lemma 3.6 *Let X be an adapted, right-continuous, τ be a stopping time. Then it holds:*

$$X_\tau \textit{ is } \mathfrak{A}_\tau \textit{ measurable on } \{\tau < \infty\}.$$

Proof To prove this, we first show that X is progressively measurable, i.e. $\forall t$ is the mapping of $[0, t] \times \Omega \longrightarrow \mathbb{R}^d$, $(s, \omega) \longrightarrow X_s(\omega)$, $\mathfrak{B}[0, t] \otimes \mathfrak{A}_t$ is measurable. This follows by discrete approximation. For $X^{(n)}(s, \omega) := X(\frac{kt}{2^n}, \omega)$, $\frac{(k-1)t}{2^n} \leq s < \frac{kt}{2^n}$ is measurable with respect to $\mathfrak{B}_t \otimes \mathfrak{A}_t$ and X is right-continuous. Therefore, convergence follows $X^{(n)}(s, \omega) \longrightarrow X(s, \omega)$ and X is as a limit of $X^{(n)}$ progressively measurable.
We now show: $X_\tau \mathbb{1}_{\{\tau \leq t\}} \in \mathcal{L}(\mathfrak{A}_t), \forall t \geq 0$. (From this follows the assertion of the lemma.)

The assertion $X_\tau \mathbb{1}_{\{\tau \leq t\}} \in \mathcal{L}(\mathfrak{A}_t)$ is equivalent to $\{X_\tau \in A\} \cap \{\tau \leq t\} \in \mathfrak{A}_t, \forall A \in \mathbb{B}^d$ and, therefore, also equivalent to $\{X_{\underbrace{\tau \wedge t}_{=:S}} \in A\} \cap \{\tau \leq t\} \in \mathfrak{A}_t, \forall A \in \mathbb{B}^d$.

Now consider the composite mapping

$$\begin{array}{ccccc} (\Omega, \mathfrak{A}_t) & \longrightarrow & \big([0, t] \times \Omega, \mathfrak{B}[0, t] \otimes \mathfrak{A}_t\big) & \longrightarrow & (\mathbb{R}^d, \mathbb{B}^d) \\ \omega & \overset{\text{measurable}}{\longmapsto} & (S(\omega), \omega), \qquad (s, \omega) & \overset{\text{measurable}}{\longrightarrow} & X_s(\omega) \end{array}$$

Compound measurable mappings are measurable with respect to $\mathfrak{A}_t$ and, therefore, the assertion of the lemma holds. □

Remark 3.7 For right-continuous martingales (submartingales), the inequalities and convergence theorems apply as in discrete time, such as Doob's theorem and the closure theorem. They are obtained by discrete approximation using the discrete convergence theorems from the proof of Theorem 3.3. For example the following results hold:

Theorem of Doob Let (X_t) be a submartingale with $\sup_t EX_t^+ < \infty$

$\Longrightarrow \lim_{t \longrightarrow \infty} X_t$ exists P-almost surely.

Closure theorem (X_t) is a uniformly integrable martingale

$\Longleftrightarrow \exists X_\infty \in L^1 : X_t = E(X_\infty \mid \mathfrak{A}_t)\,[P]$.

An important role plays the optional sampling theorem.

Theorem 3.8 (Optional Sampling Theorem)

(a) *Let X be a martingale (submartingale), right-continuous, and S, T bounded stopping times, $S \leq T$ then it follows*

$$X_S \overset{(\leq)}{=} E(X_T \mid \mathfrak{A}_S)\,[P].$$

(b) *Let $(X_t, \mathfrak{A}_t)_{0 \leq t \leq \infty}$ be a uniformly integrable right-continuous martingale (submartingale), let S, T be stopping times, with $S \leq T$, then it follows*

$$X_S \overset{(\leq)}{=} E(X_T \mid \mathfrak{A}_S)\,[P].$$

(c) *Let $X = (X_t, \mathfrak{A}_t)$ be a right-continuous supermartingale, $X \geq 0$, $S \leq T$ stopping times, then it follows*

$$X_S \geq E(X_T \mid \mathfrak{A}_S)\,[P].$$

Proof

(a) Let T be a bounded stopping time. Then there exists a sequence (T^k) of stopping times, $T^k \downarrow T$; $|T^k(\Omega)| < \infty$ (only finitely many values), e.g.: $T^k = \frac{l}{2^n}$, if $\frac{l-1}{2^n} \leq T \frac{\ell}{2^n}$. Let S^k, T^k; $S^k \downarrow S$, $T^k \downarrow T$; $S^k \leq T^k$; S^k, T^k have finitely many values. According to the optional sampling theorem in the discrete case it holds

$$\int_A X_{S^k}\,\mathrm{d}P \leq \int_A X_{T^k}\,\mathrm{d}P, \quad \forall A \in A_S \subset A_{S^k}.$$

$X_{S^k} \longrightarrow X_S$, $X_{T^k} \longrightarrow X_T$ almost surely, since X is right-continuous. Moreover, convergence holds in L^1, since both sequences are inverse submartingales with $EX_{S^k} \geq EX_0$ are. From this follows

$$\int_A X_S\,\mathrm{d}P \leq \int_A X_T\,\mathrm{d}P, \quad \forall A \in \mathfrak{A}_S.$$

(b) Let $S \leq T$ be arbitrary stopping times. Then there exist stopping times $S^n \leq T^n$, $S^n \downarrow S$ and $T^n \downarrow T$ such that S^n, T^n have countably many values. According to the discrete optional sampling theorem it follows

$$\int_A X_{S^n}\,\mathrm{d}P \le \int_A X_{T^n}\,\mathrm{d}P, \quad \forall A \in \mathfrak{A}_{S^n}$$

and, therefore, this inequality is also valid for $A \in \mathfrak{A}_{S+} = \bigcap_n \mathfrak{A}_{S^n}$.

S is a stopping time and $S \le S^n$ and, therefore, $\mathfrak{A}_S \subset \mathfrak{A}_{S^n}$. So the inequality also holds for $A \in \mathfrak{A}_S$. Further $(X_{S^n}, \mathfrak{A}_{S^n})$ is an inverse submartingale, $EX_{S^n} \downarrow$, $EX_{S^n} \ge EX_0$. From this it follows that $\{X_{S^n}\}$ is uniformly integrable.

Similarly $\{X_{T^n}\}$ are uniformly integrable, $X_S(\omega) = \lim X_{S^n}(\omega)$ and $X_T(\omega) = \lim X_T(\omega)$ almost surely and in L^1. It follows that $X_S, X_T \in L^1$ and $\int_A X_S\,\mathrm{d}P \le \int_A X_T\,\mathrm{d}P$.

(c) see, for example, Elliott (1982, p. 36) or Revuz and Yor (2005, p. 65).

□

Remark 3.9 It follows from the above proof:
If X is a uniformly integrable submartingale (martingale or closed submartingale), i.e. $X_t \overset{(=)}{\le} E(X_\infty \mid \mathfrak{A}_t)$, then it follows: $\{X_S;\ S \le T \textit{ stopping time}\}$ is uniformly integrable and it holds for stopping times $S \le T$:

$$X_S \overset{(=)}{\le} E(X_T \mid \mathfrak{A}_S) \overset{(=)}{\le} E(X_\infty \mid \mathfrak{A}_S).$$

The following application of stopping times is important for a number of uniqueness statements in the following.

Corollary 3.10 *Let X be a nonnegative right-continuous supermartingale, and let*

$$\tau := \inf\{t;\ X_t = 0 \text{ or } X_{t-} = 0\}.$$

Then it holds: $X_.(\omega) = 0$ at $[\tau(\omega), \infty)$.

Proof Let $\tau_n := \inf\{t;\ X_t \le \frac{1}{n}\}$ then it follows $\tau_{n-1} \le \tau_n \le \tau$; the sequence τ_n is increasing (Fig. 3.1).
On $\{\tau_n = \infty\} \subset \{\tau = \infty\}$ there is nothing to show.
On $\{\tau_n < \infty\}$ holds $X_{\tau_n} \le \frac{1}{n}$, since X is right-continuous. Let $q \in \mathbb{Q}$, $q > 0$. Then $\tau + q$ is a stopping time and $\tau + q > \tau_n$. Now, according to the optional sampling theorem.

$$\frac{1}{n} \ge E(X_{\tau_n} \mathbb{1}_{(\tau_n \le \infty)}) \ge E(X_{\tau+q} \mathbb{1}_{(\tau_n \le \infty)}) \ge 0.$$

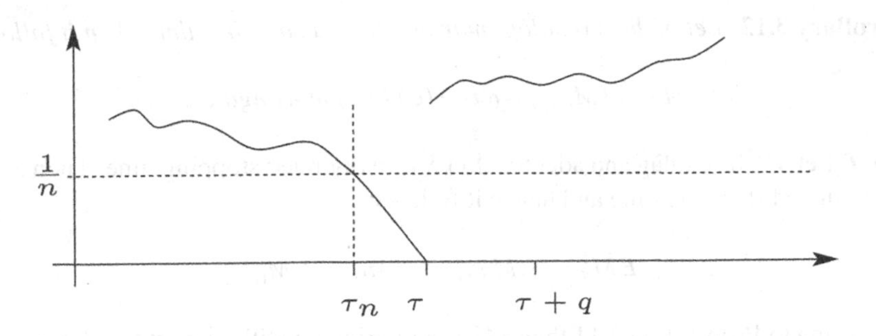

Fig. 3.1 Supermartingale

For $n \to \infty$, therefore, by the monotone convergence theorem

$$E(X_{\tau+q}\mathbb{1}_{(\tau_n<\infty,\forall n)}) = 0.$$

Because of $\{\tau < \infty\} \subset \bigcap_n \{\tau_n < \infty\}$ it follows that $X_{\tau+q} = 0$ almost surely on $\{\tau < \infty\}$, $\forall q \in \mathbb{Q}$, $q > 0$. The right-continuity of X then implies the assertion $X_\cdot(\omega) = 0$ on $[\tau, \infty)$. □

As a corollary of the optional sampling theorem, we obtain the following characterization of the martingale property by the conservation of expected values under stopping times.

Proposition 3.11 *Let X be càdlàg and adapted. Then it holds:*
X is a martingale $\Longleftrightarrow \forall$ bounded stopping times τ is $X_\tau \in L^1$ and $EX_\tau = EX_0$.

Proof "$\Longrightarrow$": Follows from the optional sampling theorem.
"$\Longleftarrow$": For $s < t$ and $A \in \mathfrak{A}_s$ define the stopping time $\tau := t\mathbb{1}_{A^c} + s\mathbb{1}_A$.
Then: $EX_0 = EX_\tau = EX_t\mathbb{1}_{A^c} + EX_s\mathbb{1}_A$.
Further:

$$EX_0 = EX_t = EX_t\mathbb{1}_{A^c} + EX_t\mathbb{1}_A,$$

because also t is a stopping time. This implies the martingale property :

$$X_s = E(X_t \mid \mathfrak{A}_s).$$

□

The martingale property of a process is preserved under stopping.

Corollary 3.12 *Let M be a (càdlàg) martingale, τ a stopping-time, then it follows*

$$M^\tau = (M_{t\wedge\tau})_{\tau\geq 0} \text{ is a (càdlàg) martingale.}$$

Proof Let M^τ be càdlàg and adapted. Let S be a bounded stopping time. Then $S\wedge\tau$ is a bounded stopping time and hence it follows:

$$EM_S^\tau = EM_{S\wedge\tau} = EM_0 = EM_0^\tau.$$

According to Proposition 3.11 then M^τ is a martingale with respect to $(\mathfrak{A}_t)$. □

A standard tool for martingales are Doob's maximal inequalities.

Theorem 3.13 (Doob's Maximal Inequality) *Let $X = (X_t)_{t\in T}$ be a right-continuous martingale or positive submartingale and let $X^* := \sup_t |X_t|$. Then it follows*

$$P(X^* \geq \lambda) \leq \sup_t \frac{E|X_t|^p}{\lambda^p} \text{ for } p \geq 1, \lambda > 0.$$

Further it holds for $1 < p < \infty$:

$$\|X^*\|_p \leq \frac{p}{p-1} \sup_t \|X_t\|_p.$$

Proof The proof follows from the discrete Doob inequality. Let $D \subset T$ be countable dense, then because of the right-continuity of X, $X^* = \sup_{t\in D} |X_t|$. So the discrete Doob inequality can be applied. □

We need an essential extension of the martingale notion, the local martingale.

Definition 3.14

(a) An adapted process M is called **local martingale** if there is a sequence of stopping times τ_n with $\tau_n \uparrow \infty$ such that $\forall n : M^{\tau_n} \in \mathcal{M}$, i.e. M^{τ_n} is a martingale. (τ_n) is called **localizing sequence**.
$\mathcal{M}_{\text{loc}} :=$ set of local martingales.

(b) $M \in \mathcal{M}_{\text{loc}}$ is called **local L^2-martingale** $\iff \exists a$ localizing sequence (τ_n) such that $M^{\tau_n} \in \mathcal{M}^2$.
$\mathcal{M}^2_{\text{loc}} :=$ set of local L^2-martingales.

Remark 3.15

(a) $\mathcal{M} \subset \mathcal{M}_{\text{loc}}$. Consider the localizing sequence $\tau_n = n$.
(b) $\mathcal{M}_{\text{loc},c} \subset \mathcal{M}^2_{\text{loc}}$,

because, e.g. $\tau_n = \inf\{t; |M_t| \geq n\} = \inf\{t; |M_t| = n\}$ is a localizing sequence with $M^{\tau_n} \in \mathcal{M}^2_{\text{loc}}$.

(c) For $M \in \mathcal{M}_{\text{loc}}$ the following characterization of the martingale property holds
$M \in \mathcal{M} \iff M$ is of class (DL), i.e. $\{M_T; T \text{ stopping time}, T \leq a\}$ is uniformly integrable, $\forall a < \infty$.

Proof

"$\Longrightarrow$" According to the optional sampling theorem, it follows

$$\sup_{T\leq a} \int_{\{M_T > K\}} M_T \,\mathrm{d}P = \sup_{T \leq a} \int_{\{M_T > K\}} M_a \,\mathrm{d}P \xrightarrow{K\to\infty} 0,$$

so $\{M_T; T \leq a\}$ is uniformly integrable.

"$\Longleftarrow$" Let (T_n) be a localizing sequence of stopping times and let $s < t$ then $M^{T_n \wedge t} \in \mathcal{M}$ and it holds for $A \in \mathfrak{A}_s$

$$\int_A M_s^{T_n} \,\mathrm{d}P = \int_A M_s^{T_n \wedge t} \,\mathrm{d}P = \int_A M_t^{T_n} \,\mathrm{d}P.$$

By the uniform integrability assumption, it follows because of $M_s^{T_n\wedge t} \longrightarrow M_s$ and $M_t^{T_n \wedge t} \longrightarrow M_t$

$$\int_A M_s \,\mathrm{d}P = \int_A M_t \,\mathrm{d}P.$$

□

For a process A define
$V_t^A := \sup_{0=t_0<\cdots<t_n=t} \sum_{i=1}^n |A_{t_i} - A_{t_{i-1}}|$, the **variation of A in $[0, t]$**.

Definition 3.16

(a) Let $\mathcal{V}^+$ is the set of all **increasing** right-continuous real, adapted **processes** with $A_0 \geq 0$. Let further $\mathcal{V}_0^+ := \{A \in \mathcal{V}^+; A_0 = 0\}$.
(b) Let $\mathcal{V} := \{A = A^1 - A^2; A^i \in \mathcal{V}^+\}$ be the set of adapted **processes of finite variation.**
(c) A is called a process of **bounded variation** if there $\exists K < \infty$ such that $V_t^A \leq K, \forall t$.

Remark 3.17

(a) Increasing processes $A^i \in \mathcal{V}^+$ also have left-sided limits and are, therefore, càdlàg. So it holds: $A \in \mathcal{V} \Longrightarrow A$ càdlàg.

(b) Hence $\mathcal{V} = \{A$ adaptiert, càdlàg; $V_t^A < \infty, \forall t < \infty\} =: FV$ is called the term "of finite variation". For the proof, note that every $A \in FV$ has a unique decomposition $A = A^+ - A^-$ with $A_0^+ = A_0$ and $A^+, A^- \in \mathcal{V}^+$ where A^+ is minimal. This decomposition is given by

$$A^+ = \frac{1}{2}(V^A + A), \quad A^- = \frac{1}{2}(V^A - A).$$

$V_t^A \uparrow$ variation measure produces a measure the **variation measure** V_A:
$V^A((s,t]) := V_t^A - V_s^A$ is a *random measure*. So $\int_0^t f(s)\,\mathrm{d}V^A(s)$ is well defined as Lebesgue–Stieltjes integral, and

$$\int_0^t f(s)\,\mathrm{d}A(s) = \int_0^t f(s)\,\mathrm{d}A^+(s) - \int_0^t f(s)\,\mathrm{d}A^-(s).$$

After these preparations, we can now introduce the central notion of semimartingale. The semimartingale is a fundamental term for modeling. It is composed additively of a trend component $A \in \mathcal{V}$ and a stochastic fluctuation $M \in \mathcal{M}_{\mathrm{loc}}$

$$X = X_0 + M + A.$$

For some purposes, an integrability condition for A is useful.

Definition 3.18

(a) An adapted process X is called **semimartingale** if a local martingale $M \in \mathcal{M}_{\mathrm{loc}}$ and a process of finite variation $A \in \mathcal{V}$ exist such that

$$X_t = X_0 + M_t + A_t, \quad t \geq 0.$$

(b) $\mathcal{A}^+ := \{A \in \mathcal{V}^+;\ EA_\infty < \infty\}$ is the set of **integrable increasing** processes.
$\mathcal{A} := \mathcal{A}^+ - \mathcal{A}^+$ is the set of **processes of integrable variation**, i.e. for $A \in \mathcal{A}$ holds $EV_\infty^A < \infty$.

Remark 3.19

(a) X is called **locally bounded process**, if there is a localizing sequence of stopping times $\tau_n \uparrow \infty$ such that X^{τ_n} is bounded.
$\mathcal{A}_{\mathrm{loc}}$ denotes the set of **locally integrable processes**, i.e., the processes of locally integrable variation.
$A \in \mathcal{A}_{\mathrm{loc}} \Longrightarrow E\sup_t |A_t^{\tau_n}| < \infty$, i.e. $(A_t^{\tau_n})$ is uniformly integrable for a localizing sequence (τ_n).
(b) According to the technically complex "**previsible section theorem**" holds:
For $\mathcal{A} \in \mathcal{V}$ "previsible" holds: $A \in \mathcal{A}_{\mathrm{loc}}$.

For the notion of previsibility used above, cf. Definition 3.21.

(c) $M \in \mathcal{M}_{\text{loc}} \cap \mathcal{V} \longrightarrow M \in \mathcal{A}_{\text{loc}}$. (cf. also Proposition 3.20)

(d) The Brownian motion B is a martingale, $B \in \mathcal{M}$ but B has non-finite variation $B \notin \mathcal{V}$.

(e) For $A \in \mathcal{V}$, X locally bounded, progressively measurable, the stochastic Stieltjes integral can be introduced

$$(X \cdot A)_t := \int_0^t X_s \, \mathrm{d}A_s.$$

It holds: $X \cdot A \in \mathcal{V}$, because: $V_t^{X \cdot A} \leq \sup_{s \leq t} |X_s| V_t^A < \infty$.

With the progressive measurability of X it follows that $X \cdot A$ is adapted.

Continuous local martingales are typically not of finite variation. So no pathwise construction of the stochastic integral is possible. For discontinuous local martingales cf. Remark 3.19, c).

Proposition 3.20 *Let $M \in \mathcal{M}_{\text{loc},c} \cap \mathcal{V}$ be a continuous local martingale of finite variation. Then there exists a $c \in \mathbb{R}^1$ such that: $M = c\,[P]$.*

Proof W.l.g. let $M_0 = 0$ and $M \in \mathcal{M}_c^2$; this can be obtained by localization.

The variational process $V^M = M^+ + M^-$ is continuous.

Let $S_n := \inf\{s;\ V_s^M \geq n\} = \inf\{s;\ V_s^M = n\}$. S_n is a stopping time, M^{S_n} is of bounded variation ($\leq n$) and $M^{S_n} \leq M_0 + n$.

So it suffices to prove the assertion for the case $|M| \leq K$ and $V^M \leq K$.

Let $\Delta = (t_i)$, $\quad 0 = t_0 < t_1 < \cdots < t_k = t$ be a decomposition of $[0, t]$. Then it holds:

$$\begin{aligned}
EM_t^2 &= E \sum_{i=0}^{k-1} (M_{t_{i+1}}^2 - M_{t_i}^2) && \textit{telescopesum} \\
&= E \sum_{i=0}^{k-1} (M_{t_{i+1}} - M_{t_i})^2, && M \in \mathcal{M}^2, s < t : EM_s M_t = EM_s^2 \\
&\leq E V_t^M \sup_i |M_{t_{i+1}} - M_{t_i}| \\
&\leq K \underbrace{E \sup_i |M_{t_{i+1}} - M_{t_i}|}_{\longrightarrow 0 \text{ for} |\Delta| \longrightarrow 0} \\
&\longrightarrow 0 && \text{because } |M| \leq K, M \in \mathcal{M}_c.
\end{aligned}$$

Here we use that continuous functions are uniformly continuous on compact intervals. It follows that $M_t = 0\,[P]$. □

On $\overline{\Omega} := [0, \infty) \times \Omega$ we define suitable σ-algebras $\mathcal{P}, \mathcal{O}, \mathcal{P}_r$ that describe measurability requirements for processes.

Definition 3.21

(a) Let $\mathcal{E}$ the class of **elementary predictable or previsible processes K** of the form
$K(t, \omega) = K_t(\omega) = K_0(\omega)\mathbb{1}_{\{0\}} + \sum_{j=1}^{m-1} K_j(\omega)\mathbb{1}_{(t_j,t_{j+1}]}(t)$, with $0 < t_0 < t_1 < \cdots < t_m$
$K_j \in B(\mathfrak{A}_{t_j}) = L_b(\mathfrak{A}_{t_j})$, i.e. K_j are bounded, $\mathfrak{A}_{t_j}$-measurable, $1 \le j \le m$,
(b) $\mathcal{P} := \mathcal{P}\big((\mathfrak{A}_t)\big) = \sigma(\mathcal{E})$ the σ-algebra of **predictable sets** in $\overline{\Omega}$, the σ-algebra generated by $\mathcal{E}$.
(c) A process Y is called **predictable** if $Y \in L(\mathcal{P})$.

Remark 3.22 For stopping times S, T we define the **stochastic interval** $[\![S, T[\![$ by $[\![S, T[\![= \{(t, \omega); S(\omega) \le t < T(\omega)\}$; the stochastic interval $[\![S, T]\!]$ is defined analogously.
T is a **predictable stopping time** if stopping times (T_n) exist with
$T_n(\omega) \uparrow T(\omega)$, $T_n(\omega) < T(\omega)$ almost surely on $\{T > 0\}$.

T is called **reachable** if there is a sequence (T_n) of predictable stopping times such that

$$[\![T]\!] \subset \bigcup [\![T_n]\!] \quad \text{and} \quad P\Big(\bigcup_n \{T_n = T\}\Big) = 1.$$

We now use the following theorem about monotone classes:

Theorem 3.23 (Theorem on Monotone Classes) *Let $(\Omega_0, \mathfrak{A}_0)$ be a measure space and $\mathcal{H} \subset B(\Omega_0)$ a vector space, closed under bounded pointwise convergence (i.e. $f_n \in \mathcal{H}$, $f_n \longrightarrow f$, $|f_n| \le K \Longrightarrow f \in \mathcal{H}$). Let $\mathcal{G} \subset \mathcal{H}$ be an algebra in $\mathcal{H}$ with $1 \in \mathcal{G}$ (or let there exist $K_n \in \mathcal{G}$ with $K_n \longrightarrow 1$ pointwise).*
Then it holds: $\mathcal{H} \supset B(\Omega_0, \sigma(\mathcal{G}))$.

Proposition 3.24 *Let $\mu \in M_e(\overline{\Omega}, \mathcal{P})$, then $\mathcal{E}$ dense in $L^2(\overline{\Omega}, \mathcal{P}, \mu)$.*

Proof $\mathcal{E}$ is an algebra of functions and there exists an approximation K^n of $\mathbb{1}_{\overline{\Omega}}$, such as $K^n := \mathbb{1}_{[0,n]} \in \mathcal{E}$, $K^n \longrightarrow \mathbb{1}_{\overline{\Omega}}$.
Using $\mathcal{G} = \mathcal{E}$ and $\mathcal{H}$ as a closure of $\mathcal{E}$ with respect to bounded pointwise convergence in $B(\overline{\Omega})$ holds according to the monotone class theorem

$$\mathcal{H} \supset B(\sigma(\mathcal{E})) = B(\mathcal{P}).$$

Since μ is finite measure, holds $B(\mathcal{P}) \subset L^2(\mathcal{P}, \mu)$ is dense.

However, it follows that according to the definition of $\mathcal{H}$ that $\mathcal{E}$ is dense in $L^2(\mathcal{P}, \mu)$. □

Proposition 3.25

$$\begin{aligned} \mathcal{P} &= \sigma(\mathcal{E}) \\ &= \sigma\ (\textit{adapted left continuous processes}) \\ &= \sigma\ (\textit{continuous processes}) \end{aligned}$$

Proof Let $\tau_1 := \sigma(\mathcal{E}) = \mathcal{P}$,
$\tau_2 := \sigma$ (adapted left continuous processes), and
$\tau_3 := \sigma$ (adapted continuous processes).
To show: $\tau_1 = \tau_2 = \tau_3$.

(1) $\tau_3 = \tau_2$, clear.
(2) $\tau_1 \subset \tau_2$ since processes in $\mathcal{E}$ are left continuous
(3) $\tau_2 \subset \tau_1$: Let K be a left continuous process on $[0, \infty)$ and let $K_t^n(\omega) := X_0(\omega)\mathbb{1}_0(t) + \sum_{k=0}^{n2^n} K_{k/2^n}(\omega)\mathbb{1}_{(\frac{k}{2^n}, \frac{k-1}{2^n}]}(t) \in \mathcal{E}$.
Then it follows that $K_t^n \longrightarrow K_t$.
So K is $\sigma(\mathcal{E}) = \tau_1$ measurable and thus $\tau_2 \subset \tau_1$.
(4) $\tau_1 \subset \tau_3$: To $s < t$ there exist $f^n \in C_K$ with support in $(s, t + \frac{1}{n})$ so that $f^n \longrightarrow \mathbb{1}_{(s,t]}$ (Fig. 3.2).

So elements from $\mathcal{E}$ allow pointwise approximation by adapted continuous processes. From this the assertion follows. □

Remark 3.26 The statement of Proposition 3.20 can be extended to predictable (rather than continuous) local martingales. Thus it holds:

$$M \in \mathcal{M}_{\text{loc}} \cap \mathcal{V} \textit{ and } M \textit{ predictable} \Longrightarrow M = c\ [P].$$

Examples of predictable processes are constructed in the following lemma using stopping times.

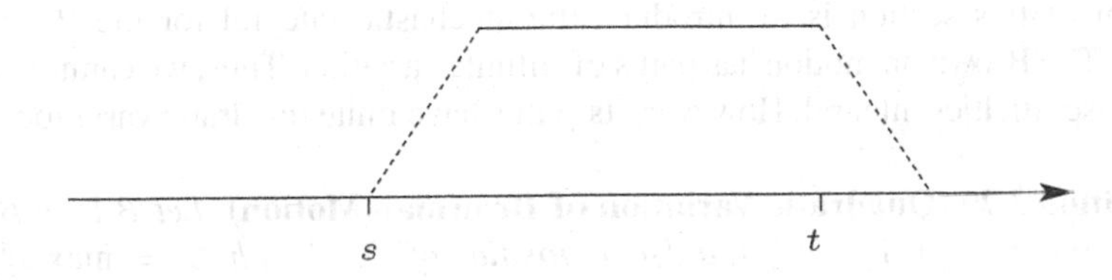

Fig. 3.2 Continuous approximation

Lemma 3.27 *Let $\sigma \leq \tau$ be stopping times and X be càdlàg, adapted. Then it holds:*

$$f_t(\omega) := \mathbb{1}_{[0,\tau(\omega)]}(t)$$
$$g_t(\omega) := \mathbb{1}_{(\sigma(\omega),\tau(\omega)]}(t)$$
$$h_t(\omega) := X_{\sigma(\omega)}(\omega)\mathbb{1}_{(\sigma(\omega),\tau(\omega)]}(t)$$

are predictable processes.

Proof f, g, h are left-continuous. To show is: "adapted".

First $\{f_t = 1\} = \{t \leq \tau\} \in \mathfrak{A}_t$, since τ is a stopping time.

Further: $\{g_t = 1\} = \{\sigma < t\} \cap \{t \leq \tau\} \in \mathfrak{A}_t$.

So f, g are adapted. Finally, we define for fixed t

$$\sigma_n(\omega) := \frac{[2^n\sigma(\omega) + 1]}{2^n} \wedge t.$$

On $\{\sigma < t\}$ holds: $\sigma_n \downarrow \sigma, \sigma_n \leq t$.
It follows that: $h_t(\omega) = \lim_n X_{\sigma_n(\omega)}\mathbb{1}_{\{\sigma(\omega)\leq t\}}\mathbb{1}_{\tau(\omega)\geq t\}}$ and $X_{\sigma_n(\omega)} \in \mathcal{L}(\mathfrak{A}_t), \forall n$. But this implies $h_t \in \mathcal{L}(\mathfrak{A}_t)$. □

Remark 3.28 $\mathcal{O} := \sigma$ (càdlàg processes) is called **optional σ-algebra** (σ-algebra of optional sets/processes)

$\mathcal{P}_r := \sigma$(progressive measurable processes) is called **progressive σ-algebra**.
Both are σ-algebras on $\overline{\Omega}$. It holds:

$\mathcal{P} \subset \mathcal{O} \subset \mathcal{P}_r \subset \mathfrak{B}_+ \otimes \mathfrak{A}$ (continuous $\rightarrow$ càdlàg $\rightarrow$ progressive $\rightarrow$ product measurable)

$\mathcal{P}$ and $\mathcal{O}$ can also be generated by suitable stochastic intervals. It holds

$\mathcal{P} = \sigma(\{[\![0, T]\!];\ T \text{ stopping times}\} \cup \widehat{\mathfrak{A}_0})$ where $\widehat{\mathfrak{A}_0} = \{\{0\} \times A;\ A \in \mathfrak{A}_0\}$

$\mathcal{O} = \sigma(\{[\![0, T[\![;\ T \text{ stopping times}\}) = \sigma(\{[\![S, T[\![;\ S, T \text{ stopping times}, S \leq T\})$.

3.2 Itô Integral for the Brownian Motion

The aim of this section is to introduce the stochastic integral for the Brownian motion. The Brownian motion has paths of infinite variation. Thus, we cannot define a pathwise Stieltjes integral. However, its paths have finite quadratic variation.

Proposition 3.29 (Quadratic Variation of Brownian Motion) *Let B be a Brownian motion, be (t_i^n), $1 \leq i \leq n$, a decomposition of $[0, T]$ with $\Delta_n = \max |t_{i+1}^n - t_i^n| \longrightarrow 0$.*

(a) For the paths of the Brownian motion

$$V^n := \sum_{i=0}^{n} |B_{t_{i+1}^n} - B_{t_i^n}| \longrightarrow \infty \text{ almost surely,}$$

i.e., the paths are P almost surely not of finite variation.

(b) The paths are P-a.s. of finite quadratic variation. It holds:

$$Q^n := \sum_{i=0}^{n} (B_{t_{i+1}^n} - B_{t_i^n})^2 \longrightarrow T \text{ almost surely.}$$

Proof

(a) is a well-known probabilistic property of a Brownian motion (cf. Rüschendorf (2016)).

(b) follows from (b): because $Q^n \leq \underbrace{\max |B_{t_{i+1}^n} - B_{t_i^n}|}_{\longrightarrow 0} V^n$.

Since $\max |B_{t_{i+1}^n} - B_{t_i^n}| \longrightarrow 0$ P a.s., it follows that $V^n \longrightarrow \infty$ P-a.s.

□

Remark 3.30 (Properties of Brownian Motion) The following properties of the Brownian motion are used frequently in the following:

- The paths of the Bronwian motion are P-almost surely not of finite variation.
- But the paths are P-almost surely of finite quadratic variation.
- The Brownian motion has nowhere differentiable paths and independent, normally distributed increments. It holds: $(B_t - B_s)_{t \geq s}, \mathfrak{A}_s$ are independent.
- A Brownian motion is a martingale.
- $(B_t^2 - t)$ is a martingale

First, we define the stochastic integral for elementary processes.

Definition 3.31 **(Stochastic Integral** (in several steps))

$$\text{Let} \quad \mathcal{E} = \left\{ f; f_t(\omega) = \sum_{j=1}^{k} U_j(\omega) \mathbb{1}_{(s_{j-1}, s_j]}(t), \right.$$
$$\left. 0 = s_0 < s_1 < \cdots < s_k, \; U_j \in B(\mathfrak{A}_{s_{j-1}}) \right\}$$

the set of elementary predictable processes.

(1) Integral for Elementary Processes *For $f \in \mathcal{E}$ define the stochastic integral*

$$\int_0^t f\,\mathrm{d}B := \sum_{j=1}^k U_j(B_{s_j \wedge t} - B_{s_{j-1} \wedge t}) = \sum_{j=1}^k U_j(B^t_{s_j} - B^t_{s_{j-1}}).$$

Basic properties of this integral are presented in the following proposition.

Proposition 3.32

(a) $f, g \in \mathcal{E} \Longrightarrow \int_0^t (f+g)\,\mathrm{d}B = \int_0^t f\,\mathrm{d}B + \int_0^t g\,\mathrm{d}B$.
(b) $f \in \mathcal{E} \Longrightarrow Y_t := \int_0^t f_s\,\mathrm{d}B_s$ *and* $Y_t^2 - \int_0^t f_s^2\,\mathrm{d}s$ *are continuous* $(\mathfrak{A}_t)$ *martingales.*
(c) $E\sup_{t \le T} \left| \int_0^t f_s\,\mathrm{d}B_s \right|^2 \le 4E(\int_0^T f_s^2\,\mathrm{d}s)$

Proof

(a) Follows directly from the definition and by using a refinement of the decomposition of the time interval.
(b) Continuity follows directly by definition.

$$\text{(1)}\quad E(Y_t \mid \mathfrak{A}_s) = E\left(\int_0^s f_u\,\mathrm{d}B_u + \int_s^t f_u\,\mathrm{d}B_u \mid \mathfrak{A}_s\right), \qquad \text{w.l.g. let } s,t \in \{s_j\}$$
$$= Y_s + \sum_{\substack{1 \le j \le n \text{ with} \\ s < s_j, s_{j+1} \le t}} E(U_{s_j}(B_{s_{j+1}} - B_{s_j}) \mid \mathfrak{A}_s).$$

Note that $E(U_{s_j}(B_{s_{j+1}} - B_{s_j}) \mid \mathfrak{A}_s) = E\,(U_{s_j}(B_{s_{j+1}} - B_{s_j}) \mid \mathfrak{A}_{s_j}) \mid \mathfrak{A}_s) = E(U_{s_j} E(B_{s_{j+1}} - B_{s_j} \mid \mathfrak{A}_{s_j}) \mid \mathfrak{A}_s) = 0$.

Therefore, by the martingale property of B, it follows that $E(Y_t \mid \mathfrak{A}_s) = Y_s$.

(2) $Y_t^2 = \sum_j U_j^2(B_{s_j \wedge t} - B_{s_{j-1} \wedge t})^2 + \sum_{j \ne k} U_j U_k (B_{s_j \wedge t} - B_{s_{j-1} \wedge t})(B_{s_k \wedge t} - B_{s_{k-1} \wedge t})$.

To show: $E(Y_t^2 \mid \mathfrak{A}_s) - E(\int_0^t f_u^2\,\mathrm{d}u \mid \mathfrak{A}_s) = Y_s^2 - \int_0^s f_u^2\,\mathrm{d}u$.
To do this, we first consider

1. **mixed term:** o.E. $s, t \in \{s_j\}$
If s_{j-1} or $s_{k-1} \ge t \Longrightarrow$ the bracket expressions are 0.
If s_{j-1} or $s_{k-1} \ge s \Longrightarrow E(\cdots \mid \mathfrak{A}_s) = 0$ because of the independence of the increments of B.

If $s_j, s_k \le s \Longrightarrow$ the bracket expressions are $\mathfrak{A}_s$ measurable $\Longrightarrow$ the term is preserved by conditioning.

Similarly, for
$\int_0^t f_u^2\,\mathrm{d}u = \sum_{s_j \le t} U_j(s_j \wedge t - s_{j-1} \wedge t) + \sum_{j \ne k} U_j U_k (s_j \wedge t - s_{j-1} \wedge t)(s_k \wedge t - s_{k-1} \wedge t)$,
it follows that for $s < s_{j-1}, s_{k-1} < t$ the terms are preserved.

2. **quadr. terms:** If $s_{j-1} \ge t$ then the quadratic term disappears.

(2a) $s \le a := s_{j-1} < b := s_j \le t$, $U = U_j \in \mathcal{L}(\mathfrak{A}_a)$. Then it holds:

$$E(U^2(B_b - B_a)^2 \mid \mathfrak{A}_s) = E(E(U^2 \underbrace{(B_b - B_a)^2 \mid \mathfrak{A}_a}_{\text{independent increments}}) \mid \mathfrak{A}_s)$$

$$= E(U^2(b-a) \mid \mathfrak{A}_s) \sim \text{fraction of } E\left(\int_o^t f_u^2 du \mid \mathfrak{A}_s\right)$$

(2b) $a < s < b \Longrightarrow$

$$\begin{aligned} E(U^2(B_b - B_a)^2 \mid \mathfrak{A}_s) &= E(U^2(B_b^2 - 2B_bB_a + B_a^2) \mid \mathfrak{A}_s) \\ &= E(U^2(B_s - B_a)^2 + (b-s) \mid \mathfrak{A}_s), \end{aligned}$$

as $B_t^2 - t$ is a martingale. This corresponds to the fraction of $E(U^2(s-a) + b - s \mid \mathfrak{A}_s)$.

(2c) $a = s_{j-1} < b = s_j \le s$. Then the term is measurable with respect to $\mathfrak{A}_s$.

From these cases statement (b) follows.

(c) (Y_t) is a continuous martingale. According to the Doob inequality then follows:

$$E \sup_{t \le T} Y_t^2 \le 4EY_T^2.$$

From (b), it follows that

$$E \sup_{t \le T} |Y_t|^2 \le 4EY_T^2 = 4E \int_0^T f_s^2 \, ds.$$

□

3.2.1 Extension of the Integral to L^2-Integrands

In the next step, the integral is extended to predictable L^2-integrands.

Definition 3.33

(a) Let $\mathcal{L}^2(B) := \left\{f \in \mathcal{L}(\mathcal{P}); E\int_0^T f_s^2 \, ds < \infty, \forall T < \infty\right\}$ be the class of **L^2-integrands** with respect to Brownian motion.

(b) Define the **Doléans measure** on $(\overline{\Omega}, \mathcal{P})$

$\mu(C) := E\int_0^\infty \mathbb{1}_C(s, \omega) \, ds$, $C \in \mathcal{P}$

$\mu_T(C) := E\int_0^T \mathbb{1}_C(s, \omega) \, ds$, $C \in \mathcal{P}$ is that on $[0, T] \times \Omega$ **restricted Doléans measure**, where the expected value with respect to P is used.

Remark 3.34

(a) $\mu_T \in M_e(\overline{\Omega}, \mathcal{P})$ is a finite measure. μ is a σ-finite measure and it holds

$$\mathcal{L}^2(\mu) = L^2(\mu, \mathcal{P}) \subset \bigcap_{T>0} L^2(\mu_T) = \mathcal{L}^2(B).$$

(b) $\mathcal{E} \subset \mathcal{L}^2(\overline{\Omega}, \mathcal{P}, \mu_T)$ is dense according to Proposition 3.24, $\forall\, T > 0$.
For $f \in \mathcal{L}^2(B) = \bigcap \mathcal{L}^2(\mu_T)$, therefore, there exist $f^T \in \mathcal{E}$ so that

$$\int_0^T |f_s^T - f_s|^2 \, d\mu_T = E \int_0^T |f_s^T - f_s|^2 \, ds \le 2^{-2T}, \quad \forall\, T.$$

Define:

$$Y_t^N := \int_0^t f^N \, dB.$$

Y^N is a continuous process.
Assertion: $\exists$ a continuous adapted process $Y = (Y_t)$ with

$$E \sup_{t \le T} |Y_t^N - Y_t|^2 \underset{N \to \infty}{\longrightarrow} 0, \quad \forall\, T > 0. \tag{3.2}$$

Proof By Proposition 3.32 holds

$$\begin{aligned} E \sup_{t \le T} |Y_t^{N+1} - Y_t^N|^2 &\le 4E \int_0^T |f_s^{N+1} - f_s^N|^2 \, ds \\ &\le 4 \cdot 2 \;\; E \int_0^T \big((f_s^{N+1} - f_s)^2 + (f_s^N - f_s)^2\big) \, ds \\ &\le 8(2^{-2(N+1)} + 2^{-2N}) \quad \le 16 \cdot 2^{-2N}. \end{aligned}$$

With the above estimate we get $E \sum_{N \ge 1} \sup_{t \le T} |Y_t^{N+1} - Y_t^N| < \infty$.
This implies the a.s. finiteness of the infinite series

$$\sum_{N \ge 1} \sup_{t \le T} |Y_t^{N+1} - Y_t^N|^2 < \infty \, [P].$$

So there is a null set A and a continuous process Y such that $\forall\, \omega \in A^c$:

$$Y_t^N(\omega) \longrightarrow Y_t(\omega) \text{ evenly on } [0, T], \quad \forall\, T < \infty.$$

It is enough to consider a sequence $T_n \uparrow \infty$ and to unite the countably many zero-sets. Thus it holds:

$$E \sup_{t\leq T} |Y_t^N - Y_t|^2 \longrightarrow 0, \quad \forall T.$$

(2) Integral for $\mathcal{L}^2(B)$ Define for $f \in \mathcal{L}^2(B) = \bigcap_{T>0} \mathcal{L}^2(\mu_T)$ the **stochastic integral**

$$\int_0^t f\,\mathrm{d}B := \lim_{N\longrightarrow\infty} \int_0^t f^N\,\mathrm{d}B = Y_t.$$

To show: the definition is independent of the approximating sequence.

Let $(g^N) \in \mathcal{E}$ be such that $E\int_0^N |g_s^N - f_s|^2\,\mathrm{d}s \longrightarrow 0$ (convergence in $\mathcal{L}^2(\mu_N)$). Then it holds:

$$\begin{aligned} E \sup_{t\leq T} \left| \int_0^t g_s^N\,\mathrm{d}B_s - \int_0^t f_s^N\,\mathrm{d}B \right|^2 &= E \sup_{t\leq T} \left| \int_0^t (g_s^N - f_s^N)\,\mathrm{d}B \right|^2 \\ &\leq 4E \int_0^T (g_s^N - f_s^N)^2\,\mathrm{d}s \quad \longrightarrow 0. \end{aligned}$$

The stochastic integral $\int_0^t f\,\mathrm{d}B$ in (2) is thus uniquely defined. □

Proposition 3.35 *The for $f \in \mathcal{L}^2(B) = \bigcap_{T>0} \mathcal{L}^2(\mu_T)$ defined process (Y_t) is called stochastic integral from f with respect to Brownian motion B,*

$$Y_t := \int_0^t f_s\,\mathrm{d}B_s.$$

The properties (a)–(c) from Proposition 3.32 hold true.

Proof Y is a martingale as L^1 limit of the martingale Y^N. The properties (a)–(c) transfer with the approximation in (3.2). □

Remark 3.36

(a) $Y_t^2 - \int_0^t f_s^2\,\mathrm{d}s \in \mathcal{M}$ is a martingale. It follows that

$$EY_t^2 = E\int_0^t f_s^2\,\mathrm{d}s = \|f\|_{2,\mu_t}^2.$$

(b) Let $\mathcal{H}^2 := \{Z \in \mathcal{M};\ \underbrace{\sup_{t<\infty} EZ_t^2}_{=EZ_\infty^2} < \infty\}$ be the set of **L^2-bounded martingales.**

Then by Doob's convergence theorem $Z_t \longrightarrow Z_\infty$ a.s. in L^2 and it holds $EZ_\infty^2 = \sup_t EZ_t^2$. We now define on $\mathcal{H}^2$ the norm

$$\|Z\|_{\mathcal{H}^2} := (EZ_\infty^2)^{1/2}.$$

Then $(\mathcal{H}^2, \|\cdot\|_{\mathcal{H}^2})$ is a Hilbert space. The stochastic integral I on $\mathcal{L}^2(\mu)$,

$$\mathcal{L}^2(\mu) \xrightarrow{I} \mathcal{H}^2, \ with\ f \longrightarrow Y := \int_0^\cdot f\,\mathrm{d}B$$

is an isometry between Hilbert spaces. This is also true for

$$\mathcal{L}^2(\mu_T) \xrightarrow{I} \mathcal{H}_T^2 := \mathcal{M}^2([0,T])\ with\ \|Y\|_{2,\mu_T} = (EY_T^2)^{1/2}.$$

This provides an alternative natural way to obtain the stochastic integral for integrands in the smaller class $\mathcal{L}^2(\mu)$ to be defined as follows:

Step 1: Define $I(f) = \int f\,\mathrm{d}B$ for $f \in \mathcal{E}$ as before. I is an isometry of $\mathcal{E} \longrightarrow \mathcal{H}^2$.

Step 2: $\mathcal{E}$ is dense in $\mathcal{L}^2(\mu)$. Therefore, I has a unique continuation as isometry on $\mathcal{L}^2(\mu)$ that **stochastic integral** on $\mathcal{L}^2(\mu)$.

For a further extension of the notion of integral, the connection of integration with stopping times is of interest.

Proposition 3.37 *Let $f \in \mathcal{L}^2(B)$, $Z_t := \int_0^t f\,\mathrm{d}B$ and σ a stopping time. Then it holds:*

$$\int_0^t f\mathbb{1}_{[0,\sigma]}\,\mathrm{d}B = Z_{t\wedge\sigma} = Z_t^\sigma = \int_0^{t\wedge\sigma} f\,\mathrm{d}B. \tag{3.3}$$

Proof

(1) It suffices to show the statement for bounded stopping times.

For $\sigma_n := \sigma \wedge n \longrightarrow \sigma$. Thus by majorized convergence $Z_{t\wedge\sigma_n} = \int_0^t \underbrace{f\mathbb{1}_{[0,\sigma_n]}}_{=:f_n}\,\mathrm{d}B$ converges, because due to $f_n \longrightarrow f1_{[0,\sigma]}$ in $\mathcal{L}^2(\mu_T)$ it follows that $\int_0^t f_n\mathrm{d}B \longrightarrow \int_0^t f1_{[0,\sigma]}\,\mathrm{d}B$ in $L^2(P)$ and $Z_{t\wedge\sigma_n} \longrightarrow Y_{t\wedge\sigma}$. Thus the representation in (3.2) for σ follows.

(2) Let $f \in \mathcal{E}$.

(a) If the stopping time σ has only finitely many values, the assertion follows by definition of the integral for $f \in \mathcal{E}$, since $f \mathbb{1}_{[0,\sigma]} = \sum f \mathbb{1}_{[0,\sigma_i]} \mathbb{1}_{\sigma=\sigma_i\}} \in \mathcal{E}$.

(b) If σ is a bounded stopping time, then define $\sigma_m := \frac{\lceil 2^m \sigma \rceil}{2^m}$. The stopping time σ_m has finitely many values and $\sigma_m \downarrow \sigma$. So (3.2) for σ_m.

Since $f^m := f \mathbb{1}_{[0,\sigma_m]} \longrightarrow f \mathbb{1}_{[0,\sigma]}$ in $\mathcal{L}^2(\mu_T)$ (3.2) also holds for σ.

(c) For $f \in \mathcal{L}^2(B)$ is $\int_0^t f \, \mathrm{d}B = \lim \int_0^t \underbrace{f^N}_{\in \mathcal{E}} \, \mathrm{d}B$ and it holds:

$$E \sup_{t \leq T} \Big| \underbrace{\int_0^t f \, \mathrm{d}B}_{=:Z_t} - \underbrace{\int_0^t f^N \, \mathrm{d}B}_{=:Z_t^N} \Big|^2 \longrightarrow 0.$$

If $\sigma \leq T$ then we get as a conclusion

$$E |Z_{t \wedge \sigma} - Z_{t \wedge \sigma}^N|^2 \longrightarrow 0.$$

Then, according to (b):

$$Z_{t \wedge \sigma}^N = \int_0^t f^N \mathbb{1}_{[0,\sigma]} \, \mathrm{d}B \longrightarrow \int_0^t f \mathbb{1}_{[0,\sigma]} \, \mathrm{d}B = Z_{t \wedge \sigma}.$$

□

For a bounded stopping time $\sigma \leq T$ we define:

(3) Integral and Stopping Times

$$\int_0^\sigma f \, \mathrm{d}B := \int_0^T f \mathbb{1}_{[0,\sigma]} \, \mathrm{d}B, \quad \forall f \in \mathcal{L}^2(B).$$

Remark 3.38

(a) Extension to adapted integrands.

With $\hat{P} := \lambda_{[0,\infty)} \otimes P$ let

$\widehat{\mathcal{L}^2}(B) := \{g$ adapted; there exists $f \in \mathcal{L}^2(B)$ such that $\widehat{P}(\{f \neq g\}) = 0\}$.

According to the **predictable projection theorem** (an elaborate measure-theoretic theorem) holds the equality

$$\widehat{\mathcal{L}^2}(B) = \Big\{ g \text{ adapted}; E \int_0^T g_s^2 \, \mathrm{d}s < \infty, \forall T \Big\}$$

$$= \widehat{\mathcal{L}^2}(\widehat{P}) \text{ the set of adapted processes in } \mathcal{L}^2(\widehat{P}).$$

This equality allows the definition of the stochastic integral for adapted integrands. For this class of functions, Itô introduced the stochastic integral in his original paper.

(b) Let $f, g \in \mathcal{L}^2(B)$ and σ be a stopping time and assume that: $P(f_t = g_t, \forall t \leq \sigma) = 1$.
Let $Y_t := \int_0^t f \, dB$, $Z_t := \int_0^t g \, dB$, then it holds:

$$Y_t^\sigma = \int_0^t f \mathbb{1}_{[0,\sigma]} \, dB = \int_0^t g \mathbb{1}_{[0,\sigma]} \, dB = Z_t^\sigma \textit{ almost surely}$$

and furthermore $P(Y^\sigma = Z^\sigma) = 1$, since the processes are continuous.

3.2.2 Construction of the Integral for $\mathcal{L}^0(B)$

We now construct the extension of the integral to the final integration class $\mathcal{L}^0(B)$.

$$\mathcal{L}^0(B) := \left\{ f \in \mathcal{L}(\mathcal{P}); \int_0^t f_s^2 \, ds < \infty \text{ almost surely, } \forall t < \infty \right\}.$$

Let $f \in \mathcal{L}^0(B)$ and $\sigma^n := \inf\{t \geq 0; \int_0^t f_s^2 \, ds \geq n\}$. Then

$$\sigma^n \uparrow \infty, \text{ the sequence of stopping times converges to } \infty.$$

With $f_t^n := f_t \mathbb{1}_{[0,\sigma^n]}(t)$ is $\int_0^T (f_t^n)^2 \, dt \leq n$ for all T, so $f^n \in \mathcal{L}^2(B)$.

Let $Z_t^n := \int_0^t f^n \, dB \in \mathcal{M}_c$. By Remark 3.38 (*b*) with $\sigma := \sigma^n$, $f := f^n$, $g := f^{n+1}$ it holds

$$P(Z_t^n = Z_{t \wedge \sigma^n}^{n+1}, \forall t) = 1.$$

Thus

$$Z_t(\omega) := \begin{cases} Z_t^n(\omega), & \sigma^{n-1}(\omega) < t \leq \sigma^n(\omega) \\ 0, & \text{other} \end{cases} \tag{3.4}$$

is well-defined and according to Proposition 3.37 it holds:

$$Z_{t \wedge \sigma^n} = Z_t^n = Z_t^{\sigma^n}.$$

Therefore, $Z = (Z_t) \in \mathcal{M}_{\mathrm{loc},c}$ with localizing sequence (σ^n).

Further

$$X_t := Z_t^2 - \int_0^t f_s^2 \, ds \in \mathcal{M}_{\mathrm{loc},c},$$

as $X_{t\wedge\sigma^n} = (Z_t^n)^2 - \int_0^t (f_s^n)^2\,\mathrm{d}s \in \mathcal{M}_c$.

For a stopping time σ such that $f\mathbb{1}_{[0,\sigma]} \in \mathcal{L}^2(B)$ then it follows according to Proposition 3.37

$$Z_{t\wedge\sigma}^n = \int_0^t f\mathbb{1}_{[0,\sigma]}\mathbb{1}_{[0,\sigma^n]}\,\mathrm{d}B$$

It follows,

$$P\Big(Z_{t\wedge\sigma} = \int_0^t f\mathbb{1}_{[0,\sigma]}\,\mathrm{d}B, \forall t\Big) = 1. \tag{3.5}$$

Because by (3.4) it holds:

$$\begin{aligned} Z_{t\wedge\sigma} &= Z_{t\wedge\sigma}^n && \text{if } \sigma^{n-1} < t\wedge\sigma \le \sigma^n, \\ &= \int_0^t f\mathbb{1}_{[0,\sigma]}\mathbb{1}_{[0,\sigma^n]}\,\mathrm{d}B && \text{as } t\wedge\sigma \le \sigma^n \\ &= \int_0^t f\mathbb{1}_{[0,\sigma]}\,\mathrm{d}B, \end{aligned}$$

independent of n and well-defined, since $f\mathbb{1}_{[0,\sigma]} \in \mathcal{L}^2(B)$.

Thus we define for $f \in \mathcal{L}^0(B)$:

(4) Integral for $\boldsymbol{\mathcal{L}^0(B)}$ The process (Z_t) defined in (3.4) is called the stochastic integral of f,

$$Z_t =: \int_0^t f\,\mathrm{d}B.$$

Proposition 3.39 *Let $f, g \in \mathcal{L}^0(B)$, then it holds:*

(a) $\int_0^t (f+g)\,\mathrm{d}B = \int_0^t f\,\mathrm{d}B + \int_0^t g\,\mathrm{d}B$

(b) $Y_t := \int_0^t f\,\mathrm{d}B \in \mathcal{M}_{\mathrm{loc},c}$ *and* $X_t := Y_t^2 - \int_0^t f_s^2\,\mathrm{d}s \in \mathcal{M}_{\mathrm{loc},c}$

(c) $\forall\lambda > 0, \varepsilon > 0$ *holds:*

$$P\Big(\sup_{t\le T}\Big|\int_0^t f\,\mathrm{d}B\Big| > \lambda\Big) \le 4\frac{\varepsilon}{\lambda^2} + P\Big(\int_0^t f_s^2\,\mathrm{d}s > \varepsilon\Big)$$

(d) If for $f^n, f \in \mathcal{L}^0(B)$, $\int_0^T |f_s^n - f_s|^2\, \mathrm{d}s \xrightarrow{P} 0$, then it follows:

$$\sup_{0\le t\le T}\left|\int_0^t f_s^n\,\mathrm{d}B_s - \int_0^t f_s\,\mathrm{d}B_s\right| \xrightarrow{P} 0.$$

Thus uniform stochastic convergence on compacta holds.

(e) Let $Y_t = \int_0^t f\,\mathrm{d}B$ and σ be a stopping time, then it holds:

$$\int_0^t f\mathbb{1}_{[0,\sigma]}\,\mathrm{d}B = Y_{t\wedge\sigma}.$$

Proof

(a), (b) follows from the definition with localizing sequence σ^n.
(c) Let $\sigma := \inf\{t \ge 0;\ \int_0^t f_s^2\,\mathrm{d}s \ge \varepsilon\}$ then $f\mathbb{1}_{[0,\sigma]} \in \mathcal{L}^2(B)$.

$$P\Bigg(\sup_{t\le T}\bigg|\underbrace{\int_0^t f\,\mathrm{d}B}_{=:Z_t}\bigg| > \lambda\Bigg) \le P(\sigma < T) + P(\sup_{t\le T\wedge\sigma}|Z_t| > \lambda, \sigma \ge T)$$

$$\le P(\sigma < T) + P(\sup_{t\le T}|Z_{t\wedge\sigma}| > \lambda)$$

$$\le P(\sigma \le T) + P\bigg(\sup_{t\le T}\left|\int_0^t f\mathbb{1}_{[0,\sigma]}\,\mathrm{d}B\right| > \lambda\bigg)$$

using the Doob inequality, Theorem 3.13

$$\le P(\sigma \le T) + \frac{1}{\lambda^2}E\sup_{t\le T}\bigg|\int_0^t \underbrace{f\mathbb{1}_{[0,\sigma]}}_{\in\mathcal{L}^2(B)}\,\mathrm{d}B\bigg|^2$$

with Proposition 3.32 or 3.35

$$\le P(\sigma \le T) + 4\frac{1}{\lambda^2}E\int_0^{T\wedge\sigma} f_s^2\,\mathrm{d}s$$

by definition of σ

$$\le P(\sigma \le T) + 4\frac{\varepsilon}{\lambda^2}$$

$$= P\bigg(\int_0^T f_s^2\,\mathrm{d}s > \varepsilon\bigg) + 4\frac{\varepsilon}{\lambda^2}.$$

(d) follows from (c).

(e) Let $Y_t^n := \int_0^t \underbrace{f \mathbb{1}_{[0,\sigma^n]}}_{\in \mathcal{L}^2(B)} \mathrm{d}B$ be an approximating sequence. Then

$$Y_{t\wedge\sigma}^n = \int_0^t \underbrace{f \mathbb{1}_{[0,\sigma]} \mathbb{1}_{[0,\sigma^n]}}_{\in \mathcal{L}^2(B)} \mathrm{d}B.$$

It holds $Y_{t\wedge\sigma}^n \longrightarrow Y_{t\wedge\sigma}$ and $\int_0^t f \mathbb{1}_{[0,\sigma]} \mathbb{1}_{[0,\sigma^n]} \mathrm{d}B \longrightarrow \int_0^t f \mathbb{1}_{[0,\sigma]} \mathrm{d}B$ according to the definition of the integral. But from this follows the assertion.

□

Lemma 3.40 *Let* $U \in L(\mathfrak{A}_u)$, $u < r$. *Define:* $h_t := U \mathbb{1}_{(u,r]}(t)$ ($\notin \mathcal{E}$, *since* U *is not bounded*).

Then

$$h \in \mathcal{L}^0(B) \text{ and } \int_0^t h \,\mathrm{d}B = U(B_{r\wedge t} - B_{u\wedge t}).$$

Proof Let $h_t^n := U \mathbb{1}_{(U\le n)} \mathbb{1}_{(u,r]}(t)$. Then $h^n \in \mathcal{E}$ and

$$\int_0^t h_t^n \,\mathrm{d}B = U \mathbb{1}_{(U\le n)}(B_{r\wedge t} - B_{u\wedge t}).$$

It holds: $\int_0^t |h_s^n - h_s|^2 \,\mathrm{d}s = U^2 \mathbb{1}_{(U>n)} \int_0^t \mathbb{1}_{(u,r]}(s) \,\mathrm{d}s \xrightarrow{P} 0$. According to Proposition 3.39 the assertion follows. □

We can now derive the following approximation result.

Theorem 3.41 (Riemann Approximation) *Let* f *be a square-integrable, left continuous, adapted process in the Riemann sense, then it holds with* $\Delta = \{0 = s_0 < s_1 < \cdots < s_k = t\}$, $|\Delta| := \max |s_{i+1} - s_i|$

$$\int_0^t f \,\mathrm{d}B = \lim_{|\Delta|\longrightarrow 0} \sum_{i=0}^{k-1} f_{s_i}(B_{s_{i+1}} - B_{s_i}) \quad \textit{(stoch.convergence)}.$$

Proof Let $f^\Delta := \sum_{i=1}^k f_{s_i} \mathbb{1}_{(s_i,s_{i+1}]}$ (the f_{s_i} are not bounded). Then by Lemma 3.40 $\int_0^t f^\Delta \,\mathrm{d}B = \sum_i f_{s_i}(B_{s_{i+1}} - B_{s_i})$. According to the properties of the usual Riemann Integrals holds $\int_0^t |f_s^\Delta - f_s|^2 \,\mathrm{d}s \xrightarrow{P} 0$ and thus according to Proposition 3.39 (d) the assertion follows. □

Example 3.42 It holds $B \in \mathcal{L}^2(B)$ and

$$\int_0^T B_s \,\mathrm{d}B_s = \frac{1}{2}(B_T^2 - T).$$

Proof Let $\Delta_n = \{0 = t_0^n < \cdots < t_k^n\}$, $k = 2^n$, $t_i^n = \frac{i}{2^n}T$. With Proposition 3.41 via the Riemann approximation holds:

$$\sum_{i=0}^{2^n-1} \underbrace{B_{t_i^n}}_{=:a} (\underbrace{B_{t_{i+1}^n}}_{=:b} - \underbrace{B_{t_i^n}}_{=:a}) \xrightarrow{P} \int_0^T B_s \,\mathrm{d}B_s.$$

Because of $a(b-a) = \frac{1}{2}(b^2 - a^2 - (b-a)^2)$ the left side is equal:

$$\frac{1}{2}\sum_{i=0}^{2^n-1} \left(B_{t_{i+1}^n}^2 - B_{t_i^n}^2 - (B_{t_{i+1}^n} - B_{t_i^n})^2\right)$$

$$= \frac{1}{2}(B_T^2 - Q_T^n) \text{ with } Q_T^n := \sum_{i=0}^{2^n-1}(B_{t_{i+1}^n} - B_{t_i^n})^2 \longrightarrow T.$$

The right side is just the quadratic variation of B: $\langle B\rangle_T = T$.

We thus obtain: $\displaystyle\int_0^T B_s \,\mathrm{d}B_s = \frac{1}{2}(B_T^2 - T)$.

As a corollary this implies:

$$B_t^2 = 2\int_0^t B_s \,\mathrm{d}B_s + t$$

is the decomposition of the submartingale B_t^2 into a martingale and a predictable process of finite variation. □

We now introduce infinite integrals to conclude this section.

Theorem 3.43 *Let $f \in \mathcal{L}(\mathcal{P})$, $\int_0^\infty f_s^2 \,\mathrm{d}s < \infty$ almost surely. Then there exists a random variable Z_∞ such that*

$$Z_t := \int_0^t f \,\mathrm{d}B \xrightarrow{P} Z_\infty \quad \textit{for } t \longrightarrow \infty.$$

We define the infinite integral by

(5) Integral for Infinite Intervals $\int_0^\infty f\,\mathrm{d}B := Z_\infty.$

Proof Let $t \geq u$ and $g := f\mathbb{1}_{[u,\infty)}$, then

$$\int_0^t g\,\mathrm{d}B = Z_t - Z_u = \int_0^t f\,\mathrm{d}B - \int_0^u f\,\mathrm{d}B.$$

From Proposition 3.39 (c) we get

$$P(|Z_t - Z_u| > \delta) \leq 4\frac{\varepsilon}{\delta^2} + P\Big(\int_u^t f_s^2\,\mathrm{d}s > \varepsilon\Big).$$

To $\varepsilon > 0, \delta > 0$ be $\eta > 0$ such that $\frac{\varepsilon}{\delta^2} < \frac{\eta}{8}$. Then choose $t_0 > 0$ such that $\forall u$ where $t_0 < u < t$:

$$P\Big(\int_u^t f_s^2\,\mathrm{d}s > \varepsilon\Big) \leq \frac{\eta}{2}.$$

Then it follows

$$P(|Z_t - Z_u| > \delta) \leq \eta, \quad \forall u \text{ with } t_0 < u \leq t.$$

Then by the Cauchy criterion for stochastic convergence there exists a limit Z_∞,

$$Z_t \xrightarrow{P} Z_\infty \text{ for } t \longrightarrow \infty.$$

□

This allows the integral to be introduced for arbitrary stopping boundaries.

Remark 3.44 Let τ be a stopping time, $f \in \mathcal{L}(\mathcal{P})$, $\int_0^\tau f_s^2\,\mathrm{d}s < \infty$, then it holds:

$$\int_0^t f\mathbb{1}_{[0,\tau]}\,\mathrm{d}B \xrightarrow{P} \int_0^\infty f\mathbb{1}_{[0,\tau]}\,\mathrm{d}B.$$

So for a stopping time τ and for $f \in \mathcal{L}(\mathcal{P})$ such that

$$\int_0^\tau f_s^2\,\mathrm{d}s < \infty \tag{3.6}$$

we now define

(6) Integral for Pathwise $\mathcal{L}^2$-Processes $\int_0^\tau f\,\mathrm{d}B := \int_0^\infty f\mathbb{1}_{[0,\tau]}\,\mathrm{d}B.$

The condition (3.6) holds for $f \in \mathcal{L}^0(B)$ if τ is a finite stopping time, or if $\int_0^\tau f_s^2 \,\mathrm{d}s < \infty$ and τ is any stopping time.

We now obtain the optional sampling theorem for stochastic integrals as a consequence of the above considerations.

Theorem 3.45 (Optional Sampling Theorem for Local Martingales) *Let $f \in \mathcal{L}^0(B)$, $Y_t := \int_0^t f_s \,\mathrm{d}B_s$ and τ be a stopping time with $E \int_0^\tau f_s^2 \,\mathrm{d}s < \infty$. Then it follows:*

$$EY_\tau = 0, \qquad EY_\tau^2 = E \int_0^\tau f_s^2 \,\mathrm{d}s < \infty.$$

If in particular $E\tau < \infty$ then it holds with $f \equiv 1$:

$$EB_\tau = 0, \qquad EB_\tau^2 = E\tau.$$

Proof Let $Z_t := \int_0^t \underbrace{f \mathbb{1}_{[0,\tau]}}_{\in \mathcal{L}^2(B)} \,\mathrm{d}B$, then by Theorem 3.43 and by the definition in (6) the following holds:

$$I := \int_0^\tau f \,\mathrm{d}B = \lim_{t \longrightarrow \infty} Z_t \quad \text{(stochastic limit)}.$$

It suffices to show: $Z_t \xrightarrow{L^2} I$, because:

$$EZ_t = 0, \qquad EZ_t^2 = 1 = E \int_0^{t \wedge \tau} f_s^2 \,\mathrm{d}s.$$

With L^2-convergence it follows:

$$EI = 0, \qquad EI^2 = \int_0^\tau f_s^2 \,\mathrm{d}s$$

thus the assertion. It holds with majorized convergence:

$$E(Z_t - Z_s)^2 = E \int_s^t f_u^2 \mathbb{1}_{[0,\tau]}(u) \,\mathrm{d}u = E \int_{s \wedge \tau}^{t \wedge \tau} f_u^2 \,\mathrm{d}u \overset{s,t \to \infty}{\longrightarrow} 0.$$

So (Z_t) is a Cauchy net in L^2 and, therefore, there exists a limit in I of Z_t in L^2,

$$Z_t \xrightarrow{L^2} I.$$

□

Remark 3.46 (Integral Construction) For the construction of the stochastic integral with respect to Brownian motion, the maximal inequality for martingales and the martingale property of (B_t) as well as of $(B_t^2 - t)$ are used. For continuous martingales $M \in \mathcal{M}_c$ this process can be carried out analogously.

Let $M \in \mathcal{M}_c^2$ and let $(M_t^2 - A_t) \in \mathcal{M}_c$ with $A \in \mathcal{V}_+$ then the stochastic integral $\int_0^t f \,\mathrm{d}M$ can be constructed analogously to that for the Brownian motion if $E \int_o^T f_s^2 \,\mathrm{d}A_s < \infty$. For this class of processes, the integral was introduced by Kunita and Watanabe (1967). We will generalize the definition of the stochastic integral to this class of integrators in the following section.

3.3 Quadratic Variation of Continuous Local Martingales

For processes $A \in \mathcal{V}$ and adapted locally bounded progressively measurable integrands X, the integral $(X \cdot A)_t := \int_0^t X_s \,\mathrm{d}A_s$ is defined as a stochastic Stieltjes integral and it holds:

$$X \cdot A \text{ is adapted and } X \cdot A \in \mathcal{V}.$$

If X is càdlàg, then the above conditions are satisfied. X is locally bounded and progressively measurable. Our goal is to find the integral $X \cdot M$ for continuous martingales M. As for Brownian motion, this requires finiteness of the quadratic variation of M.

Definition 3.47 $X = (X_t)$ is called a process of **finite quadratic variation**, if $\exists\, A \in \mathcal{V}^+$ such that $\forall t > 0$

$$T_t^\Delta := T_t^\Delta(X) = \sum_i (X_{t_{i+1}\wedge t} - X_{t_i \wedge t})^2 \xrightarrow{P} A_t,$$

for all decompositions (t_i) of $\mathbb{R}_+$ with $|\Delta| \longrightarrow 0$.

$A_t := [X]_t$ is called quadratic variation process of X.

Remark 3.48 For the Brownian motion, the quadratic variation is

$$[B]_t = t \qquad \textit{and it holds} \qquad B_t^2 - t \in \mathcal{M}.$$

B is not of finite variation but of finite quadratic variation.

The following basic theorem ensures finite quadratic variation of continuous bounded martingales.

Theorem 3.49

(a) Every continuous bounded martingale M is of finite quadratic variation and $[M] \in \mathcal{V}_c^+$, i.e., the variation process is nonnegative, increasing, and continuous.

(b) There exists a unique continuous adapted process $A \in \mathcal{V}^+$ with $A_0 := 0$ such that

$$M^2 - A \in \mathcal{M}_c.$$

$A =: \langle M \rangle$ is called the ***predictable quadratic variation*** *of M. Uniqueness also holds in the class $\mathcal{P} \cap \mathcal{V}^+$ of predictable and increasing processes.*

(c) It holds: $[M] = \langle M \rangle$.

Proof We first prove uniqueness in (b).

(b) **Uniqueness**

Let $A, B \in \mathcal{V}_c^+$, $A_0 = B_0 = 0$ and $M^2 - A,\ M^2 - B \in \mathcal{M}_c$. Then it follows $A - B \in \mathcal{M}_c \cap \mathcal{V}$, and $A_0 = B_0 = 0$. According to Proposition 3.20 it follows that $A = B$. Uniqueness also holds in the class of predictable elements in $\mathcal{V}^+$.

(a) **Existence and finiteness of quadratic variation:**

The proof is divided into several steps.

1st step:

$M_t^2 - T_t^\Delta(M) \in \mathcal{M}_c$

proof Let $t_i < s \le t_{i+1}$, then it follows :

$E\big((M_{t_{i+1}} - M_{t_i})^2 \mid \mathfrak{A}_s\big) = E\big((M_{t_{i+1}} - M_s)^2 \mid \mathfrak{A}_s\big) + (M_s - M_{t_i})^2.$

The mixed term is dropped because of the martingale property.

Let $s < t$, $t_i < s \le t_{i+1} < \cdots < t_{i+k} < t \le t_{i+k+1}$. Then it follows

$$\begin{aligned}
&E\big(T_t^\Delta(M) - T_s^\Delta(M) \mid \mathfrak{A}_s\big) \\
&= \sum_{j=1}^{\infty} E\big((M_{t_j\wedge t} - M_{t_{j-1}\wedge t})^2 - (M_{t_j\wedge s} - M_{t_{j-1}\wedge s})^2 \mid \mathfrak{A}_s\big) \\
&= \sum_{l=0}^{k} E\big((M_{t_{i+l+1}\wedge t} - M_{t_{i+l}\wedge t})^2 - (M_{t_{i+l+1}\wedge s} - M_{t_{i+l}\wedge s})^2 \mid \mathfrak{A}_s\big) \\
&= \sum_{l=1}^{k} E\big((M_{t_{i+l+1}\wedge t} - M_{t_{i+l}\wedge t})^2 \mid \mathfrak{A}_s\big) \\
&\qquad + E\big((M_{t_{i+1}} - M_{t_i})^2 - (M_s - M_{t_i})^2 \mid \mathfrak{A}_s\big) \\
&= E(M_t^2 - M_{t_{i+1}}^2 + (M_{t_{i+1}}^2 - M_s^2) \mid \mathfrak{A}_s) \\
&\qquad \text{because of the martingale property} \\
&= E\big(M_t^2 - M_s^2 \mid \mathfrak{A}_s\big).
\end{aligned} \tag{3.7}$$

From this follows $M_t^2 - T_t^\Delta(M) \in \mathcal{M}_c$.

2. Second step:

In the second step, our goal is to show that for $a > 0$ and a decomposition Δ_n of $[0, a]$ with $|\Delta_n| \longrightarrow 0$ it holds that $T_a^{\Delta_n}$ has a limit in L^2:

$$T_a^{\Delta_n} \xrightarrow{L^2} [M]_a.$$

Assertion: $(T_a^{\Delta_n})$ is a Cauchy sequence in L^2.
For this let Δ, Δ' be decompositions. Then according to (3.7) $X := T^\Delta - T^{\Delta'} \in \mathcal{M}_c$. From this it follows by the first step applied to X

$$EX_a^2 = E(T_a^\Delta - T_a^{\Delta'})^2 = ET_a^{\Delta\Delta'}(X),$$

where $\Delta\Delta'$ denotes the joint refinement of Δ and Δ'. Thus it holds:

$$EX_a^2 \le 2 \cdot E\big(T_a^{\Delta\Delta'} T^\Delta + T_a^{\Delta\Delta'} T^{\Delta'}\big),$$

using $(x + y)^2 \le 2(x^2 + y^2)$. So it suffices to show:

(a) $ET_a^{\Delta\Delta'} T^\Delta \longrightarrow 0$ for $|\Delta| + |\Delta'| \longrightarrow 0$.

Let $s_k \in \Delta\Delta'$ and $t_l \in \Delta$ be the largest point in Δ before s_k such that $\underbrace{t_l}_{\in\Delta} \le s_k \le s_{k+1} \le \underbrace{t_{l+1}}_{\in\Delta}$. Then it holds:

$$\begin{aligned} T_{s_{k+1}}^\Delta - T_{s_k}^\Delta &= (M_{s_{k+1}} - M_{t_l})^2 - (M_{s_k} - M_{t_l})^2 \\ &= (M_{s_{k+1}} - M_{s_k})(M_{s_{k+1}} + M_{s_k} - 2M_{t_l}). \end{aligned}$$

It follows that:

$$T_a^{\Delta\Delta'} T^\Delta \le \sup_k |M_{s_{k+1}} + M_{s_k} - 2M_{t_l}|^2 T_a^{\Delta\Delta'}(M).$$

By Cauchy–Schwarz it follows that:

$$ET_a^{\Delta\Delta'} T^\Delta \le \underbrace{\big(E \sup_k |M_{s_{k+1}} + M_{s_k} - 2M_{t_l}|^2\big)^{\frac{1}{2}}}_{=:a_{\Delta\Delta'}} \big(E(T_a^{\Delta\Delta'}(M))^2\big)^{\frac{1}{2}}$$

It holds: $a_{\Delta\Delta} \longrightarrow 0$ if $|\Delta| + |\Delta'| \longrightarrow 0$. M is, therefore, uniformly continuous on compacta.

Now it is to show in the next step:

(b) $(E(T_a^{\Delta\Delta'}(M)^2)^{\frac{1}{2}}$ is bounded independent of Δ, Δ'.

W.l.g., let: $a := t_n$, and we write Δ instead of $\Delta\Delta'$. It holds:

$$(T_a^\Delta)^2 = \left(\sum_{k=1}^n (M_{t_k} - M_{t_{k-1}})^2\right)^2$$

$$= 2\sum_{k+1}^n \underbrace{(T_a^\Delta - T_{t_k}^\Delta)}_{\sum_{l>k}(M_{t_l}-M_{t_{l-1}})^2} \underbrace{(T_{t_k}^\Delta - T_{t_{k-1}}^\Delta)}_{(M_{t_k}-M_{t_{k-1}})^2} + \sum_k (M_{t_k} - M_{t_{k-1}})^4$$

By the 1st step it follows $E(T_a^\Delta - T_{t_k}^\Delta \mid \mathfrak{A}_{t_k}) = E\big((M_a - M_{t_k})^2 \mid \mathfrak{A}_k\big)$ as M is a martingale.

Thus, by conditioning under $\mathfrak{A}_{t_k}$

(c) $E\ (T_a^\Delta)^2$
$= 2\sum_{k=1}^n E[(M_a - M_{t_k})^2(T_{t_k}^\Delta - T_{t_{k-1}}^\Delta)] + \sum_{k+1}^n E(M_{t_k} - M_{t_{k-1}})^4$
$\leq E\big[\big(2\sup_k \underbrace{| M_a - M_{t_k}|^2}_{\leq 4C^2} + \sup \underbrace{| M_{t_k} - M_{t_{k-1}}|^2}_{\leq 4C^2}\big) \underbrace{T_a^\Delta}_{\text{increasing}}\ \big]$
$\leq 12C^2 ET_a^\Delta \leq 12C^4$ as $ET_a^\Delta \overset{1.)}{=} EM_a^2 \leq C^2$.
Thus, (b) follows.

Using the Doob inequality for martingales applied to $T^{\Delta_n} - T^{\Delta_m}$ it follows:

(d) $E\sup_{t\leq a} |T_t^{\Delta_n} - T_t^{\Delta_m}|^2 \leq 4E(T_a^{\Delta_n} - T_a^{\Delta_m})^2 \longrightarrow 0$ for $|\Delta_n| \longrightarrow 0$ after 2.), 2.a), 2.c). From this assertion 2. follows:

$(T_a^{\Delta_n})$ is a Cauchy sequence and has a limit in L^2 independent of the decomposition, $T_a^{\Delta_n} \xrightarrow{L^2} A_a = [M]_a$.

Step 3: Properties of $[M]$:

(a) $M^2 - [M] \in \mathcal{M}$, is a martingale, because $T_t^{\Delta_n} - T_t^{\Delta_m} \in \mathcal{M}$ and $T_t^{\Delta_m} \xrightarrow{L^2} [M]_t$ for $|\Delta_m| \longrightarrow 0$. From this follows for $|\Delta_m| \longrightarrow 0$ convergence in L^1 and thus

$$T_t^{\Delta_n} - [M]_t \in \mathcal{M}.$$

So after the 1st step $M^2 - [M] \in \mathcal{M}$.

(b) $[M]$ is continuous.

This is because, according to the Doob inequality.
$E\sup_{t\leq a} |T_t^{\Delta_n} - [M]_t|^2 \leq 4E|T_a^{\Delta_n} - [M]_a|^2 \longrightarrow 0.$

So there exists a subsequence (n_k) : $\sup_{t\leq a} |T_t^{\Delta_{n_k}} - [M]_t| \longrightarrow 0$ almost surely (uniform convergence).

From this follows that $[M]_t$ is continuous.

(c) $[M]_t \subset \mathcal{V}_c^+$ is increasing.

For without restriction Δ_{n+1} is a refinement of Δ_n *and* $\bigcup \Delta_n$, is dense in $[0, a]$. Then $T_s^{\Delta_n} \le T_t^{\Delta_n}$, $s \le t$, $s, t \in \Delta_n$. Thus it follows:

$[M]_t$ is increasing on $\bigcup \Delta_n$, so also on $[0, a]$, since it is continuous.

(3a), (b), (c) $\Longrightarrow$ The quadratic variation process $[M]$ is the unique continuous process A such that $M^2 - A \in \mathcal{M}_c$. Thus it holds that $[M]$ is equal to the predictable quadratic variation,

$$[M] = \langle M \rangle.$$

□

For non-continuous martingales, it will be shown later that quadratic variation and predictable quadratic variation are different processes.

To extend Theorem 3.49 to non-bounded processes, we need the following connection with stopping times.

Proposition 3.50 *If τ is a stopping time, and if $M \in \mathcal{M}_c$ is bounded, then*

$$\langle M^\tau \rangle = \langle M \rangle^\tau.$$

Proof It holds that $M^2 - \langle M \rangle \in \mathcal{M}_c$. The martingale property is preserved by stopping. Thus, according to Corollary 3.12

$M^2_{t\wedge\tau} - \langle M \rangle_{t\wedge\tau} \in \mathcal{M}_c$ thus $(M^\tau)^2 - \langle M \rangle^\tau \in \mathcal{M}_c$.

Because of the uniqueness of the predictable quadratic variation from Theorem 3.49 it follows:

$$\langle M^\tau \rangle = \langle M \rangle^\tau.$$

□

Remark 3.51 (Localization)

(a) It holds: $X \in \mathcal{M}_{\text{loc}} \Longleftrightarrow \exists(\tau_n)$ a sequence of stopping times with $\tau_n \uparrow \infty$ such that X^{τ_n} is a uniformly integrable martingale.

Let (τ_n) be a localizing sequence. Then $\sigma_n = \tau_n \wedge n$ defines a localizing sequence such that X^{σ_n} is a uniformly integrable martingale.

For $M \in \mathcal{M}_{\text{loc},c}$, let be $S_n := \inf\{t : |M_t| \ge n\}$. By the transition from τ_n to $\tau_n \wedge S_n$, one obtains a localizing sequence (τ_n) so that (M^{τ_n}) is even a uniformly integrable bounded martingale. So by localizing one can assume w.l.g. that $M \in \mathcal{M}_{\text{loc},c}$ is uniformly integrable and bounded.

(b) There are examples of uniformly integrable local martingales that are not martingales.
(c) Let $M = (M_n)$ be a discrete martingale and $\varphi = (\varphi_n)$ be predictable. Then the martingale transformation

$$\varphi \cdot M = \big((\varphi \cdot M)_n\big) \text{ with } (\varphi \cdot M)_n = \sum_{k=1}^{n} \varphi_k (M_k - M_{k-1})$$

is a local martingale, $\varphi \cdot M \in \mathcal{M}_{\text{loc}}$.
This is in analogy to the stochastic integral in continuous time:
If $f \in \mathcal{L}^0(B)$, then $(f \cdot B)_t = \int_0^t f \, \mathrm{d}B_s \in \mathcal{M}_{\text{loc}}$ is a local martingale.

The following definition describes the difference between the local martingale property and the martingale property.

Definition 3.52 Let X be an adapted process

(a) X is a process of class (D) (**Dirichlet class**),
if $\{X_\tau;\ \tau \text{ a finite stopping time}\}$ is uniformly integrable.
(b) X is called process of class (DL) (**local Dirichlet class**),
if $\forall a > 0 : \{X_\tau;\ \tau \text{ stopping time}, \tau \leq a\}$ is uniformly integrable.

Theorem 3.53 *If X is a local martingale, then it holds:*

(a) $X \in \mathcal{M} \Longleftrightarrow X$ is of class (DL)
(b) X is a uniformly integrable martingale $\Longleftrightarrow X$ is of the class (D)

Proof

(a) "$\Longrightarrow$": Be $X \in \mathcal{M}$ and $\tau \in \gamma^a$, i.e. τ is a stopping time and $\tau \leq a$. Then, according to the optional sampling theorem, it follows

$$X_\tau = E(X_a \mid \mathfrak{A}_\tau).$$

But this implies the uniform integrability of $\{X_\tau;\ \tau \in \gamma^a\}$.
"$\Longleftarrow$": Let $X \in DL$ and (τ_n) be a localizing sequence for X such that X^{τ_n} is a uniformly integrable martingale. Then it holds

$$X_{s \wedge \tau_n} = X_s^{\tau_n} = E(X_t^{\tau_n} \mid \mathfrak{A}_s) = E(X_{t \wedge \tau_n} \mid \mathfrak{A}_s).$$

Since $\tau_n \uparrow \infty$ it follows: $X_{s \wedge \tau_n} \longrightarrow X_s$ a.s. and $\{X_{s \wedge \tau_n}\}_n$ is uniformly integrable by condition (DL). So also convergence holds in L^1.

Similarly $X_{t \wedge \tau_n} \longrightarrow X_t$ a.s. and in L^1. It follows:

$$X_{s \wedge \tau_n} = E(X_{t \wedge \tau_n} \mid \mathfrak{A}_s) \longrightarrow E(X_t \mid \mathfrak{A}_s)$$

and, therefore, $X_s = E(X_t \mid \mathfrak{A}_s)$ i.e. $X \in \mathcal{M}$.

(b) "$\Longleftarrow$": If $X \in \mathcal{M}_{\text{loc}} \cap (D)$, then the set

$$\{X_t;\ 0 \le t \le \infty\} \subset \{X_\tau;\ \tau \text{ a finite stopping time}\},$$

is, therefore, uniformly integrable. The assertion then follows from a).

"$\Longrightarrow$": If X is a uniformly integrable martingale, and τ is a finite stopping time, then the closure theorem yields $X_t \longrightarrow X_\infty$ almost surely and in L^1 and $(X_t)_{t \le \infty}$ is a martingale.

By the optional sampling theorem, it follows that $X_\tau = E(X_\infty \mid \mathfrak{A}_\tau)$. Thus it follows that:

$$\{X_\tau;\ \tau \text{ a finite stopping time}\} = \{E(X_\infty \mid \mathfrak{A}_\tau);\ \tau \text{ a finite stopping time}\}$$

is uniformly integrable.

□

The following theorem shows the existence of predictable quadratic variation $\langle M \rangle$ for continuous local martingales. As for bounded local martingales, it agrees with the quadratic variation $[M]$.

Theorem 3.54 (Quadratic Variation of Continuous Local Martingales) *Let $M \in \mathcal{M}_{\text{loc},c}$ then it holds:*

(a) There exists exactly one process $\langle M \rangle \in \mathcal{V}_c^+$ the predictable quadratic variation of M such that

$$M^2 - \langle M \rangle \in \mathcal{M}_{\text{loc},c}.$$

(b) $\forall t$, $\forall \Delta_n$ decompositions of $\mathbb{R}^+$, with $|\Delta_n| \longrightarrow 0$, holds $\sup_{s \le t} |T_s^{\Delta_n} - \langle M \rangle_t| \xrightarrow{P} 0$, i.e. $\langle M \rangle$ is identical with the quadratic variation $[M]$.

Proof

(a) $\exists$ a sequence of stopping times $(T_n) \uparrow \infty$ such that with $X_n = M^{T_n} \in \mathcal{M}_c$ is a bounded martingale. According to Theorem 3.49 there exists a unique process

$(A_n) \subset \mathcal{V}_c^+$, $A_n(0) = 0$ such that

$$X_n^2 - A_n \in \mathcal{M}_c.$$

Thus it follows

$$(X_{n+1}^2 - A_{n+1})^{T_n} = X_n^2 - A_{n+1}^{T_n} \in \mathcal{M}_c.$$

The uniqueness implies that $A_{n+1}^{T_n} = A_n$ almost surely. It follows that the process $\langle M \rangle := A_n$ at $[0, T_n]$ is well-defined. It holds:

$$(M^{T_n})^2 - \langle M \rangle^{T_n} \in \mathcal{M}_c, \quad \text{d.h. } M^2 - \langle M \rangle \in \mathcal{M}_{\text{loc},c}.$$

Thus the existence of predictable quadratic variation holds. The uniqueness of $\langle M \rangle$ follows from the uniqueness on $[0, T_n]$, $\forall\, n$.

(b) Let $\delta, \varepsilon > 0$; to $t > 0$ there exists a stopping time S such that M^S is bounded and $P(S \leq t) \leq \delta$.
To show this let (τ_n) be a localizing sequence and define

$$\nu_n := \inf\{t;\, |M_t^{\tau_n}| \geq a_n\}.$$

Then it holds $\nu_n \uparrow \infty$ if $a_n \uparrow \infty$ sufficiently fast.
For $P(\nu_n \leq t) = P(\sup_{s \leq t} |M_s^{\tau_n}| \geq a_n) \leq \frac{E(M_t^{\tau_n})^2}{a_n} \longrightarrow 0$ if $a_n \uparrow \infty$ sufficiently fast. So choose, for example $S = \nu_{n_0} \wedge \tau_{n_0}$.
It holds: $T^\Delta(M) = T^\Delta(M^S)$ and $\langle M \rangle = \langle M^S \rangle$ to $[0, S]$.
From this follows

$$P(\sup_{s \leq t} |T_s^\Delta(M) - \langle M^S \rangle| > \varepsilon) \leq \delta + P\Big(\underbrace{\sup_{s \leq t} |T_s^\Delta(M^S) - \langle M^S \rangle| > \varepsilon}_{\longrightarrow 0 \text{ for } |\Delta| \longrightarrow 0} \Big).$$

□

Remark 3.55 From Proposition 3.50 and the proof of Proposition 3.54 it follows for $M \in \mathcal{M}_{\text{loc},c}$ and a stopping time τ that

$$\langle M^\tau \rangle = \langle M \rangle^\tau.$$

The Doob–Meyer decomposition is an important general version of Theorem 3.54, which also holds for non-continuous processes, which we state without proof.

Theorem 3.56 (Doob–Meyer Decomposition) *Let X be a submartingale of class (DL). Then it holds:*

(a) $\exists M \in \mathcal{M}$ *and* $\exists A \in \mathcal{V}_0^+ \cap \mathcal{P}$, *i.e., there exist a martingale M and an increasing predictable process A with* $A_0 = 0$ *such that*

$$X = M + A.$$

(b) The decomposition is unique with respect to A in the class of increasing and predictable processes starting in 0. □

For a proof, compare Karatzas and Shreve (1991, pp. 22–28). A describes the predictable drift of X.

Remark 3.57 By localization, the Doob–Meyer decomposition can be extended to other classes of processes. Here are some variants of such extensions:

(a) To $M \in \mathcal{M}^2_{\text{loc}}$ there exists a predictable process A of finite variation, $A \in \mathcal{V}_0^+ \cap \mathcal{P}$ such that

$$M^2 - A \in \mathcal{M}_{\text{loc}}.$$

(b) If $X \in \mathcal{A}^+ = \{A \in \mathcal{V}^+ \,;\, EA_\infty < \infty\}$ then X is a submartingale of class (D).
It follows that X has a Doob–Meyer decomposition.
A consequence of this is:

(c) If $A \in \mathcal{A}_{\text{loc}}(\mathcal{A}^+_{\text{loc}})$ then there exists exactly one local martingale $M \in \mathcal{M}_{\text{loc}}$ and $A^p \in \mathcal{A}_{\text{loc}} \cap \mathcal{P} \quad (\mathcal{A}^+_{\text{loc}} \cap \mathcal{P})$ such that $A = M + A^p$.
A^p is called **predictable compensator (predictable compensator)** of A or also **dual predictable projection**. Equivalent to the definition of the predictable compensator A^p is the validity of the projection equations

$$EHA^p_\infty = EHA_\infty, \quad \forall H \in \mathfrak{L}_+(\mathcal{P})$$

(d) In contrast to the "dual predictable projection" A^p is the predictable projection **(predictable projection)** ${}^pX$ of a process $X \in \mathfrak{L}(\mathfrak{A} \otimes \mathfrak{B}_+)$ uniquely defined by the following properties:
(d.1) ${}^pX \in \mathcal{P}$
(d.2) For all predictable stopping times T holds:
${}^pX_T = E(X_T \mid \mathfrak{A}_{T-})$ on $\{T < \infty\}$.
For integrable processes X the predictable projection exists and is unique except for equivalent versions. It holds:

$${}^p(X^T) = {}^pX1_{[\![0,T]\!]} + X^T 1_{]\!]T,\infty[\![}.$$

For a Poisson process N with intensity $\lambda > 0$ it holds:

$${}^pN_t = N_{t-} \textit{ and } N_t^p = \lambda t.$$

(d.3) A process X is called **semimartingale**, if X has a decomposition of the form

$$X = X_0 + \underbrace{M}_{\in \mathcal{M}_{\text{loc}}} + \underbrace{A}_{\in \mathcal{V} \cap \mathcal{O}}.$$

So X is decomposable into a local martingale and an optional process of finite variation. Let $\mathcal{S}$ be the set of all semimartingales (cf. Definition 3.18).

In general, the decomposition of a semimartingale X is not unique. However, if there exists a decomposition with $A \in \mathcal{P}$, i.e. A is predictable, then the decomposition is unique in the class $A \in \mathcal{V} \cap \mathcal{P}$.

X is called **special semimartingale** if A is selectable in $\mathcal{V} \cap \mathcal{P}$.

The decomposition $X = X_0 + M + A$, $\quad A \in \mathcal{V} \cap \mathcal{P}$, then is called **canonical decomposition of X**. Let $\mathcal{S}_p$ be the set of special semimartingales.

It follows from the Doob–Meyer decomposition: If X is a submartingale in the class (DL), then $X \in \mathcal{S}_p$.

The property of being a special semimartingale is essentially an integrability condition. It means that the jumps do not become too large. E.g. α-stable processes are special semimartingales for $\alpha > 1$. The Cauchy process with $\alpha = 1$ is not a special semimartingale.

To describe the relationship between two processes, we define the predictable quadratic covariation. The predictable quadratic variation corresponds to the variance, the predictable quadratic covariation to the covariance.

Proposition 3.58 *Let M and N be continuous local martingales, $M, N \in \mathcal{M}_{\text{loc},c}$, then it holds:*

(a) There exists a unique continuous process $\langle M, N \rangle$ with finite variation such that $MN - \langle M, N \rangle$ is a continuous local martingale, i.e.
$\exists!\ \langle M, N \rangle \in \mathcal{V}_c$, $\langle M, N \rangle_0 = 0$, $MN - \langle M, N \rangle \in \mathcal{M}_{\text{loc},c}$
*$\langle M, N \rangle$ is the **predictable quadratic covariation** resp. **bracket process**.*

(b) Let

$$\widetilde{T}_s^{\Delta_n} := \sum_i \left(M_{t_{i+1}}^s - M_{t_i}^s\right)\left(N_{t_{i+1}}^s - N_{t_i}^s\right);$$

then $\widetilde{T}^{\Delta_n}$ converges uniformly to the predictable covariation process $\langle M, N \rangle$.

$$\sup_{s \le t} \left\| \widetilde{T}_s^{\Delta_n} - \langle M, N \rangle_s \right\| \xrightarrow{P} 0, \quad \textit{if} \qquad |\Delta_n| \longrightarrow 0,$$

i.e., the quadratic covariation $[M, N]$ (cf. Proposition 3.78) exists and is identical to the predictable quadratic covariation $\langle M, N\rangle$.

Proof

(a) **Existence**
Consider the local continuous martingales $M + N, M - N \in \mathcal{M}_{\text{loc},c}$. By Theorem 3.54 there exist associated to them unique predictable quadratic variation processes
$\langle M + N\rangle \in \mathcal{V}_c^+, \langle M - N\rangle \in \mathcal{V}_c^+$. Thus it holds
$(M + N)^2 - \langle M + N\rangle \in \mathcal{M}_{\text{loc},c}$ and $(M - N)^2 - \langle M - N\rangle \in \mathcal{M}_{\text{loc},c}$.
By polarization $MN = \frac{1}{2}((M + N)^2 - (M - N)^2)$ it follows
$MN - \underbrace{\tfrac{1}{2}(\langle M + N\rangle - \langle M - N\rangle)}_{=:\langle M,N\rangle} \in \mathcal{M}_{\text{loc},c}$.
Uniqueness follows as in Theorems 3.54 and 3.49. Localization reduces the problem to the case of bounded martingales. For $M \in \mathcal{M}_{\text{loc},c}$ it holds $\langle M^\tau\rangle = \langle M\rangle^\tau$. This provides consistency in the above definition of covariation and, as a consequence, uniqueness.

(b) follows by polarization as in (a) of the corresponding quadratic increments.

□

Remark 3.59 It follows from (b): The (predictable) quadratic covariation is a bilinear form on the set of continuous local martingales

$$\langle \alpha M_1 + \beta M_2, N\rangle = \alpha\langle M_1, N\rangle + \beta\langle M_2, N\rangle.$$

That quadratic covariation has similar properties to the covariance is also seen in the following proposition; part (c).

Proposition 3.60 *Let M be a continuous local martingale $M \in \mathcal{M}_{\text{loc},c}$.*

(a) Let τ be a stopping time, then

$$\langle M^\tau, N^\tau\rangle = \langle M, N^\tau\rangle = \langle M, N\rangle^\tau$$

(b) $\langle M\rangle_a = 0$ a.s. $\Longleftrightarrow M_t = M_0$ a.s. $\forall t \le a$
(c) In general for $a < b$ it holds:

$$\{\omega;\ M_t(\omega) = M_a(\omega),\ \forall t \in [a, b]\} = \{\omega;\ \langle M\rangle_b(\omega) = \langle M\rangle_a(\omega)\}\ a.s.,$$

i.e., if the quadratic variation on an interval is constant, then the process is also constant there.

Proof

(a) follows from the construction of the quadratic covariation and the corresponding property of the quadratic variation.
(b) (c) By (a), it suffices to show (b) for the case of bounded martingales M. Then

$$M_t^2 - \langle M \rangle_t \in \mathcal{M}_c.$$

It follows for all $t \leq a$:
$E(M_t - M_0)^2 = E\langle M \rangle_t$ and analogously $E(M_b - M_a)^2 = E\langle M \rangle_b - E\langle M \rangle_a = 0$.
From this, (b) and (c) follow.

□

The predictable quadratic variation describes the growth of the process. Like covariance, quadratic covariation describes the dependence behavior. The quadratic covariation process induces a measure

$$\mathrm{d}|\langle M, N \rangle| := \mathrm{d}V^{\langle M, N \rangle} \ll \mathrm{d}\langle M \rangle.$$

This measure is absolutely continuous with respect to the variation process $\langle M \rangle$ or also with respect to $\langle N \rangle$. This continuity is a consequence of the following inequality.

Proposition 3.61 *Let $M, N \in \mathcal{M}_{\mathrm{loc},c}$, $H, K : (\Omega \times \mathbb{R}_+, \mathcal{A} \otimes \mathfrak{B}_+) \longrightarrow (\mathbb{R}, \mathfrak{B})$*

$$\int_0^t |H_s K_s| \, d|\langle M, N \rangle|_s \leq \left(\int_0^t H_s^2 \, \mathrm{d}\langle M \rangle_s \right)^{1/2} \cdot \left(\int_0^t K_s^2 \, \mathrm{d}\langle N \rangle_s \right)^{1/2}.$$

Proof It suffices to consider the case where the integrands are bounded and ≥ 0. We first show the statement for elementary processes. Then the statement follows for bounded positive integrands with the monotone class theorem.

To this end let K, H be processes of the form $K = K_0 \mathbb{1}_{\{0\}} + K_1 \mathbb{1}_{[0,t_1]} + \cdots + K_n \mathbb{1}_{[t_{n-1}, t_n]}$, $H = H_0 \mathbb{1}_{\{0\}} + H_1 \mathbb{1}_{[0,t_1]} + \cdots + H_n \mathbb{1}_{[t_{n-1}, t_n]}$ with H_i, $K_i \in \mathcal{L}(\mathcal{A})$ and H_i, K_i bounded. With $\langle M, N \rangle_s^t := \langle M, N \rangle_t - \langle M, N \rangle_s$ is then for $r \in \mathbb{Q}$:

$$\langle M + rN, M + rN \rangle_s^t = \underbrace{\langle M \rangle_s^t}_{=:c} + 2r \underbrace{\langle M, N \rangle_s^t}_{=:b} + r^2 \underbrace{\langle N \rangle_s^t}_{=:a} \geq 0.$$

Therefore, the above quadratic form is also positive for $r \in \mathbb{R}$ and as a consequence the discriminant $D = b^2 - ac \leq 0$, i.e.

$$|\langle M, N\rangle_s^t| \leq (\langle M\rangle_s^t)^{1/2}(\langle N\rangle_s^t)^{1/2} \quad \text{a.s.}$$

But from this, according to Cauchy–Schwarz,it follows.

$$\begin{aligned}
\int_0^t |H_s K_s| \, \mathrm{d}|\langle M, N\rangle|_s &\leq \sum_i |H_i K_i| |\langle M, N\rangle_{t_i}^{t_{i+1}}| \\
&\leq \sum_i |H_i||K_i|(\langle M\rangle_{t_i}^{t_{i+1}})^{1/2}(\langle N\rangle_{t_i}^{t_{i+1}})^{1/2} \\
&\leq \Big(\sum_i H_i^2 \langle M\rangle_{t_i}^{t_{i+1}}\Big)^{1/2}\Big(\sum_i K_i^2 \langle N\rangle_{t_i}^{t_{i+1}}\Big)^{1/2} \\
&= \Big(\int_0^t H_s^2 \, \mathrm{d}\langle M\rangle_s\Big)^{1/2}\Big(\int_0^t K_s^2 \, \mathrm{d}\langle N\rangle_s\Big)^{1/2}
\end{aligned}$$

In particular $V^{\langle M,N\rangle}(\mathrm{d}s)$ is continuous with respect to the measure $\mathrm{d}\langle M\rangle_s$. The statement of the theorem now follows from monotone convergence. □

The variation measure of the covariation is, therefore, continuous with respect to the variation measure of each component. The proposition is central to the extension of the stochastic integral to L^2-martingales. With the help of Hölders inequality it follows from Proposition 3.61

Corollary 3.62 (Kunita–Watanabe Inequality) *Let be $p, q \geq 1$, $\frac{1}{p} + \frac{1}{q} = 1$, and H, K, M, N be as in Proposition 3.61. Then it holds:*

$$E \int_0^\infty |H_s K_s| \, \mathrm{d}|\langle M, N\rangle|_s \Big\| \Big(\int_0^\infty H_s^2 \, \mathrm{d}\langle M\rangle_s\Big)^{1/2}\Big\|_p \cdot \Big\|\Big(\int_0^\infty K_s^2 \, \mathrm{d}\langle N\rangle_s\Big)^{1/2}\Big\|_q$$

The Kunita–Watanabe inequality induces an important bilinear form on the set of L^2-bounded martingales.

Definition 3.63 Let

$$\mathcal{H}^2 := \{M \in \mathcal{M};\ \sup_t EM_t^2 < \infty\}$$

be the class of $\boldsymbol{L^2}$**-restricted martingale.**

Remark 3.64 By Doob's convergence theorem, the following holds for $M \in \mathcal{H}^2$:

$$M_t \longrightarrow M_\infty, \ \text{a.s. and in } L^2.$$

Moreover the **closure theorem** holds: If we add the limit to the martingale, the martingale property is preserved,

$$(M_t)_{t\leq\infty} \in \mathcal{M}^2.$$

In particular, the set $(M_t)_{t\leq\infty}$ is uniformly integrable and it holds

$$M_t = E(M_\infty \mid \mathcal{A}_t), \qquad M_\infty \in L^2.$$

The Brownian motion is not in $\mathcal{H}^2$ because it is not uniformly bounded in L^2, but only after localization.

Two important norms on $\mathcal{H}^2$ are the following:

Norms for the Class $\mathcal{H}^2$
Define $M^*_\infty := \sup_t |M_t| \in L^2$ for $M \in \mathcal{H}^2$. Let

$$\|M\|_{\mathcal{M}_2^*} := \|M^*_\infty\|_2 = \left(E \sup_t M_t^2\right)^{1/2} \qquad \textit{the } \mathcal{M}_2^*\textit{-norm}$$
$$\|M\|_{\mathcal{H}^2} := \|M_\infty\| = \left(EM_\infty^2\right)^{1/2} = \lim \left(EM_t^2\right)^{1/2} \quad \textit{the } \mathcal{H}^2\textit{-norm}$$

The $\mathcal{M}_2^*$-norm is the $\mathcal{L}^2$-norm of the supremum M^*_∞. The $\mathcal{H}^2$-norm, on the other hand, is the $\mathcal{L}^2$-norm of M_∞, i.e. the limit of $\left(EM_t^2\right)^{1/2}$. It holds according to the maximal inequality of Doob

$$\|M\|_{\mathcal{H}^2} \leq \|M^*_\infty\|_2 \overset{\text{Doob}}{\leq} 2 \cdot \|M\|_{\mathcal{H}^2},$$

i.e. the norms are equivalent. $\mathcal{H}^2$ provided with $\|\cdot\|_{\mathcal{H}^2}$ is a Hilbert space. $\mathcal{H}^2$ with $\|\cdot\|_{\mathcal{M}_2^*}$ it is not a Hilbert space. However, this norm is useful for approximation arguments.

Proposition 3.65

(a) $\left(\mathcal{H}^2, \|\cdot\|_{\mathcal{H}^2}\right)$ *is a Hilbert space.*
(b) The mapping $\mathcal{H}^2 \longrightarrow L^2(\Omega, \mathcal{A}_\infty, P)$, $M \longrightarrow M_\infty$ *is an isometry.*
(c) $\mathcal{H}_c^2 \subset \mathcal{H}^2$ *is a closed subspace.*

Proof

(a), (b) $M \longrightarrow M_\infty$ is bijective, because one can specify the inverse directly using the representation $M_t = E[M_\infty \mid \mathcal{A}_t]$. The isometry property follows from the definition, because the $\mathcal{H}^2$-norm is defined via the L^2-norm of M_∞. Each $Y \in L^2(\Omega, \mathcal{A}_\infty, P)$ generates an L^2-martingale $M_t = E(Y \mid \mathcal{A}_t)$ such that $M_t \longrightarrow Y$. The L^2-bounded martingales can, therefore, be identified with the L^2-space.

(c) Closedness of the set of elements in $\mathcal{H}^2$: We consider a sequence of continuous L^2-bounded martingales which converges in $\mathcal{H}^2$:
Let $\{M^n\} \subset \mathcal{H}^2_c$, $\quad M^n \longrightarrow M \quad$ in $\quad \mathcal{H}^2$. Then it follows:

$$E \sup_t |M^n_t - M_t|^2 \leq 4 \cdot \|M^n - M\|^2_{\mathcal{H}^2} = 4 \cdot \|M^n_\infty - M_\infty\|^2_2 \longrightarrow 0.$$

Thus there exists a subsequence such that $\sup_t |M^{n_k}_t - M_t| \longrightarrow 0$ almost surely. Thus $M \in \mathcal{H}^2_c$, because the limit of continuous functions under uniform convergence is continuous. The closure property for continuous L^2-bounded martingales is thus obtained using the equivalence of the two norms introduced above.

□

For continuous local martingales, the property to be an element in $\mathcal{H}^2_c$ can be written via the quadratic variation.

Proposition 3.66 *Let $M \in \mathcal{M}_{\text{loc},c}$. Then it holds:*

$$M \in \mathcal{H}^2_c \Longleftrightarrow M_0 \in L^2 \quad \textit{and} \quad \langle M \rangle \in \mathcal{A}^+,$$

$\langle M \rangle \in \mathcal{A}^+$ means: the variation process is an integrable increasing process, i.e. $E\langle M \rangle_\infty < \infty$. Further it holds for $M \in \mathcal{H}^2_c$:

(a) $M^2 - \langle M \rangle$ is a uniformly integrable martingale
(b) $\forall$ stopping times $S \leq T$ it holds:
$E(M^2_T - M^2_S \mid \mathcal{A}_S) = E((M_T - M_S)^2 \mid \mathcal{A}_S) = E(\langle M \rangle^T_S \mid \mathcal{A}_S)$,
i.e., the expected conditional quadratic increment is described by the conditional variation process between stopping times.

Proof "$\Longrightarrow$": Let $T_n \uparrow \infty$ be stopping times such that M^{T_n} is a bounded martingale. Then, according to the optional sampling theorem,

$$\begin{aligned} E\left(\left(M^{T_n}\right)^2_t - \langle M^{T_n} \rangle_t\right) &= EM^2_{T_n \wedge t} - E\langle M \rangle_{T_n \wedge t} \\ &= EM^2_0 < \infty \text{ so } M_0 \in L^2. \end{aligned} \tag{3.8}$$

Further $M^*_\infty \in L^2$ and $E\langle M \rangle_{T_n \wedge t} \longrightarrow E\langle M \rangle_\infty$ (monotone convergence). It follows that the right-hand side in (3.8) for $n \longrightarrow \infty, t \to \infty$ converges to $EM^2_\infty - E\langle M \rangle_\infty$.

So it holds: $0 \leq EM_0^2 = EM_\infty^2 - E\langle M\rangle_\infty$ and thus $E\langle M\rangle_\infty < \infty$.

"$\Longleftarrow$": By (3.8) it is $EM_{T_n\wedge t}^2 \leq E\langle M\rangle_\infty + EM_0^2 =: K < \infty$. It follows by the Lemma of Fatou:

$$EM_t^2 \leq \underline{\lim} EM_{T_n\wedge t}^2 \leq K.$$

So $\{M_t\} \subset L^2$ is bounded.

In particular, therefore, L^1-convergence holds. To prove the martingale property: With (T_n) as introduced above,it holds for $s \leq t$:

$$E(M_{t\wedge T_n} \mid \mathfrak{A}_s) = M_{s\wedge T_n}.$$

From this it follows by means of uniform integrability $E(M_t \mid \mathfrak{A}_s) = M_s$.

Because $EM_\infty^2 \leq \underline{\lim} EM_t^2 \leq K < \infty$ (M_t) is a L^2-bounded martingale and thus $M \in \mathcal{H}_c^2$.

For the proof of (a) and (b):

$M^2 - \langle M\rangle$ is a uniformly integrable martingale, because

$$\sup_t |M_t^2 - \langle M\rangle_t| \leq (M_\infty^*)^2 + \langle M\rangle_\infty \in L^1 \text{ as } \langle M\rangle \in \mathcal{A}^+.$$

Therefore, $M^2 - \langle M\rangle$ is uniformly majorized by a L^1-function and, therefore, uniformly integrable.

By Theorem 3.53 is $M^2 - \langle M\rangle$ of the class (D), i.e., the set $\{M_\tau^2 - \langle M\rangle_\tau;\ \tau$ finite stopping time$\}$ is uniformly integrable.

Hence, by the optional sampling theorem, (b) it follows. □

A consequence of the martingale property of $M^2 - \langle M\rangle$ is the following corollary:

Corollary 3.67 *Let $M \in \mathcal{H}_c^2$, $M_0 = 0$. Then it follows:*

$$\|M\|_{\mathcal{H}^2} = \|\langle M\rangle_\infty^{1/2}\|_2 = \big(E\langle M\rangle_\infty\big)^{1/2}.$$

The Hilbert space norm on $\mathcal{H}^2$, is identical to the L^2-norm of $\langle M\rangle_\infty^{1/2}$ i.e. with $\big(E\langle M\rangle_\infty\big)^{1/2}$.

The following proposition gives an application of the variation process to the characterization of the convergence set of continuous local martingales. This is an extension of Doob's convergence theorems to non L^1-bounded local martingales.

Proposition 3.68 (Convergence Sets) *Let M be a continuous local martingale $M \in \mathcal{M}_{\mathrm{loc},c}$, then it holds:*

$$\{\langle M\rangle_\infty < \infty\} \subset \Big\{\lim_{t\to\infty} M_t \text{ exists } P\text{-almost surely}\Big\} \quad \text{almost surely.}$$

Proof Let w.l.g. $M_0 = 0$, and let $\tau_n := \inf\{t; \langle M\rangle_t \geq n\}$. Then $\langle M^{\tau_n}\rangle = \langle M\rangle^{\tau_n} \leq n$. It follows that (M^{τ_n}) is bounded in $\mathcal{L}^2$ since $\left(M^{\tau_n}\right)^2 - \langle M^{\tau_n}\rangle$ is a martingale. Thus, according to the martingale convergence theorem, there exists $\lim_{t\to\infty} M_t^{\tau_n}$ a.s. and in L^2.
On $\{\langle M\rangle_\infty < \infty\}$ it holds $\tau_n = \infty$ for $n \geq n_0(\omega)$. But this implies

$$\lim_{t\to\infty} M_t^{\tau_n} = \lim_{t\to\infty} M_t, \qquad n \geq n_0.$$

□

On the set where the variation is finite at infinity, the martingale converges. Even the inverse holds, i.e. a.s. equality of the sets holds.

3.4 Stochastic Integral of Continuous Local Martingales

We construct the stochastic integral in a series of steps.

3.4.1 Stochastic Integral for Continuous L^2-Martingales

We define the stochastic integral first for the class $\mathcal{M}_c^2 = \{M \in \mathcal{M}_c; EM_t^2 < \infty, \forall t\}$ of continuous L^2 martingales. Important examples for continuous L^2-martingales can be found in the class of diffusion processes. These processes are of importance in stochastic analysis. They are the stochastic analogue of elliptic partial differential equations.

First, as before we introduce the integral for elementary predictable processes.

Let $f \in \mathcal{E}$, $f_t(\omega) = U_0 1\!\!1_{\{0\}} + \sum_{j=0}^m U_j \cdot 1\!\!1_{(s_j,s_{j+1}]}, 0 = s_0 < s_1 < \cdots < s_m, U_j \in B(\mathcal{A}_{s_{j-1}})$.

For these integrands, we define the integral as for the Brownian motion:

$$\int_0^t f\,\mathrm{d}M := U_0M_0 + \sum_{j=0}^{m-1} U_j\big(M_{s_{j+1}}^t - M_{s_j}^t\big) =: (f\cdot M)_t.$$

By analogy with the stochastic integral for Brownian motion, and using predictable quadratic variation, we obtain the basic properties.

Proposition 3.69 *Let $f, g \in \mathcal{E}$, $M, N \in \mathcal{M}_c^2$, then it holds:*

(a) $\displaystyle\int (f+g)\,\mathrm{d}M = \int f\,\mathrm{d}M + \int g\,\mathrm{d}M$

(b) $\displaystyle\int f d(M+N) = \int f\,\mathrm{d}M + \int f\,\mathrm{d}N$

(c) $f \cdot M \in \mathcal{M}_c^2$ *and* $\langle f \cdot M, N\rangle_t = \int_0^t f_s \,\mathrm{d}\langle M, N\rangle_s = \big(f \cdot \langle M, N\rangle\big)_t$

(d) $E \sup_{t \le T} \left| \int_0^t f \,\mathrm{d}M \right|^2 \le 4 \cdot E \int_0^t g^2 \,\mathrm{d}\langle M\rangle, \quad \forall T < \infty.$

Proof

(a), (b) analogous to Proposition 3.32

(c) The space of L^2-martingales $\mathcal{M}_c^2$ is a vector space.
The mapping $X \mapsto \langle X, N\rangle$ is linear. So it suffices to prove (c) for the case $f := U\mathbb{1}_{(s,u]}(t), \quad U \in B(\mathcal{A}_s)$. In this case:
$(f \cdot M)_t = U\big(M_{u\wedge t} - M_{s\wedge t}\big) = U\big(M_t^u - M_t^s\big) \in \mathcal{M}_c^2$. Next it is

$$\begin{aligned}
\langle f \cdot M, N\rangle_t &= \langle U\big(M^u - M^s\big), N\rangle_t \\
&= U\big(\langle M^u, N\rangle_t - \langle M^s, N\rangle_t\big) \\
&= U\big(\langle M, N\rangle_t^u - \langle M, N\rangle_t^s\big)
\end{aligned}$$

connection between quadratic variation and stopping times

$$= \big(f \cdot \langle M, N\rangle\big)_t$$

(d) follows from the maximal inequality of Doob.

□

The predictable covariation of $(f \cdot M)$ with another process N is given by the Lebesgue–Stieltjes integral of f with respect to the covariation process $\langle M, N\rangle$. In particular it holds:

$$\langle f \cdot M, f \cdot M\rangle = \langle f \cdot M\rangle = f^2 \cdot \langle M\rangle$$

For the extension of the stochastic integral, a measure-theoretic argument for the Doléans measure introduced in the following definition is important.

Definition 3.70

(a) Let M be a continuous L^2-martingale, $M \in \mathcal{M}_c^2$, then we define

$$\mathcal{L}^2(M) := \left\{ f \in \mathcal{L}(\mathcal{P}); \quad E \int_0^T f^2 \,\mathrm{d}\langle M\rangle < \infty, \quad \forall T < \infty \right\}$$

(b) We define the **restricted Doléans measure** μ_T on the σ-algebra $\mathcal{P}$ in $\overline{\Omega} = [0, \infty) \times \Omega$

$$\mu_T(C) = \int_\Omega \int_0^T \mathbb{1}_C \, \mathrm{d}\langle M\rangle_t \otimes P = E \int_0^T \mathbb{1}_C(t, \omega) \, \mathrm{d}\langle M\rangle_t.$$

(c) As **Doléans measure** we denote the measure μ at $(\overline{\Omega}, \mathcal{P})$ defined by

$$\mu(C) = E \int_0^\infty \mathbb{1}_C(t, \omega) \, \mathrm{d}\langle M\rangle_t.$$

It holds:

$$\mathcal{L}^2(M) = \bigcap_{T>0} \mathcal{L}^2(\mu_T, \mathcal{P}).$$

μ_T is a finite measure on $\overline{\Omega}$ since $\mu_T(\overline{\Omega}) = EM_T^2 < \infty$, $M \in \mathcal{M}_c^2$. For $M \in \mathcal{H}_c^2$ the Doléans measure μ is finite on $\overline{\Omega}$ and it holds that $\mathcal{L}^2(\mu, \mathcal{P}) \subseteq \mathcal{L}^2(M)$.

For L^2-martingales M, the restricted Doléans measures are finite measures. Because of the finiteness of the measures, the elementary predictable processes are dense in $\mathcal{L}^2(\overline{\Omega}, \mathcal{P}, \mu_T)$, $\forall\, T > 0$. Using this measure-theoretic denseness statement, the extension of the stochastic integral to integrands in $\mathcal{L}^2(M)$ is as follows:

To $g \in \mathcal{L}^2(M)$ there exists a sequence of elementary processes $(g^N) \in \mathcal{E}$ such that:

$$\int (g^N - g)^2 \, \mathrm{d}\mu_N = E \int_0^N (g^N - g)^2 \, \mathrm{d}\langle M\rangle \le 2^{-2N}.$$

For elementary integrands we had already introduced the stochastic integral. We now define

$$Y_t^N := \int_0^t g^N \, \mathrm{d}M \in \mathcal{M}_c^2.$$

This sequence is a Cauchy sequence. As in Chap. 3.2 for the Brownian motion, there exists a limiting process $Y \in \mathcal{M}_c^2$ so that

$$E \sup_{t \le T} \left(Y_t^N - Y_t\right)^2 \longrightarrow 0.$$

For each further elementary process $f \in \mathcal{E}$ holds:

$$E \sup_{t \le T} |Y_t^N - (f \cdot M)_t|^2 \le 4 \cdot E \int_0^T |g_s^N - f_s|^2 \, \mathrm{d}\langle M\rangle_s.$$

For $N \longrightarrow \infty$ it follows

$$E \sup_{t \le T} \left| Y_t - \int_0^t f \, \mathrm{d}M \right|^2 \le 4E \int_0^T (g_s - f_s)^2 \, \mathrm{d}\langle M \rangle_s = 4 \int (g_s - f_s)^2 \, \mathrm{d}\mu_T.$$

So the limit Y is defined independently of the approximating sequence (g^N) and we get

Theorem 3.71 *Let M be a continuous L^2-martingale and let $g \in \mathcal{L}^2(M)$. Then the associated process $Y = (Y_t)$ is called* ***stochastic integral of g with respect to M***

$$Y_t := \int_0^t g \, \mathrm{d}M = (g \cdot M)_t.$$

The properties (a)–(d) from Proposition 3.69 hold true.

This part of the definition process is as for Brownian motion. The following characterization of the stochastic integral using the covariation process gives an alternative and somewhat more abstract way to introduce the stochastic integral. This construction route is taken in Revuz and Yor (2005).

Theorem 3.72 (Characterization of the Stochastic Integral) *Let be $M \in \mathcal{M}_c^2$ and $g \in \mathcal{L}^2(M)$ then it holds: $g \cdot M$ is the is the uniquely determined element $\Phi \in \mathcal{M}_c^2$ which solves the following system of equations:*

$$\langle \Phi, N \rangle = g \cdot \langle M, N \rangle, \quad \forall N \in \mathcal{M}_c^2. \tag{3.9}$$

Proof By Theorem 3.71 it holds:

$$\langle g \cdot M, N \rangle = g \cdot \langle M, N \rangle, \quad \forall N \in \mathcal{M}_c^2,$$

i.e. the stochastic integral $g \cdot M$ satisfies the system of equations (3.9).

To prove uniqueness, we assume that (3.9) holds for $\Phi \in \mathcal{M}_c^2$. Then it follows:

$$\langle \Phi - g \cdot M, N \rangle = 0 \, [P]\text{-a.s.} \quad \forall N \in \mathcal{M}_c^2.$$

If one chooses N specifically $N := \Phi - g \cdot M$, then it follows:

$$\langle N, N \rangle = \langle N \rangle = 0, \quad N_0 = 0.$$

With Proposition 3.60 then follows $N = 0$, i.e. $\Phi = g \cdot M$. □

Remark 3.73 **(Stochastic Integral and Theorem of Riesz)**

(a) Equation (3.9) leads to a an alternative construction possibility for the stochastic integral: For $M \in \mathcal{H}_c^2 \subset \mathcal{M}_c^2$ and $f \in \mathcal{L}^2(\mu) \subset \mathcal{L}^2(M)$, μ the Doléans measure, we define on the Hilbert space $\big(\mathcal{H}_c^2, \|\cdot\|_{\mathcal{H}_c^2}\big)$ the functional

$$\mathcal{H}_c^2 \xrightarrow{T} \mathbb{R}, \quad N \longrightarrow E\big(f \cdot \langle M, N\rangle\big)_\infty. \tag{3.10}$$

The mapping T is linear and continuous in the Hilbert space norm $\|\cdot\|_{\mathcal{H}_c^2}$. This follows from the Kunita–Watanabe inequality. The norm of the image can be estimated by the norm of the preimage. By the Theorem of Riesz it follows that: $\exists!\Phi \in \mathcal{H}_c^2$ so that

$$(\Phi, N)_{\mathcal{H}_c^2} = E\Phi_\infty N_\infty = E\big(f \cdot \langle M, N\rangle\big)_\infty = T(N). \tag{3.11}$$

We now define the stochastic integral: $f \cdot M$ as this solution $\Phi = \Phi_f$:

$$\Phi = \Phi_f =: f \cdot M.$$

By Theorem 3.72 it remains to show that Φ is a solution of (3.9) in order to obtain as a consequence the equality of both constructions of the stochastic integral. Note that for all stopping times τ:

$$\begin{aligned}
E(\Phi_f N)_\tau &= EE(\Phi_f \mid \mathfrak{A}_\tau)N_\tau = E(\Phi_f)_\infty N_\tau \\
&= E(\Phi_f)_\infty N_\infty^\tau = E(f \cdot \langle M, N^\tau\rangle)_\infty \\
&= E(f \cdot \langle M, N\rangle^\tau)_\infty \\
&= E(f \cdot \langle M, N\rangle)_\tau.
\end{aligned}$$

It follows that

$$\Phi_f N - f \cdot \langle M, N\rangle \in \mathcal{M}.$$

This implies:

$$\langle \Phi_f, N\rangle = f \cdot \langle M, N\rangle. \tag{3.12}$$

Thus, according to (3.9) the stochastic integral as defined by Riesz's representation theorem Φ_f is identical to the earlier introduced stochastic integral.

(a) **Stochastic Integral as an Isometry**
The stochastic integral for $M \in \mathcal{H}_c^2$ is an isometry of $\mathcal{L}^2(\mu)$ to $\mathcal{H}_c^2$, i.e.

$$\mathcal{L}^2(\mu) \longrightarrow \mathcal{H}_c^2, \quad f \longrightarrow \Phi_f = f \cdot M \text{ is an isometry.} \tag{3.13}$$

For by definition of the Doléans measure it holds:

$$\|f\|^2_{\mathcal{L}^2(\mu)} = E\int f^2 \,\mathrm{d}\langle M\rangle = E\Phi^2_\infty = \|\Phi\|^2_{\mathcal{H}^2_c} \tag{3.14}$$

by the characterization in (3.9). Because of this isometry, the stochastic integral for $\mathcal{H}^2_c$ and integrands in $\mathcal{L}^2(\mu)$ is most simply introduced as an isometric continuation of the stochastic integral for elementary integrands $\mathcal{E} \subset \mathcal{L}^2(\mu)$, since these lie dense in $\mathcal{L}^2(\mu)$.

3.4.2 Extension to the Set of Continuous Local Martingales

In this section we extend the construction of the stochastic integral to the set of continuous local martingales. Analogous to the Brownian motion case, we denote by $\mathcal{M}_{\text{loc},c}$, the class of continuous local martingales.

For $M \in \mathcal{M}_{\text{loc},c}$ we introduce as class of integrands similar to Brownian motion the class $\mathcal{L}^0(M)$ as a class of integrands:

$$\mathcal{L}^0(M) := \left\{ f \in \mathcal{L}(\mathcal{P});\quad \int_0^T f_s^2 \,\mathrm{d}\langle M\rangle_s < \infty \quad \text{a.s.,} \quad \forall T < \infty \right\}.$$

The difference with Brownian motion here is the integrator. Instead of the Lebesgue measure, we use the variation process. Let $\tau_n' \nearrow \infty$ be a localization such that $M^{\tau_n'} \in \mathcal{M}$, $n \in \mathbb{N}$. For the localization τ_n with

$$\tau_n := \inf\left\{ t \geq 0;\quad \langle M\rangle_t \geq n \text{ or } \int_0^t f_s^2 \,\mathrm{d}\langle M\rangle_s \geq n \right\} \wedge \tau_n'$$

it follows that the second moments of the stopped processes are bounded:

$$E\big(M_t^{\tau_n}\big)^2 = E\langle M\rangle_t^{\tau_n} \leq n.$$

It holds: $M^{\tau_n} \in \mathcal{M}^2_c$. The stopped martingale is even a uniformly bounded martingale and for $f_t^n := f_t \mathbb{1}_{[0,\tau_n]}(t)$ is $f^n \in \mathcal{L}^2\big(\mu_{M^{\tau_n}}, \mathcal{P}\big)$, i.e. f^n is an element of the $\mathcal{L}^2$ with respect to the Doléans measure. f^n is predictable, $f^n \in \mathcal{L}(\mathcal{P})$ since $\mathbb{1}_{[0,\tau_n](t)}$ is left-continuous on. Thus we define the stochastic integral of f^n with respect to dM^{τ_n}

$$X_t^n := \int_0^t f^n \,\mathrm{d}M^{\tau_n}$$

These integrals are consistently defined. Analogously to the argument for the Brownian motion it holds

$$P\big(X^n_{t\wedge\tau_n} = X^{n+1}_{t\wedge\tau_n}, \quad \forall t\big) = 1,$$

since

$$P\big(f^n_{t\wedge\tau_n} = f^{n+1}_{t\wedge\tau_n}, \quad \forall t\big) = 1.$$

Let Ω^0 be the set of all ω for which equality holds:

$$\Omega^0 := \Big\{\omega; \quad X^n_{t\wedge\tau_n(\omega)}(\omega) = X^{n+1}_{t\wedge\tau_n(\omega)}(\omega), \quad \forall t, \forall n\Big\}$$

then it follows $P(\Omega^0) = 1$ and we can consistently define X_t:

$$X_t(\omega) := \begin{cases} X^n_t(\omega), \; \tau_{n-1}(\omega) \le t < \tau_n(\omega), & \omega \in \Omega_0 \\ \qquad 0 & \text{other.} \end{cases}$$

X_t is well-defined and agrees before time τ_n with X^n_t. So X_t is continuous. Thus we can define the stochastic integral as this process

$$X_t := \int_0^t f \, \mathrm{d}M =: [f \cdot M]_t.$$

With the help of this localization, the basic integral properties carry over from the case $\mathcal{M}^2_c$, Proposition 3.69. The argument in (3.12) yields the basic relation $\langle f \cdot M, N\rangle = f \cdot \langle M, N\rangle$ for $M, N \in \mathcal{M}_{\mathrm{loc},c}$.

Theorem 3.74 *Let $M, N \in \mathcal{M}_{\mathrm{loc},c}$, then it holds for $f \in \mathcal{L}^0(M)$*

(a) $X := f \cdot M \in \mathcal{M}_{\mathrm{loc},c}$, *$f \cdot M$ is a continuous local martingale*
(b) $\langle X\rangle = f^2 \cdot \langle M\rangle$
(c) $\langle X, N\rangle = f \cdot \langle M, N\rangle$
(d) $(f+g) \cdot M = f \cdot M + g \cdot M, \quad f, g \in \mathcal{L}^0(M)$
(e) $f \cdot (M+N) = f \cdot M + f \cdot N, \quad f \in \mathcal{L}^0(M) \cap \mathcal{L}^0(N).$

The construction of the stochastic integral is based on localizations. The following lemma gives us a maximal inequality of the integral on $[0, \sigma]$ for finite stopping times σ.

Lemma 3.75 *Let $M \in \mathcal{M}_{\text{loc},c}$, $f \in \mathcal{L}^0(M)$ and let σ be a finite stopping time. Then it holds:*

$$E \sup_{t \le \sigma} \left\| \int_0^t f \, \mathrm{d}M \right\|^2 \le 4 \cdot E \int_0^\sigma f_s^2 \, \mathrm{d}\langle M \rangle_s .$$

Proof Let w.l.g. $E \int_0^\sigma f_s^2 \, \mathrm{d}\langle M \rangle_s < \infty$. We define

$$\sigma_k := \inf \left\{ t; \int_0^t f_s^2 \, \mathrm{d}\langle M \rangle_s \ge k \right\} \text{ and } f_t^k := f_t \mathbb{1}_{[0,\sigma_k]}(t).$$

For $X := f \cdot M$, $X^k := f^k \cdot M$ then holds:

$$X_{t \wedge \sigma_k} = X_{t \wedge \sigma_k}^k .$$

Since $E \int_0^\infty (f_s^k)^2 \, \mathrm{d}\langle M \rangle_s \le k$; we get $f^k \in \mathcal{L}^2(M)$. Thus $X^k \in \mathcal{M}_c^2$.

Then, according to the Doob inequality, it follows:

$$\begin{aligned} E \sup_{t \le \sigma \wedge \sigma_k} (X_s^k)^2 \le 4E[X_{\sigma \wedge \sigma_k}]^2 &= 4E(X_{\sigma \wedge \sigma_k}^k)^2 \\ &= 4E \int_0^{\sigma \wedge \sigma_k} (f_s^k)^2 \, \mathrm{d}\langle M \rangle_s \\ &= 4E \int_0^{\sigma \wedge \sigma_k} f_s^2 \, \mathrm{d}\langle M \rangle_s \le 4E \int_0^\sigma f_s^2 \, \mathrm{d}\langle M \rangle_s . \end{aligned}$$

By the monotone convergence theorem it follows

$$\begin{aligned} E \sup_{t \le \sigma} X_t^2 &= \lim_{k \longrightarrow \infty} E \sup_{t \le \sigma \wedge \sigma_k} X_{t \wedge \sigma_k}^2 \\ &= \lim_{k \longrightarrow \infty} E \sup_{t \le \sigma} (X_t^k)^2 \le 4E \int_0^\sigma f_s^2 \, \mathrm{d}\langle M \rangle_s . \end{aligned}$$

□

There is also a theorem on dominated convergence for stochastic integrals.

Theorem 3.76 (Dominated Convergence) *Let M be a continuous local martingale and let f^n, f, and g be predictable processes from $\mathcal{L}^0(M)$. Let f^n be uniformly bounded $|f^n| \le g$ and let the sequence f^n converge pointwise, $f^n \longrightarrow f$. Then it follows:*

$$\sup_{t \le T} \left| \int_0^t f^n \, \mathrm{d}M - \int_0^t f \, \mathrm{d}M \right| \xrightarrow{P} 0, \quad \forall T < \infty.$$

Proof The sequence of stopping times $\tau_k := \inf\{t;\ \int_0^t g_s^2 \,\mathrm{d}\langle M\rangle_s \geq k\}$ converges to ∞, $\tau_k \nearrow \infty$ since $g \in \mathcal{L}^0(M)$. Then it holds

$$\int_0^{\tau_k} g_s^2 \,\mathrm{d}\langle M\rangle_s \leq k.$$

According to Lemma 3.75 it holds using the theorem on majorized convergence:

$$\begin{aligned} b_{k,n} &:= E \sup_{t\leq\tau_k} \left| \int_0^t f^n \,\mathrm{d}M - \int_0^t f \,\mathrm{d}M \right|^2 \\ &\leq 4E \int_0^{\tau_k} \underbrace{|f_s^n - f_s|^2}_{\leq 2(g+|f|)^2 \leq 4(g^2+f^2)} \mathrm{d}\langle M\rangle_s \longrightarrow 0 \quad \forall k. \end{aligned}$$

To $\varepsilon > 0, \delta > 0$: there exists k such that: $P(\tau_k < T) < \delta/2$, since $\tau_k \nearrow \infty$. It follows:

$$P\left(\sup_{t\leq T} \left| \int_0^t f^n \,\mathrm{d}M - \int_0^t f \,\mathrm{d}M \right| > \varepsilon \right) \leq \frac{b_{k,n}}{\varepsilon^2} + P(\tau_k < T) < \delta \text{ for } n \geq n_0.$$

□

3.4.3 Extension to the Case of Continuous Semimartingales

Let X be a **continuous semimartingale**, i.e. X has a decomposition of the form

$$X_t = X_0 + M_t + A_t, \text{ with } M \in \mathcal{M}_{\mathrm{loc},c}, \quad A \in \mathcal{V}_c, \text{ i.e. } V_t^A < \infty, \quad \forall t < \infty.$$

With $\mathcal{S}$ resp. $\mathcal{S}_c$ we denote the set of all semimartingales or continuous semimartingales.

For the case of continuous semimartingales, the procedure is the same as for continuous local martingales. First we define the classes of integrands:

$$\mathcal{L}^0(X) := \left\{ f \in \mathcal{L}(\mathcal{P});\quad \int_0^t f_s^2 \,\mathrm{d}\langle M\rangle_s < \infty \quad \text{and} \quad \int_0^t |f_s| \,\mathrm{d}A_s < \infty, \quad \forall t \right\}.$$

Remark 3.77 (Càdlàg Processes) If the process Y is a càdlàg process, then we define Y_- by

$$(Y_-)_0 := 0, \qquad (Y_-)_t := \lim_{s\uparrow t} Y_s$$

Then it holds:

Y_- is left continuous, in particular predictable,

Y, Y_- are locally bounded, i.e. bounded on finite intervals.

For $X \in \mathcal{S}_c$ is then $\int_0^t (Y_-)_s^2 \, d\langle M\rangle_s < \infty$ and $\int_0^t (Y_-)_s \, d|A_s| < \infty$, $\forall t$. So $Y_- \in \mathcal{L}^0(X)$ for all càdlàg processes Y.

For the class $\mathcal{L}^0(X)$ we define the stochastic integral

$$\int_0^t f \, dX := \int_0^t f \, dM + \int_0^t f \, dA. \tag{3.15}$$

The properties of stochastic integrals for the class of locally continuous martingales carry over to the class of continuous semimartingales: linearity, majorized convergence, Riemann sum approximation.

Proposition 3.78 (Quadratic Variation) *Let X be a continuous semimartingale and let $\Delta^n = (\tau_i^n)$ be sequences of decompositions of $\mathbb{R}^+$ with stopping times τ_i^n, such that $|\Delta^n| = \sup\left(\tau_{i+1}^n - \tau_i^n\right) \longrightarrow 0$ a.s. and let*

$$Q_t^n = \sum_i \left(X_{\tau_{i+1}^n}^t - X_{\tau_i^n}^t\right)^2.$$

Then it follows

(a) There exists a process $[X] \in \mathcal{V}_c^+$ such that $Q_t^n \xrightarrow{P} [X]_t$.
*This process is called **quadratic variation process** (square bracket process) (angular bracket process).*
(b) The quadratic variation is defined independently of the decomposition and it holds:

$$X_t^2 = X_0^2 + 2\int_0^t X_{s-} \, dX_s + [X]_t.$$

(c) Uniform stochastic convergence holds:

$$\sup_{t \le T} \left|Q_t^n - [X]_t\right| \xrightarrow{P} 0.$$

Proof For a continuous local martingale M $[M]$ denotes the quadratic variation and $\langle M\rangle$ the predictable quadratic variation, i.e., it holds that $M^2 - \langle M\rangle \in \mathcal{M}_{\mathrm{loc},c}$. Let Y be a càdlàg process and let

$$W_t^n := \sum_i Y_{\tau_i^n}^t (X_{\tau_{i+1}^n}^t - X_{\tau_i^n}^t) = \int_0^t Y^n \, dX$$

with $Y^n := \sum Y_{\tau_i^n} \mathbb{1}_{(\tau_j^n, \tau_{j+1}^n]} \in \mathcal{L}(\mathcal{P})$. Y^n converges pointwise to Y_-,

$$Y^n \longrightarrow Y_-.$$

$h_t := \sup_{s \le t} |Y_s|$ is locally bounded and, therefore, $h \in \mathcal{L}^0(X)$ and $|Y_t^n| \le h_t$. Thus it follows according to Theorem 3.76 on dominated convergence

$$\sup_{t \le T} \left| \int_0^t Y^n \,\mathrm{d}X - \int_0^t Y_- \,\mathrm{d}X \right| \xrightarrow[P]{} 0. \tag{3.16}$$

In particular for $Y = X$, therefore, it follows:

$$\sup_{t \le T} \left| W_t^n - \int_0^t X_s \,\mathrm{d}X_s \right| \xrightarrow[P]{} 0. \tag{3.17}$$

With $a := X_{\tau_i^n}^t$, $b := X_{\tau_{i+1}^n}^t$ and $b^2 - a^2 = 2a(b-a) + (b-a)^2$ it then follows after summation over i

$$X_t^2 - X_0^2 = 2W_t^n + Q_t^n.$$

Thus by (3.17) it holds:

$$Q_t^n \xrightarrow[P]{} [X]_t := X_t^2 - X_0^2 - 2\int_0^t X_{s-} \,\mathrm{d}X_s.$$

For $\tau_i^n \le t$ is $Q_{\tau_i^n}^n \le Q_t^n$. Therefore, $[X]_t$ is increasing and $[X] \in V_c^+$. The convergence is uniform on $[0, T]$. As a corollary, the quadratic and the predictable quadratic variation are equal. □

Corollary 3.79 *Let $M \in \mathcal{M}_{\mathrm{loc},c}$, $M_0 := 0$, then it holds:*

$$M_t^2 - [M]_t = 2\int_0^t M_{s-} \,\mathrm{d}M_s \in \mathcal{M}_{\mathrm{loc},c}.$$

This implies

$$[M] = \langle M \rangle.$$

Proof What remains to be shown is the uniqueness of the predictable quadratic variation. This follows analogously to Theorem 3.54. □

The quadratic covariation process can now be introduced by polarization.

Definition 3.80 (Quadratic Covariation) Let X, Y be continuous semimartingales. Then

$$[X, Y]_t := \frac{1}{4} \cdot \big([X + Y]_t - [X - Y]_t\big)$$

is called **quadratic covariation** of X and Y.

Proposition 3.81 (Partial Integration Formula) *For two continuous semimartingales X and Y it holds*

$$X_t Y_t = X_0 Y_0 + \int_0^t X_{s-} \, \mathrm{d}Y_s + \int_0^t Y_{s-} \, \mathrm{d}X_s + [X, Y]_t.$$

Proof By Proposition 3.78 it holds:

$$[X + Y]_t = (X + Y)_t^2 - (X_0 + Y_0)^2 - 2 \int_0^t (X_- + Y_-)_s \, \mathrm{d}(X + Y)_s$$

$$[X - Y]_t = (X - Y)_t^2 - (X_0 - Y_0)^2 - 2 \int_0^t (X_- - Y_-)_s \, \mathrm{d}(X - Y)_s$$

By $a \cdot b = \frac{1}{4}\big((a + b)^2 - (a - b)^2\big)$ and the bilinearity of the integral follows the assertion. □

Remark 3.82

(a) The above partial integration formula also applies to general semimartingales (cf. Proposition 3.94).
(b) If $f = f(t)$ is a deterministic continuous function, then (cf. Proposition 3.86)

$$f \in \mathcal{S}_c \iff f \in \mathcal{V}_c.$$

Thus, the local martingale part is identically zero. For $f, g \in \mathcal{S}_c$ holds $[f, g] = 0$ and the above partial integration formula includes the partial integration formula for (continuous) functions of finite variation.

3.5 Integration of Semimartingales

For the integration of non-continuous martingales and more generally of semimartingales, some decomposition theorems are needed. These are compended below (partly without proof). A càdlàg process X is called a **semimartingale** $X \in \mathcal{S}$ if it can be decomposed into a local martingale and a process of finite variation.

$$X = X_0 + M + A \tag{3.18}$$

with $M \in \mathcal{M}_{\text{loc}}$ and $A \in \mathcal{V}$. Examples of semimartingales are jump processes with countable state space such as generalized Poisson processes, Markovian jump processes, (exponential) Lévy processes with general state space, and diffusion processes with jumps.

The **special semimartingales** $X \in \mathcal{S}_p$ are the semimartingales for which A in the above decomposition can be chosen predictable, i.e. $A \in \mathcal{V}_p := \mathcal{V} \cap \mathcal{P}$. For special semimartingales this decomposition is unique (canonical decomposition), but this is not true for general semimartingales. The special semimartingales are exactly those semimartingales for which the part of finite variation is locally integrable.

More precisely, the following equivalence holds:

Lemma 3.83 *For a semimartingale X it holds:*

$$\begin{aligned} X \in \mathcal{S}_p &\iff \exists \text{ a decomposition of } X \text{ with } A \in \mathcal{A}_{\text{loc}}, \text{ the class of processes} \\ &\qquad \text{of locally integrable finite variation} \\ &\iff \forall \text{ decompositions of } X \text{ holds } A \in \mathcal{A}_{\text{loc}} \\ &\iff Y_t = \sup_{s \le t} |X_s - X_0| \text{ is locally integrable.} \end{aligned}$$

The following lemma concretizes the above statement.

Lemma 3.84 *Let $X \in \mathcal{S}$ and $|\Delta X| \le a < \infty$, then $X \in \mathcal{S}_p$.*

For the canonical decomposition $X = X_0 + M + A$ of X it holds:

$$|\Delta A| \le a, \quad |\Delta M| \le 2a.$$

For $X \in \mathcal{S}_{p,c}$ it follows that $M \in \mathcal{M}_{\text{loc},c}$ and $A \in \mathcal{V}_{p,c}$.

$\mathcal{S}$ and $\mathcal{S}_p$ are stable under stopping.

Proposition 3.85 (Continuous Martingale Part) *To $X \in \mathcal{S}$ there is a unique decomposition of the form*

$$X = X_0 + X^c + M + A \; with \; X^c \in \mathcal{M}_{\text{loc},c}, M \in \mathcal{M}_{\text{loc}}, A \in \mathcal{V} \tag{3.19}$$

$X_0^c = 0$, such that for all decompositions $M = M_1 + M_2$ with $M_1 \in \mathcal{M}_{\text{loc},c}$, $M_2 \in \mathcal{M}_{\text{loc}}$ holds $M_1 = 0$.

X^c is called ***continuous martingale part*** *of X.*

Proposition 3.86 *Let $f = f(t)$ be a deterministic càdlàg function. Then*

$$f \in \mathcal{S} \Longleftrightarrow f \text{ is of finite variation}, \ f \in \mathcal{V}.$$

Proof Only the direction "$\Longrightarrow$" is to show. By Lemma 3.84 it holds

$$f = f(0) + M + A \in \mathcal{S}_p \text{ is a special semimartingale.}$$

There is a localization $\tau_n \uparrow \infty$:

$$f^{\tau_n} = f(0) + M^{\tau_n} + A^{\tau_n} \text{ with } EA_t^{\tau_n} < \infty.$$

It follows

$$Ef_{t\wedge\tau_n} = f(0) + \underbrace{EA_{t\wedge\tau_n}}_{\in\mathcal{V}} = \underbrace{f(t)P(\tau_n > t) + \int_0^t f(y)\,\mathrm{d}P^{\tau_n}(y)}_{\in\mathcal{V}}.$$

For all $T > 0$ and $t \le T$ it holds for $n \ge n_0$, $P(\tau_n > t) > 0$.

The above representation then implies that $f \in \mathcal{V}$. □

We now treat some decomposition theorems for general semimartingales and describe their variation and covariation processes.

3.5.1 Decomposition Theorems

Let $M \in \mathcal{M}^2_{\mathrm{loc}}$ with $M_0 = 0$, i.e. there exists a localization $\tau_n \uparrow \infty$ such that $M^{\tau_n} \in \mathcal{M}^2$ is an L^2-martingale. Then $(M^{\tau_n})^2$ is a submartingale of class (DL).

By the Doob–Meyer's theorem, it follows:
There exists exactly one predictable increasing process $A \in \mathcal{V}_+ \cap \mathcal{L}(\mathcal{P}) = (\mathcal{V}_p)_+$, $A_0 = 0$, such that $M^2 - A \in \mathcal{M}_{\mathrm{loc}}$.

(Definition on $[0, \tau_n]$, consistent continuation on $[0, \infty)$.)
The process $A =: \langle M \rangle$ is called the **predictable quadratic variation** of M.

Using the polarization formula, we can then introduce for $M, N \in \mathcal{M}^2_{\mathrm{loc}}$ the **predictable quadratic covariation** of M, N $\langle N, M \rangle \in \mathcal{L}(\mathcal{P})$ and characterize it by the property

$$MN - \langle M, N \rangle \in \mathcal{M}_{\mathrm{loc}}. \tag{3.20}$$

Since $\mathcal{M}_{\mathrm{loc},c} \subset \mathcal{M}^2_{\mathrm{loc}}$ this generalizes the predictable variation process of continuous local martingales.

As in the continuous case, we get an inequality for $N, M \in \mathcal{M}^2_{\text{loc}}$ given by:

$$|\langle M, N\rangle_t| \leq \left(\langle M\rangle_t \langle N\rangle_t\right)^{1/2} \quad \text{a.s.} \quad \forall t,$$

i.e., the root of the product of the variations is a majorant for the absolute value of the covariation.

The reason that the square-integrable processes play a special role is the following decomposition theorem:

Proposition 3.87 (Special Decomposition Theorem) *Let be $X \in \mathcal{S}$, then there exists a decomposition of X of the form*

$$X = X_0 + M + A \quad \textit{with } M \in \mathcal{M}^2_{\text{loc}} \textit{ and } A \in \mathcal{V}.$$

The above decomposition is for $A \in \mathcal{V}$ unique. The argument for the above decomposition is to transfer the large jumps of M in a decomposition to the process A without losing the finite variation.

Applying this special decomposition theorem to a local martingale, we obtain as corollary

Corollary 3.88 *Let $M \in \mathcal{M}_{\text{loc}}$, $M_0 = 0$, then there exists a process $M' \in \mathcal{M}^2_{\text{loc}}$ and $A \in \mathcal{V}$ with $M = M' + A$.*

As a consequence of this, one can restrict the definition of the stochastic integral for semimartingales to the case where the martingale component lies in $\mathcal{M}^2_{\text{loc}}$.

A useful property of predictable increasing processes (as in the definition of special semimartingales) is formulated by the following lemma.

Lemma 3.89 *Let A be a predictable process $A \in \mathcal{V}_p$ of finite variation, $A_0 = 0$, then $A \in \mathcal{A}_{\text{loc}}$, A has locally integrable variation, i.e., there exists a localization (τ_n) such that A^{τ_n} has integrable variation $\forall\, n \in \mathbb{N}$.*

As a corollary of this, we have the following integrability statement.

Lemma 3.90 *Let $f \in \mathcal{L}(\mathcal{P})$ be a predictable process, and let $A \in \mathcal{A}_+ \cap \mathcal{L}(\mathcal{P})$ be predictable and increasing, and further let $\int_0^t |f_s|\, \mathrm{d}A_s < \infty$, $\forall\, t$ a.s. Then it holds:*

$$B_t := \int_0^t f_s\, \mathrm{d}A_s \in \mathcal{L}(\mathcal{P}) \cap \mathcal{A}_{\text{loc}},$$

i.e., the integral $B = f \cdot A$ is predictable and locally integrable.

Proof Let $f \in \mathcal{E}$ and let $\mathcal{H} := \{X \text{ càdlàg}; \int_0^t f\,\mathrm{d}X \in \mathcal{L}(\mathcal{P})\}$, then $\mathcal{H} \subset$ {continuous processes}, because $f \cdot X$ is continuous. For all càdlàg processes f the integral $f \cdot X$ is defined.

By the monotone class theorem, it follows:

$$\mathcal{H} \text{ contains all constrained predictable processes.}$$

Then by localization it follows:

$$B = f \cdot A \in \mathcal{L}(\mathcal{P}) \cap \mathcal{A}_{\text{loc}}, \quad \forall A \in \mathcal{V}_+ \cap \mathcal{L}(\mathcal{P}), f \in \mathcal{E}.$$

From this, again using the monotone class theorem, the assertion for $f \in \mathcal{L}(\mathcal{P})$, f bounded, follows.

For general predictable functions, the statement is reduced to the bounded case by localization: Let $f^k := (f \wedge k) \vee (-k)$ then the sequence f^k converges pointwise: $f^k \longrightarrow f$. By assumption the sequence $|f^k|$ is majorized by an integrable function:

$$|f^k| \le |f| \text{ and } \int_0^t |f|_s\,\mathrm{d}A_s < \infty.$$

The assertion now follows with the theorem of majorized (dominated) convergence. □

3.5.2 Stochastic Integral for $\mathcal{M}^2_{\text{loc}}$

The stochastic integral is now introduced in the next step in a similar way as for $\mathcal{M}^2_c$.

Let $M \in \mathcal{M}^2_{\text{loc}}$, and $\mathcal{L}^0(M) := \left\{f \in \mathcal{L}(\mathcal{P}); \int_0^t f_s^2\,\mathrm{d}\langle M\rangle_s < \infty, \text{a.s. } \forall t < \infty\right\}$.
Then, by Lemma 3.90 $B_t := \int_0^t f_s^2\,\mathrm{d}\langle M\rangle_s \in \mathcal{P} \cap \mathcal{A}_{\text{loc}}$, for $f \in \mathcal{L}^0(M)$.

So there exists a localizing sequence $\sigma_n \uparrow \infty$: $\quad EB_t^{\sigma_n} < \infty$.
For $M \in \mathcal{M}^2$ the stochastic integral $X_t := \int_0^t f\,\mathrm{d}M$ is defined as for martingales $M \in \mathcal{M}^2_c$. Then one extends the class of integrators to $\mathcal{M}^2_{\text{loc}}$ by localization as before:
Let (σ'_n) be a localizing sequence such that $M^{\sigma'_n} \in \mathcal{M}^2$ and let $\tau_n := \sigma_n \wedge \sigma'_n$.

We define on $[0, \tau_n]$ the stochastic integral by

$$\int_0^t f\,\mathrm{d}M := \int_0^t f\,\mathbb{1}_{[0,\tau_n]}\,\mathrm{d}M^{\tau_n} \text{ on } [0, \tau_n].$$

To identify the predictable quadratic covariation, we use the Doob–Meyer decomposition:

Let $N \in \mathcal{M}_{\text{loc}} \cap \mathcal{P}$, $N_0 = 0$ and $N \in \mathcal{V}$, then

$$P(N_t = 0, \forall t) = 1$$

because of the uniqueness in Doob–Meyer's decomposition. This implies the uniqueness of $\langle X, N\rangle$.

We get $X = f \cdot M \in \mathcal{M}^2_{\text{loc}}$ and the integral properties hold (cf. Proposition 3.69) as well as the theorem on dominated convergence for the stochastic integral. In particular, the covariation formula also holds.

3.5.3 Stochastic Integral for Semimartingales

Let $X \in \mathcal{S}$ be a semimartingale. Then according to the decomposition theorem one can decompose X:

$$X = M + A, \qquad M \in \mathcal{M}^2_{\text{loc}}, A \in \mathcal{V}.$$

This decomposition is not unique.

Our goal is to construct an integral that is independent of the decomposition. Suppose

$$X = M + A = N + B, \qquad M, N \in \mathcal{M}^2_{\text{loc}}, A, B \in \mathcal{V}$$

are two decompositions of X. For an elementary functions $f \in \mathcal{E}$ then, according to the definition of the integral

$$\int_0^t f \, \mathrm{d}M + \int_0^t f \, \mathrm{d}A = \int_0^t f \, \mathrm{d}N + \int_0^t f \, \mathrm{d}B \tag{3.21}$$

The set of f satisfying this condition is closed with respect to pointwise convergence. By the monotone class theorem and the dominated convergence theorem, it follows:

Equation (3.21) holds also for bounded predictable processes $f \in \mathcal{P}$.

With a truncation argument, it follows: (3.21) holds for $f \in \mathcal{P}$ under the conditions

$$\begin{aligned} &\int_0^t f_s^2 \, \mathrm{d}\langle M\rangle_s < \infty \quad \text{a.s.,} \qquad \int_0^t |f_s| \, \mathrm{d}|A|_s < \infty \quad \text{a.s.,} \\ \text{and } &\int_0^t f_s^2 \, \mathrm{d}\langle N\rangle_s < \infty \quad \text{a.s.,} \qquad \int_0^t |f|_s \, \mathrm{d}|B|_s < \infty \quad \text{a.s.} \quad \forall t, \end{aligned} \tag{3.22}$$

i.e. equality holds in the class of all f with (3.22).

We, therefore, define the integration class

$$\mathcal{L}^0(X) := \{f \in \mathcal{L}(\mathcal{P});\ \exists \text{ decomposition } X = M + A,$$
$$\text{with } M \in \mathcal{M}^2_{\text{loc}},\ A \in \mathcal{V},\ \text{such that (3.22) holds}\Big\}.$$

For this class the definition of the integral is unambiguous and the integral properties hold as in Chap. 3.4.2, since essentially only the theorem of majorized convergence is used.

The statement on the quadratic variation for continuous semimartingales carries over as follows:

Proposition 3.91 (Quadratic Variation) *Let be $X \in \mathcal{S}$, be a càdlàg semimartingale and let (τ_i^n) be a sequence of decompositions with $|\Delta_n = | \longrightarrow 0$. Then*

$$Q_t^n := \sum \left(X_{\tau_{i+1}^n \wedge t} - X_{\tau_i^n \wedge t}\right)^2 \xrightarrow{P} [X]_t$$

and the representation formula holds

$$X_t^2 = X_0^2 + 2 \cdot \int_0^t X_{s-}\, \mathrm{d}X_s + [X]_t. \tag{3.23}$$

The convergence is uniform on $[0, t]$.

The proof of this proposition is similar to the proof in Proposition 3.78 using the theorem of dominated convergence. The essential step is the proof of the following

Proposition 3.92 *Let $X \in \mathcal{S}$ be càdlàg, Y càdlàg, and let (τ_i^n) be a sequence of decompositions with $\Delta_n \longrightarrow 0$. Let further $Z_t^n := \sum Y_{\tau_i^n}(X_{\tau_{i+1}^n \wedge t} - X_{\tau_i^n \wedge t})$. Then it holds*

$$\sup_{t \le T} \left| Z_t^n - \int_0^t Y_-\, \mathrm{d}X \right| \xrightarrow{P} 0.$$

With the quadratic variation, we also obtain the quadratic covariation by polarization from quadratic variation.

Definition 3.93 (Quadratic Covariation) For $X, Y \in \mathcal{S}$ we define the quadratic covariation

$$[X, Y] := \frac{1}{4}\big([X + Y] - [X - Y]\big)$$

The partial integration formula also carries over.

Proposition 3.94 (Partial Integration Formula) *For $X, Y \in \mathcal{S}$ it holds:*

$$X_t Y_t = X_0 Y_0 + \int_0^t X_- \,\mathrm{d}Y + \int_0^t Y_- \,\mathrm{d}X + [X, Y]_t.$$

In particular, Proposition 3.94 yields in the case of deterministic functions of finite variation, the classical partial integration formula as a special case (cf. Proposition 3.81).

Proposition 3.95 *Let $M, N \in \mathcal{M}_{\mathrm{loc}}$, and let $f \in \mathcal{L}(\mathcal{P})$ be locally bounded, then it holds:*

(a) $Y_t := \int_0^t f \,\mathrm{d}M = (f \cdot M)_t \in \mathcal{M}_{\mathrm{loc}}$
(b) $MN - [M, N] \in \mathcal{M}_{\mathrm{loc}}$.

Proof

(1) From the decomposition theorem, it follows that there exists a decomposition $M = M^* + A$ with $M^* \in \mathcal{M}^2_{\mathrm{loc}}$ and $A \in \mathcal{V}$. $A = M - M^*$ is as the difference of two local martingales itself a local martingale, $A \in \mathcal{M}_{\mathrm{loc}}$. By construction, the stochastic integral is $f \cdot M^* \in \mathcal{M}^2_{\mathrm{loc}}$.
It remains to show: $f \cdot A \in \mathcal{M}_{\mathrm{loc}}$.
Let $\tau_n^1 := \inf\{t; |A|_t \geq n\}$ and (τ_n^2) be a localizing sequence with $A^{\tau_n^2} \in \mathcal{M}$ and define $\tau_n := \inf\{\tau_n^1, \tau_n^2\}$. Then it holds for $t < \tau_n$ that $|A|_t \leq n$. Thus it follows $|A|_{\tau_n} \leq 2n + |A_{\tau_n}|$ and $|A|^{\tau_n} \in \mathcal{A}$.
Let $g = f \mathbb{1}_{[0,\tau_n]}$, $B = A^{\tau_n}$ then $(f \cdot A)^{\tau_n} = g \cdot B$ and $|B| \in \mathcal{A}$.
Define $\mathcal{H} := \{h \in \mathcal{L}(\mathcal{P}); h \text{ bounded}, h \cdot B \in \mathcal{M}\}$, then it holds:

$$\mathcal{H} \supset \mathcal{E}, \mathcal{H} \text{ is closed with respect to bounded pointwise convergence;}$$

so $\mathcal{H}$ is a monotone class.
Thus it follows: $\mathcal{H} \supset B(\mathcal{P})$ and, therefore, $f \cdot A \in \mathcal{M}_{\mathrm{loc}}$.

(2) follows from the characterization property a) and Proposition 3.91. □

Remark 3.96 For $M, N \in \mathcal{M}^2_{\mathrm{loc}}$ it holds

$$MN - \langle M, N \rangle \in \mathcal{M}_{\mathrm{loc}},$$

$\langle M, N \rangle \in \mathcal{V}_p$ is of finite variation and predictable.

The quadratic covariation $[M, N]$ is in general not predictable. By Proposition 3.95 it follows

$$[M, N] - \langle M, N \rangle \in \mathcal{M}_{\mathrm{loc}}$$

i.e. $\langle M, N\rangle$, the **predictable covariation** from M and N, is the predictable quadratic variation of the covariation process $[M, N]$.

The predictable variation process of the covariation is identical to the predictable covariation of M and N.

We now describe some properties of the **jumps of semimartingales**.

Let $(\Delta X)_t := X_t - X_{t-}$ be the **jump process** at t. If the process X is càdlàg, then there are at most countably many jumps.

Proposition 3.97 *Let $X \in \mathcal{S}$, $Y \in \mathcal{P}$ be locally bounded, $Z := Y \cdot X$, then it holds:*

(a) The jump process of Z is the integral $Y \cdot \Delta X$, $\Delta Z = Y \cdot \Delta X$
(b) $\Delta [X] = (\Delta X)^2$
(c) $\sum_{0 \le s \le t} (\Delta X_s)^2 \le [X]_t$.

Proof For the case that $Y = Y_-$ is left-continuous:

(a) Let $Y = Y_-$ and

$$Y_t^n := \sum Y_{\tau_i^n} \mathbb{1}_{(\tau_i^n, \tau_{i+1}^n]}(t),$$

$$Z_t^n := \sum Y_{\tau_i^n} \left(X_{\tau_{i+1}^n}^t - X_{\tau_i^n}^t\right) \quad \text{the integral of the elementary process}$$

Y_t^n w.r.t. X.

Then, because of the left-continuity of Y

$$Y_t^n \xrightarrow{P} Y_t.$$

By the dominated convergence theorem for semimartingales it follows analogously to Proposition 3.78:

$$\sup_{t \le T} \left| Z_t^n - \int_0^t Y \, \mathrm{d}X \right| \xrightarrow{P} 0.$$

It holds $\Delta Z^n \longrightarrow \Delta Z$ and by the dominated convergence theorem

$$Y^n \cdot \Delta X \longrightarrow Y \cdot \Delta X$$

because of the uniform convergence on compact intervals.

Because of $\Delta Z^n = Y^n \cdot \Delta X$, $\Delta Z = Y \cdot \Delta X$, therefore, it follows $\Delta Z = Y \cdot \Delta X$.

(b) Using the representation formula from Proposition 3.91 for quadratic variation, it follows:

$$X_t^2 = X_0^2 + 2 \cdot \int_0^t X_- \, \mathrm{d}X + [X]_t.$$

If we form the increment on both sides, then we get from (a):

$$\Delta X^2 = 2X_- \cdot \Delta X + \Delta \, [X].$$

On the other hand it holds:

$$\begin{aligned}(\Delta X)_t^2 &= (X_t - X_{t-})^2 \\ &= X_t^2 - X_{t-}^2 - 2 \cdot X_{t}\text{-}(X_t - X_{t}\text{-})(\Delta X^2)_t - 2X_{t-}(\Delta X)_t.\end{aligned}$$

Thus it follows $\Delta \, [X] = (\Delta X)^2$.

(c) follows by definition of $[X]$ and according to b), because the continuous component also contributes to the quadratic variation. □

Remark 3.98 The sum of the quadratic jumps of X is smaller than the variation process at t. In particular, this sum is finite. Typically, equality does not hold. In addition to the part of the quadratic variation that comes from the jumps, there is also a part that comes from the continuity regions of the process. The quadratic variation grows due to the jumps, but also due to the continuity regions.

A useful computational rule for quadratic covariation follows:

Lemma 3.99 *Let $X \in \mathcal{S}$, $A \in \mathcal{V}$, i.e. X is a semimartingale and A is of finite variation, then*

(a) If A or X is continuous, then $[X, A] = 0$.

(b) The covariation of X and A is the sum of the products of the jumps of the increments

$$[X, A] = \sum_{0<s\leq t} (\Delta X)_s (\Delta A)_s.$$

Proof

(b) Let $t_i^n = \frac{i}{2^n}t, n \geq 1, i \leq 2^n$ be a partition of the real axis, then it holds for $A \in \mathcal{V}$:

$$\sum_{i=1}^{2^n} X_{t_{i+1}^n}\left(A_{t_{i+1}^n} - A_{t_i^n}\right) \xrightarrow{P} \int_0^t X_s \, dA_s, \quad \text{because } X \text{ is càdlàg. Furtheron}$$

$$\sum_{i=1}^{2^n} X_{t_i^n}\left(A_{t_{i+1}^n} - A_{t_i^n}\right) \xrightarrow{P} \int_0^t X_{s-} \, dA_s \quad \text{(cf. Proposition 3.78).}$$

From this follows:

$$\begin{aligned}
[X, A]_t &= \lim \sum_{i=1}^{2^n} \left(X_{t_{i+1}^n} -, X_{t_i^n}\right)\left(A_{t_{i+1}^n} - A_{t_i^n}\right), \text{ stochastic limit} \\
&= \int_0^t X_s \, dA_s - \int_0^t X_{s-} \, dA_s \\
&= \int_0^t (\Delta X)_s \, dA_s \\
&= \sum_{s \leq t} (\Delta X)_s (\Delta A)_s, \text{ since } X \text{ has only a countable number of jumps.}
\end{aligned}$$

(a) follows from (b). □

Corollary 3.100 *Let X be a continuous semimartingale $X = X_0 + M + A \in \mathcal{S}_c$ with $M \in \mathcal{M}_{\mathrm{loc},c}$, $A \in \mathcal{V}_c$ then it follows:*

$$[X] = [M] = \langle M \rangle.$$

Proof

$$\begin{aligned}
[X] &= [M + A, M + A] \\
&= [M] + [A] + [M, A] + [A, M] \qquad \text{since [ ] is a bilinear form} \\
&= [M] + 0 + 0 + 0 \\
&= [M].
\end{aligned}$$

$[M]$ is continuous hence in particular predictable.

Further M is continuous, so $M \in \mathcal{M}_{\mathrm{loc}}^2$. Using the representation formula in Corollary 3.79 it follows:

$$M^2 - [M] \in \mathcal{M}_{\mathrm{loc},c}.$$

Because of the uniqueness of the predictable variation, it follows:

$$[M] = \langle M \rangle \quad \text{(cf. Theorem 3.54).}$$

□

Remark 3.101

(a) One can decompose general local martingales $M \in \mathcal{M}_{\text{loc}}$ in an unambiguous way

$$M = M_0 + M^c + M^d$$

into a purely continuous part $M^c \in \mathcal{M}_{\text{loc},c}$ and a purely discontinuous part M^d which is perpendicular to any continuous local martingale,

$$M^d \perp X, X \in \mathcal{M}_{\text{loc},c}, \quad \textit{i.e. it holds } M^d X \in \mathcal{M}_{\text{loc}}.$$

A process M^d is called purely discontinuous if for all $X \in \mathcal{M}_{\text{loc},c}$ it holds $M^d \perp X$, i.e. $M^d X$ is a local martingale, $M^d X \in \mathcal{M}_{\text{loc}}$:
The orthogonality relation $\perp$ on the set of local martingales has properties similar to the orthogonality relation in a Hilbert space. A continuous local martingale that is orthogonal to itself is zero.
For $X \in \mathcal{S}$ the decomposition $X = X_0 + M + A$ is in general not unique, but the continuous martingale part of X is independent of the representation.
Notation: $X^c := M^c$ is called the **continuous martingale part of** $\boldsymbol{X}$.
The general formula for the quadratic covariation of two semimartingales X, Y is

$$[X, Y]_t = \langle X^c, Y^c \rangle + \sum_{s \leq t} \Delta X_s \Delta Y_s.$$

$\langle X^c, Y^c \rangle$ is the quadratic covariation of the continuous part and is identical to the predictable quadratic covariation.

(b) The continuous martingale part of a semimartingale is independent of the decomposition. Based on this, one defines the semimartingale characteristic.

The **semimartingale characteristic of** $X \in \mathcal{S}_p$ has three components $(A, \langle X^c \rangle, \mu)$. Here $A \in \mathcal{V}$ is the "drift" of X, $\langle X^c \rangle$ is the predictable variation of the continuous martingale component X^c of X and μ is the compensator of the jump process.

In the case of general semimartingales $X \in \mathcal{S}$ one defines with help of a truncation function h with $h(x) = x$ in an neighborhood of 0, h bounded, e.g. $h(x) = x 1_{\{|x| \le 1\}}$,

$$\overline{X}_t(h) := \sum_{s \le t} (\Delta X_s - h(\Delta X_s)) = (x - h(x)) * \mu^X,$$

the sum of the large jumps, μ^X the jump measure of X. Then

$$X(h) := X - \overline{X}(h) \in \mathcal{S}_p$$

is a special semimartingale. The semimartingale characteristic of $X(h)$ then is called the characteristic of X.

For a Lévy process we obtain $A_t = a \cdot t$, $\langle X^c \rangle = \sigma^2 t$ and $\mu = \lambda\!\!\!\lambda^1 \otimes L$, L the Lévy measure. In this case, the characteristic is determined by the 'characteristic triplet' (a, σ, L) and it is deterministic.

Proposition 3.102 *Let $X, Y \in \mathcal{S}$, $f \in \mathcal{L}^0(X)$, then it holds:*

$$[f \cdot X, Y] = f \cdot [X, Y]$$

where $f \cdot [X, Y]$ is the Stieltjes integral with respect to $[X, Y]$.

Proof Using the partial integration formula, the above statement is equivalent to:

$$\begin{aligned}(f \cdot X)_t Y_t \\ &= (f \cdot X)_0 Y_0 + \int_0^t (f \cdot X)_- \, \mathrm{d}Y + \underbrace{\int_0^t (f \cdot Y)_- \, \mathrm{d}X}_{=\int_0^t Y_- \, \mathrm{d}(f \cdot X)} + \underbrace{\int_0^t f \, \mathrm{d}[X, Y]}_{=(f \cdot [X,Y])_t}.\end{aligned}$$

This equality holds for $f = U \cdot \mathbb{1}_{(r,s]}$, $U \in B(\mathcal{A}_r)$ by definition.
Because of the linearity of the integral, the statement follows for $f \in \mathcal{E}$.
Bounded predictable processes for which the equation holds are closed under bounded pointwise convergence. Therefore, by the monotone class theorem, equality follows for bounded predictable functions f. $\square$

Elements of Stochastic Analysis

4

The second main part of the textbook is devoted to stochastic analysis, its importance for model building and its analysis. Essential building blocks are the partial integration formula and the Itô formula with numerous applications, e.g., to Lévy's characterization of Brownian motion. The stochastic exponential is a solution of a fundamental stochastic differential equation and shows its importance in the characterization of equivalent measure changes (Girsanov's theorem). Together with the martingale representation theorem (completeness theorem) and the Kunita–Watanabe decomposition theorem, these form the foundation for arbitrage-free pricing theory in financial mathematics in the third part of the textbook.

A detailed section is devoted to stochastic differential equations. In particular, a comparison of several different approaches to diffusion processes is presented, namely (1) the Markov process (semigroup) approach, (2) the PDE approach (Kolmogorov equations), and (3) the approach by stochastic differential equations (Itô theory). Moreover, the particularly fruitful connection between stochastic and partial differential equations (Dirichlet problem, Cauchy problem, Feynman–Kac representation) is elaborated.

4.1 Itô Formula

The Itô formula is an important tool for studying functions of semimartingales. A fundamental question is closely related to the Itô formula: Let X be a semimartingale. For which functions f is $f(X)$ also a semimartingale and, if so, how do we obtain the decomposition into martingale and drift components?

L. Rüschendorf, *Stochastic Processes and Financial Mathematics*, Mathematics Study Resources 1, https://doi.org/10.1007/978-3-662-64711-0_4

Example 4.1 **(Semimartingale Decompositions)** For the Brownian motion and $f(x) = x^2$ it holds:

$$B_t^2 = 2 \underbrace{\int_0^t B_s \,\mathrm{d}B_s}_{\text{martingale}} + t.$$

In general, according to the formula for quadratic variation (Proposition 3.91) for a semimartingale X and $f(x) = x^2$ the representation

$$X_t^2 = X_0^2 + 2\int_0^t X_- \,\mathrm{d}X + [X]_t \text{ holds.}$$

For semimartingales $X, Y \in \mathcal{S}$ the partial integration formula (Proposition 3.102) states:

$$X_t Y_t = X_0 Y_0 + \int_0^t X_- \,\mathrm{d}Y + \int_0^t Y_- \,\mathrm{d}X + [X, Y]$$

The Itô formula provides a corresponding decomposition for general functions.

Definition 4.2 Let $X = (X^1, \dots, X^d) \in \mathcal{S}^1$ be a d-dimensional (semi-)martingale, i.e. X^i is a (semi-)martingale $\forall\, i$. If $X^i \in \mathbb{C}$ (complex-valued), then X is a (semi-) martingale if Re X^i and Im X^i are (semi-)martingales and we use the terms $\in \mathcal{S}^{\mathbb{C}}$, $\mathcal{M}^{\mathbb{C}}$, analogously for local martingales $\mathcal{M}_{\mathrm{loc}}^{\mathbb{C}}$ etc.

Theorem 4.3 (Itô Formula) *Let $X = (X^1, \dots X^n) \in \mathcal{S}_c^d$ be a continuous semimartingale, let $f \in C^2(\mathbb{R}^d, \mathbb{R}) = C^2$ be two times continuously differentiable, then it follows: $f(X) \in \mathcal{S}_c$ is a continuous semimartingale, and*

$$f(X_t) = f(X_0) + \sum_{i=1}^d \int_0^t \frac{\partial f}{\partial x_i}(X_s)\,\mathrm{d}X_s^i + \frac{1}{2}\sum_{i,j}\int_0^t \frac{\partial^2 f}{\partial x_i \partial x_j}(X_s)\,\underbrace{\mathrm{d}\langle X^i, X^j\rangle_s}_{=\mathrm{d}[X^i, X^j]_s}. \tag{4.1}$$

Proof In the first step of the proof, we show: If (4.1) holds for f holds, then also for $g(x) = x_i \cdot f(x)$, where x_i is the i-th component of x. Then, by induction, it follows that the formula is valid for all polynomials. We first consider the case ($d = 1$). Suppose that the Itô formula (4.1) is valid for f, i.e.

$$f(X) = f(X_0) + f'(X) \cdot X + \frac{1}{2} f''(X) \cdot \langle X \rangle.$$

Let $g(x) = xf(x)$, then $g'(x) = f(x) + xf'(x)$ and $g''(x) = f'(x) + f'(x) + xf''(x) = 2f'(x) + xf''(x)$. Thus, using the partial integration formula and (4.1) for f:

$$\begin{aligned} g(X) = Xf(X) &= X_0 f(X_0) + X \cdot f(X) + f(X) \cdot X + \langle X, f(X) \rangle \\ &= X_0 f(X_0) + (Xf'(X)) \cdot X + \frac{1}{2}\big(f''(X)X\big) \cdot \langle X \rangle \\ &\quad + f(X) \cdot X + \langle X, f'(X) \cdot X \rangle + \underbrace{\Big\langle X, \frac{1}{2} f''(X) \cdot \langle X \rangle \Big\rangle}_{=0} \\ &= g(X_0) + g'(X) \cdot X + \frac{1}{2} g''(X) \cdot \langle X \rangle, \end{aligned}$$

because the second variation term of the penultimate line is 0.

Similarly, for $d \geq 1$:

$$\begin{aligned} X_t^i f(X_t) &= X_0^i f(X_0) + \int_0^t X_s^i \,\mathrm{d} f(X_s) + \int_0^t f(X_s) \,\mathrm{d}X_s^i + \langle X^i, f(X) \rangle_t \\ &= g(X_0) + \int_0^t X_s^i \sum_j \frac{\partial f}{\partial x_j}(X_s) \,\mathrm{d}X_s^j \\ &\quad + \frac{1}{2} \sum_{j,k} \int_0^t X_s^i \frac{\partial^2 f}{\partial x_j \partial x_k}(x_s) \mathrm{d}\langle X^i, X^k \rangle_s \\ &\quad + \int_0^t f(X_s) \,\mathrm{d}X_s^i + \langle X^i, f(X) \rangle_t \\ &= g(X_0) + \sum_j \int_0^t \frac{\partial g}{\partial x_j}(X_s) \,\mathrm{d}X_s^j \\ &\quad + \frac{1}{2} \sum_{j,k} \int_o^t \frac{\partial^2 g}{\partial x_j \partial x_k}(X_s) \mathrm{d}\langle X^i, X^j \rangle_s. \end{aligned}$$

It follows: Itô's formula holds for polynomials. With stopping it is sufficient to prove (4.1) for processes X with values in a compact interval K. However, according to Weierstrass, any $f \in C^2$ can be uniformly approximated on K by a sequence of polynomials (P_n) in $C^2(K)$ so that $\| P_n - f \|_{C^2(K)} \to 0$, i.e. $P_n(X) \to f(X)$, $P_n^{(\alpha)}(X) \to f^{(\alpha)}(X)$ uniformly for $|\alpha| \leq 2$ (i.e., the first two derivatives converge). According to the theorem on majorized convergence, it follows that $P_n^{(i)}(X) \cdot X \to f^{(i)}(X) \cdot X$ and $\frac{\partial^2}{\partial x_i \partial x_j} P_n(X) \to \frac{\partial^2}{\partial x_i \partial x_j} f(X) \cdot \langle X \rangle$. This implies the assertion. $\square$

Remark 4.4

(a) Itô formula in differential notation

$$\mathrm{d}f(X_t) = \sum_i \frac{\partial f}{\partial x_i}(X_t)\,\mathrm{d}X_t^i + \frac{1}{2}\sum \frac{\partial^2 f}{\partial x_i \partial x_j}(X_t)\,\mathrm{d}\langle X^i, X^j\rangle_t.$$

(b) For $X \in \mathcal{S}_c$, $A \in \mathcal{V}$ we consider the continuous semimartingale $(X, A) \in \mathbb{R}^{d+1}$. Then it holds for $f \in C^{2,1}(\mathbb{R}^d \times \mathbb{R})$, $f = f(x, y)$

$$f(X_t, A_t) = f(X_0, A_0) + \sum_i \int_0^t \frac{\partial f}{\partial x_i}(X_s, A_s)\,\mathrm{d}X_s^i + \int_0^t \frac{\partial f}{\partial y}(X_s, A_s)\,\mathrm{d}A_s + \frac{1}{2}\sum_{i,j}\int_0^t \frac{\partial^2 f}{\partial x_i \partial x_j}(X_s, A_s)\mathrm{d}\langle X^i, X^j\rangle$$

(c) Let $d = 1$, $\Phi \in C^1$, $\operatorname{supp}\Phi \subset (0, t)$. Then with the partial integration formula it follows

$$X_t \underbrace{\Phi(t)}_{=0} = X_0 \underbrace{\Phi(0)}_{=0} + \int_0^t \Phi(s)\,\mathrm{d}X_s + \int_0^t X_s \Phi'(s)\,\mathrm{d}s + \underbrace{\langle X, \Phi\rangle}_{=0},$$

since Φ is continuous. From this follows

$$\int_0^t \Phi(s)\,\mathrm{d}X_s = -\int_0^t X_s \Phi_s'\,\mathrm{d}s.$$

This identity can be taken as the starting point for the definition of the stochastic integral, which is then extended to more general functions $\Phi(s)$ in the sense of distributions. However, the general stochastic integral also allows stochastic processes Φ as integrands.

(d) **General Itô formula** The Itô formula can also be extended to general non-continuous semimartingales. Let $X \in \mathcal{S}^d$, $f \in C^{2,1}$, $f = f(x, s)$, then it holds:

$$f(X_t, t) = f(X_0, 0) + \int_0^t f_s(X_s, s)\,\mathrm{d}s + \sum_j \int_0^t f_j(X_{s-}, s)\,\mathrm{d}X_s^j + \frac{1}{2}\sum_{j,k}\int_0^t f_{j,k}(X_{s-}, s)\mathrm{d}[X^j, X^k]_s$$

$$+ \sum_{0 \le s \le t} \Big\{ f(X_s, s) - f(X_{s-}, s) - \sum_j f_j(X_{s-}, s) \Delta X_s^j$$
$$- \frac{1}{2} \sum_{j,k} f_{j,k}(X_{s-}, s) \Delta X_s^j \Delta X_s^k \Big\}.$$

Remark 4.5 The formula is also valid for $f : U \longrightarrow \mathbb{R}$, $U \subset \mathbb{R}^d$ open, if $X, X_- \in U$, i.e. the formula is also valid for mappings of open subsets of $\mathbb{R}^d$ to $\mathbb{R}$ if X and the left-side limit X_- lie in U. The above formula gives a semimartingale representation of $f(X)$.

The Itô formula is a chain rule for stochastic integration. In particular, it states that for a C^2-function f and a semimartingale X, $f(X)$ is again a semimartingale. From the Itô formula, one can read of the semimartingale representation. In particular, for a martingale X and a convex function f, $f(X)$ is a submartingale. In this case, the Itô formula yields the Doob–Meyer decomposition of this submartingale.

The following are some direct applications of the Itô formula.

Example 4.6 Let B be the Brownian motion and $f(x) = x^m$, then according to the Itô formula

$$B_t^m = m \int_0^t B_s^{m-1} \,\mathrm{d}B_s + \frac{m(m-1)}{2} \int_0^t B_s^{m-2} \,\mathrm{d}s.$$

This leads for $m = 2$ to

$$B_t^2 = 2 \int_0^t B_s \,\mathrm{d}B_s + t \quad \textit{also} \quad \int_0^t B_s \,\mathrm{d}B_s = \frac{1}{2}(B_t^2 - t).$$

For $m = 3$ we obtain with the help of the case $m = 2$ and partial integration

$$\begin{aligned} B_t^3 &= 3 \int_0^t B_s^2 \,\mathrm{d}B_s + 3 \int_0^t B_s \,\mathrm{d}s \\ &= 6 \int_0^t \Big(\int_0^s B_u \,\mathrm{d}B_u \Big) \mathrm{d}B_s + 3 \int_0^t s \,\mathrm{d}B_s + 3 \int_0^t B_s \,\mathrm{d}s \\ &= 3t B_t + 6 \int_0^t \Big(\int_0^s B_u \,\mathrm{d}B_u \Big) \mathrm{d}B_s. \end{aligned}$$

So one gets a formula with an iterated stochastic integral.

In the following theorem, the Itô formula is used to construct an important class of local martingales, the stochastic exponentials.

Theorem 4.7 (Exponential Martingale) *Let $M \in \mathcal{M}_{\text{loc},c}$ and let $f \in C^{2,1}(\mathbb{R} \times \mathbb{R}_+)$, $f = f(x, y)$ be a solution of the partial differential equation*

$$\frac{\partial f}{\partial y} + \frac{1}{2}\frac{\partial^2 f}{\partial x^2} = 0.$$

Then $f\big(M_t, \langle M\rangle\big)$ is a continuous local martingale,

$$f\big(M_t, \langle M\rangle\big) \in \mathcal{M}_{\text{loc},c}.$$

In particular for $\lambda \in \mathbb{C}$ it holds

$$\mathcal{E}^\lambda(M)_t := \exp\left(\lambda M_t - \frac{\lambda^2}{2}\langle M\rangle_t\right) \in \mathcal{M}_{\text{loc},c}.$$

For $\lambda = 1$, $\mathcal{E}^1(M) := \mathcal{E}(M)$ is called exponential martingale or stochastic exponential.

Proof Let $f(x, y) := e^{x-\frac{1}{2}y}$. Then it follows: $\frac{\partial f}{\partial x} = e^{x-\frac{1}{2}y} = f$, $\frac{\partial f}{\partial y} = -\frac{1}{2}e^{x-\frac{1}{2}y}$, $\frac{\partial^2 f}{\partial x^2} = e^{x-\frac{1}{2}y}$.

So it holds: $\frac{\partial f}{\partial y} + \frac{1}{2}\frac{\partial^2 f}{\partial x^2} = 0$, that is, this function f satisfies the differential equation from the first part of the theorem. So the second part is a special case of the first part (complex martingale) noting that $\langle \lambda M\rangle = \lambda^2\langle M\rangle$.

For the proof of the first part, we consider the semimartingale $\big(M_t, \langle M\rangle_t\big) \in \mathcal{S}_c$. Then it follows from the Itô-formula using the above assumption on f

$$\begin{aligned} f\big(M_t, \langle M\rangle_t\big) &= f\big(M_0, \langle M_0\rangle\big) + \int_0^t \frac{\partial f}{\partial y}(M_s, \langle M\rangle_s)\mathrm{d}\langle M\rangle_s \\ &\quad + \int_0^t \frac{\partial f}{\partial x}\big(M_s, \langle M\rangle_s\big)\,\mathrm{d}M_s + \int_0^t \frac{1}{2}\frac{\partial^2 f}{\partial x^2}(M_s, \langle M\rangle)\mathrm{d}\langle M\rangle_s \\ &= f(M_0, \langle M\rangle_0) + \int_0^t \frac{\partial f}{\partial x}\big(M_s, \langle M\rangle_s\big)\,\mathrm{d}M_s \in \mathcal{M}_{\text{loc},c}. \end{aligned} \tag{4.2}$$

Since this is a stochastic integral with respect to a local martingale, the result is a local martingale.

In the special case of the stochastic exponential the choice $f(x, y) = e^{x-\frac{1}{2}y}$ (4.2) results in:

$$\mathcal{E}^\lambda(M)_t = 1 + \int_0^t \mathcal{E}^\lambda(M)_s\,\mathrm{d}M_s.$$

□

The exponential martingale is the solution of a basic stochastic differential equation, which explains the name *stochastic exponential.*

Corollary 4.8 (Stochastic Exponential) *Let $M \in \mathcal{M}_{\text{loc},c}$, then*

$$\mathcal{E}(M) = e^{M-\frac{1}{2}\langle M\rangle}$$

is the unique continuous solution of the integral equation

$$X_t = 1 + \int_0^t X_s \, dM_s. \tag{4.3}$$

In differential notation: $dX_t = X_t \, dM_t$ with the initial condition $X_0 = 1$.

Proof

(a) **Existence** $X = \mathcal{E}(M)$ is a solution of the differential equation (4.3) by the proof of Theorem 4.7.
(b) **Uniqueness** $X_t := \mathcal{E}(M)_t$ is solution of (4.3). Let Z be another solution of the integral equation (4.3). We set $X := \mathcal{E}(M)$ and $V := \frac{1}{X} = e^{-(M-\frac{1}{2}\langle M\rangle)} = e^{-M} \cdot e^{\frac{1}{2}\langle M\rangle}$.
V is the product of two processes. According to the partial integration formula:

$$dV = e^{-M} d\big(e^{\frac{1}{2}\langle M\rangle}\big) + e^{\frac{1}{2}\langle M\rangle} d\big(e^{-M}\big) + d\big(\underbrace{\langle e^{-M}, e^{\frac{1}{2}\langle M\rangle}\rangle}_{=0}\big).$$

□

For $e^{-M} = f(M)$ it follows with the Itô formula:

$$\begin{aligned} dV &= \frac{1}{2} e^{-M} e^{\frac{1}{2}\langle M\rangle} d\langle M\rangle + e^{\frac{1}{2}\langle M\rangle}\Big(- e^{-M} dM + \frac{1}{2} e^{-M} d\langle M\rangle\Big) \\ &= V d\langle M\rangle - V \, dM. \end{aligned} \tag{4.4}$$

We define $\xi_t := Z_t V_t = \frac{Z_t}{X_t}$. To show: $\xi_t = 1, \forall t$. For the proof we obtain with the formula for partial integration

$$\begin{aligned} \xi_t = Z_t V_t &= 1 + \int_0^t (Z \, dV + V \, dZ) + [V, Z]_t \\ &= 1 + \int_0^t \{(-ZV) \, dM + VZ d\langle M\rangle + VZ \, dM\} + [V, Z] \\ &= 1 + \int_0^t \xi d\langle M\rangle + [V, Z]. \end{aligned}$$

All processes are continuous. Thus it follows by (4.4) using $Z = 1 + Z \cdot M$

$$\begin{aligned}[V, Z] &= [V \cdot \langle M \rangle - V \cdot M, Z] \\ &= -V \cdot [Z, M] = -(VZ) \cdot [M] \\ &= -\xi \cdot [M],\end{aligned}$$

because $[V \cdot \langle M \rangle, Z] = 0$ by Lemma 3.99. But from this it follows

$$\xi = 1 + \xi \cdot \langle M \rangle - \xi \cdot \langle M \rangle = 1.$$

□

Remark 4.9

(a) **Doléans–Dade formula** We have considered the stochastic integral equation (4.3) only for continuous local martingales. The equation is also relevant for arbitrary semimartingales. Let $X \in \mathcal{S}$, then the generalized stochastic integral equation

$$Y = 1 + Y_- \cdot X.$$

in the differential notation $dY = Y_- dX$, $Y_0 = 1$, has as unique solution $\mathcal{E}(X)_t$, the **generalized stochastic exponential**

$$\mathcal{E}(X)_t := \exp\Big(X_t - X_0 - \frac{1}{2}\langle X^c \rangle_t\Big) \prod_{s \le t} \big(1 + \Delta X_s\big) e^{-\Delta X_s}$$

and this is called **Doléans–Dade formula**. Here $\langle X^c \rangle$ is the predictable quadratic variation of the continuous martingale part X^c of X. The multiplicative part $\prod_{s \le t} \big(1 + \Delta X_s\big) e^{-\Delta X}$ is equal to the jump part.

(b) **Stochastic logarithm** Let $Z \in \mathcal{S}$, $Z > 0$, $(Z, Z_- \neq 0)$ be a positive semimartingale. In real analysis, the inverse function of the exponential is the logarithm. By analogy, the 'inverse function' of the stochastic exponential is called the stochastic logarithm. Let

$$X := \frac{1}{Z_-} \cdot Z,$$

then it holds:

X is the unique element (semimartingale) $X \in \mathcal{S}$ with $X_0 = 0$ and $Z = Z_0 \mathcal{E}(X)$, i.e. Z is the stochastic exponential of the stochastic logarithm X. Notation: $\mathfrak{L}(Z) := X$.

Proof Let $X = \mathfrak{L}(Z)$ then $X_0 = 0$ and

$$\begin{aligned} Z &= Z_0 + 1 \cdot Z \\ &= Z_0 + Z_- \cdot \Big(\frac{1}{Z_-} \cdot Z\Big) \\ &= Z_0 + Z_- \cdot X, \end{aligned}$$

i.e., Z is a solution of this equation. Thus it follows: Z is a solution of the stochastic exponential equation (4.3), $Z = Z_0\mathcal{E}(X)$.

Uniqueness Let $X \in \mathcal{S}$, $X_0 = 0$ and $Z := Z_0\mathcal{E}(X)$. Assertion: $X = \mathfrak{L}(Z)$.

It holds: Z is a solution of the equation for the stochastic exponential:

$$Z = Z_0 + Z_- \cdot X.$$

From this it follows

$$\begin{aligned} \frac{1}{Z_-} \cdot Z &= \frac{1}{Z_-} \cdot (Z_- \cdot X) \\ &= \Big(\frac{1}{Z_-} Z_-\Big) \cdot X \\ &= 1 \cdot X = \int 1 \, \mathrm{d}X = X - X_0 = X. \end{aligned}$$

So the stochastic logarithm is the unique solution of the basic stochastic differential equation in (4.3). The stochastic logarithm $\mathfrak{L}(Z)$ is also well-defined for $Z \in \mathcal{S}$ such that $Z, Z_- \in \mathbb{R} \setminus \{0\}$. □

Corollary 4.10 *Let $\varphi \in \mathcal{L}^0(B)$, then the stochastic differential equation*

$$\mathrm{d}L = L\varphi \, \mathrm{d}B, \qquad L_0 = 1$$

has a unique (continuous) solution.

$$L_t = \exp\Big(\int_0^t \varphi_s \, \mathrm{d}B_s - \frac{1}{2}\int_0^t \varphi_s^2 \, \mathrm{d}s\Big).$$

Proof The equation is a special case of Remark 4.9, with $M := \varphi \cdot B \in \mathcal{M}_{\mathrm{loc},c}$. Then it holds

$$\langle M \rangle_t = \Big\langle \int_0^t \varphi_s \, \mathrm{d}B_s \Big\rangle = \langle \varphi \cdot B \rangle_t = (\varphi^2 \cdot \langle B \rangle)_t = \int_0^t \varphi_s^2 \, \mathrm{d}s.$$

According to Remark 4.9 therefore

$$L_t = \exp\Big(\int_0^t \varphi_s \, \mathrm{d}B_s - \frac{1}{2}\int_0^t \varphi_s^2 \, \mathrm{d}s\Big) = \exp\big(M_t - \frac{1}{2}\langle M\rangle_t\big) = \mathcal{E}(M)_t$$

is a unique solution of the equation $\mathrm{d}L = L\,\mathrm{d}M = L\varphi\,\mathrm{d}B$. □

In Sect. 4.4 general statements about the solution of stochastic differential equations are derived. Stochastic differential equations, like deterministic differential equations, are important for modeling. The reason is that often only local quantities, such as drift and diffusion, are needed for the formulation of these differential equations. Such local quantities are known in many applications. The solution of a (stochastic) differential equation then gives information about the global behavior of the described process. The following example, geometric Brownian motion, is the standard model used in financial mathematics.

Remark 4.11 **(Geometric Brownian Motion)** Let $\widehat{S}_t := e^{\sigma B_t - \frac{1}{2}\sigma^2 t}$, then $\widehat{S}_t$ is an exponential martingale, so it solves the differential equation (cf. Corollary 4.10)

$$\mathrm{d}\widehat{S}_t = \widehat{S}_t \sigma \, \mathrm{d}B_t.$$

$\sigma > 0$ is a positive constant. It is called in financial mathematics **volatility**. We define the product

$$S_t := e^{\mu t}\widehat{S}_t = e^{(\mu - \frac{1}{2}\sigma^2)t + \sigma B_t},$$

then it follows from the partial integration formula:

$$\begin{aligned} \mathrm{d}S_t &= e^{\mu t} d\widehat{S}_t + \mu e^{\mu t}\widehat{S}_t \, \mathrm{d}t = \mu S_t \, \mathrm{d}t + \sigma S_t \, \mathrm{d}B_t \\ &= S_t(\mu \, \mathrm{d}t + \sigma \, \mathrm{d}B_t). \end{aligned} \tag{4.5}$$

(S_t) is called a **geometric Brownian motion** with drift μ and volatility σ. (S_t) is an exponential process with independent logarithmic increases.
(S_t) is a standard model for the evolution of stock prices. The reason is the following: If one writes the above differential equation in discretized form, then the relative increments (returns) are given by

$$r_t := \frac{S_{t+h} - S_t}{S_t}.$$

These are often approximately i.i.d. and normally distributed in empirical data. Equivalently, the logarithmic increments are independent and normally distributed.

This is also a justification for the fact that important models in financial mathematics are exponential Lévy models. It are not the price processes themselves that are Lévy processes, but the logarithmic price processes that are Lévy processes. The most important example of exponential Lévy processes is the geometric Brownian motion introduced by Samuelson. The Black–Scholes formula is based on this model. With the differential equation (4.5) one can understand this process well: The local change is proportional to the process itself. We have a drift term and a volatility term. Using the Itô formula, we obtain the solution of this differential equation in explicit form.

Some remarks concerning the martingale property of the stochastic exponential follow.

Proposition 4.12

(a) Let M be a nonnegative local martingale $M \in \mathcal{M}_{\mathrm{loc},c}$, $M \geq 0$, then M is a supermartingale.
(b) Let $M \in \mathcal{M}_{\mathrm{loc}}$, $M_0 = 0$, then it holds:
  (1) $N := \mathcal{E}(M)$ is a supermartingale with $EN_t \leq 1 \quad \forall t$
  (2) $N \in \mathcal{M} \Longleftrightarrow EN_t = 1 \quad \forall t$
  (3) Each of the following conditions

$$Ee^{\frac{1}{2}\langle M\rangle_t} < \infty \quad \forall t \leq T \qquad \textbf{\textit{Novikov condition}}$$

$$E\exp\left(\frac{1}{2}M_t\right) < \infty \quad \forall t \leq T \qquad \textbf{\textit{Kazamaki condition}}$$

is sufficient for $\mathcal{E}(M)$ being a martingale on $[0, T]$.

Proof

(a) Let τ_n be a localizing sequence, then it follows

$$E(M_{t\wedge\tau_n} \mid \mathcal{A}_s) = M_{s\wedge\tau_n}, \qquad s \leq t. \tag{4.6}$$

Because of $M \geq 0$ there exists a lower bound for M. Using Fatou's lemma, it follows for $n \to \infty$:

$$\begin{aligned} M_s = \liminf M_{s\wedge\tau_n} &= \liminf E(M_{t\wedge\tau_n} \mid \mathcal{A}_s) \\ &\geq E(\liminf M_{t\wedge\tau_n} \mid \mathcal{A}_s) = E(M_t \mid \mathcal{A}_s). \end{aligned}$$

We obtain the characteristic equation for a supermartingale.

(b) (1) follows from (a)
  (2) Only the direction "$\Leftarrow$" is to be shown. By assumption N is a nonnegative local martingale, so it is also a supermartingale. According to (a) it,

therefore, follows $E(N_t \mid \mathcal{A}_s) \le N_s$. This implies

$$EN_t = EE(N_t \mid \mathcal{A}_s) \le EN_s.$$

Equality holds if and only if N is a martingale.

(3) To show: the conditions imply that one can choose a sequence of localizations such that the sequence of stopped processes $\{N_{t\wedge\tau_n}\}$ is uniformly integrable. We dispense with the somewhat involved proof. In the case M is a bounded martingale, this condition is trivially satisfied.

□

An auxiliary statement about the covariation of independent processes follows.

Lemma 4.13 *Let $M, N \in \mathcal{M}^2_{\mathrm{loc}}$. If M and N be stochastically independent, then it follows:*

$$\langle M, N\rangle = 0.$$

Proof We consider the special case that M and N are local martingales with respect to a common filtration $(\mathcal{A}_t)_{t\ge 0}$.

The condition $\langle M, N\rangle = 0$ is equivalent to the condition that $MN \in \mathcal{M}_{\mathrm{loc}}$, because

$$MN - 0 \in \mathcal{M}_{\mathrm{loc}} \quad \Longleftrightarrow \quad \langle M, N\rangle = 0;$$

The predictable quadratic covariation $\langle M, N\rangle$ is the uniquely determined process of finite variation that subtracted from MN makes it a local martingale.

Without restriction (via localization), we can assume that both processes are square-integrable martingales, $M, N \in \mathcal{M}^2$.

Let $\widetilde{\mathcal{A}}_s := \sigma\left(\mathcal{A}_s, \sigma(N)\right)$, be the σ-algebra in which the σ-algebra generated by the process N is added to the filtration. Then it follows

$$\begin{aligned} E(M_t N_t \mid \mathcal{A}_s) &= E\left(E(M_t N_t \mid \widetilde{\mathcal{A}}_s) \mid \mathcal{A}_s\right) \\ &= E\left(N_t E(M_t \mid \widetilde{\mathcal{A}}_s) \mid \mathcal{A}_s\right) \end{aligned}$$

It holds $E(M_t \mid \widetilde{\mathcal{A}}_s) = E(M_t \mid \mathcal{A}_s)$, since M, N are independent. It follows that

$$\begin{aligned} E(M_t N_t \mid \mathcal{A}_s) &= E\left(N_t E(M_t \mid \mathcal{A}_s) \mid \mathcal{A}_s\right) \\ &= M_s E(N_t \mid \mathcal{A}_s) = M_s N_s. \end{aligned}$$

□

Remark 4.14 Two processes $M, N \in \mathcal{M}^2_{\text{loc}}$ are called **orthogonal** if $MN \in \mathcal{M}_{\text{loc}}$. The notion was already introduced in the decomposition of local martingales into a continuous and an absolutely discontinuous part. The property that M is absolutely discontinuous is equivalent to saying that M is orthogonal to any continuous local martingale $N \in \mathcal{M}_{\text{loc},c}$.

The following proposition gives a characterization of harmonic functions from analysis using Brownian motion.

Proposition 4.15 (Harmonic Functions) *Let B be a Brownian motion, let $f \in C^{2,1}(\mathbb{R}^d, \mathbb{R}_+)$ and let the differential operator $\mathcal{L}$ be defined by*

$$\mathcal{L}f := \frac{1}{2}\Delta f + \frac{\partial f}{\partial t} = \frac{1}{2}\sum \frac{\partial^2 f}{\partial x_i^2} + \frac{\partial f}{\partial t}.$$

Then

$$M_t^f := f(B_t, t) - \int_0^t \mathcal{L}f(B_s, s)\,\mathrm{d}s \in \mathcal{M}_{\text{loc},c}.$$

In particular:
A function $f = f(x)$ is **harmonic** in $\mathbb{R}^d$ (i.e. $\Delta f = 0$) if and only if $f(B) \in \mathcal{M}_{\text{loc},c}$.

Proof Analogous to the proof of Theorem 4.7 this statement follows from the Itô formula and Lemma 4.13. In the Itô formula, the covariation terms occur. For independent processes these covariation terms are zero. Therefore, only the diagonal terms remain here:

$$\langle B^i, B^j \rangle_t = \partial_{ij} t.$$

Thus for $f = f(x)$

$$\begin{aligned} f(B_t) \in \mathcal{M}_{\text{loc},c} &\iff \int_0^t \Delta f(B_s)\,\mathrm{d}s = 0, \qquad \forall t \\ &\iff \Delta f(B_t) = 0, \qquad \forall t, \qquad \text{as } \Delta f \text{ is continuous} \\ &\iff \Delta f = 0 \qquad \text{i.e. } f \text{ is harmonic.} \end{aligned}$$

□

Remark 4.16 **(Infinitesimal Generator)**

(a) More generally it holds: For continuous Markov processes X there exists an operator $\mathcal{L}$ (differential operator, difference operator, or in general a linear continuous operator) such that.

$$f(X_t, t) - \int_0^t \mathcal{L}f(X_s, s)\mathrm{d}\langle X\rangle_s \in \mathcal{M}_{\mathrm{loc},c}. \tag{4.7}$$

The linear continuous operator $\mathcal{L}$ is called **infinitesimal generator** of the Markov process. This equation characterizes the Markov process X. For the Brownian motion $\mathcal{L}$ is the partial differential operator of second order, $\mathcal{L}f = \frac{1}{2}\Delta f + \frac{\partial f}{\partial t}$, $f \in C^{2,1}$; $\mathcal{L}$ is associated to the heat equation. The elliptic operators (second order partial differential operators) are associated with a subclass of Markov processes, the **diffusion processes**. The problem of finding a process by means of a martingale property as in (4.7) is called the **martingale problem.**

(b) **Uniqueness of the Dirichlet problem**

Let $D \subset \mathbb{R}^d$ be a bounded domain and let h be a continuous function on the boundary of D, $h \in C(\partial D)$. In the Dirichlet problem the search is for a function $u \in C^2(D) \cap C(\overline{D})$ that is twice continuously differentiable in the interior of the domain and continuous up to the boundary, such that two properties hold:

(1) $u = h$ on ∂D, i.e. on the boundary of D u coincides with h.
(2) In the interior of D u is harmonic, i.e. in the interior the partial differential equation $\Delta u = 0$ holds.

Using the strong Markov property of the Brownian motion, it follows for domains with regular boundary that

$$u(x) := E_x h(B_{\tau_D})$$

is a solution of the Dirichlet problem. Here τ_D is the first exit time from D. One usually infers the uniqueness of the solution using an analytic result, the maximum principle for harmonic functions. We now show that uniqueness of solutions can also be inferred directly from the Itô formula.

Uniqueness proof of the Itô formula

Let u be a solution of the Dirichlet problem on D. Then it follows by the Itô formula

$$u(B_t) = u(B_0) + \underbrace{\int_0^t \nabla u(B_s) \cdot \mathrm{d}B_s}_{\text{product in } \mathbb{R}^d} + \underbrace{\frac{1}{2}\int_0^t \Delta u(B_s)\, \mathrm{d}s}_{=0}$$

Here we use $\langle B^i, B^j\rangle_t = 0, \forall i \neq j$, since the components of B are independent and, therefore, only the diagonal terms remain, and further that u is harmonic. From this we obtain

$$u\big(B_{t\wedge\tau_D}\big) = u(B_0) + \underbrace{\int_0^{t\wedge\tau_D} \nabla u(B_s)\cdot \mathrm{d}B_s}_{M_t\in\mathcal{M}}\,.$$

The gradient of u is continuous and B_s moves up to time τ_D in the bounded domain D, i.e. M is bounded and hence M is a martingale, $M \in \mathcal{M}_c$. Using the optional sampling theorem, it now follows that

$$E_x u\big(B_{t\wedge\tau_D}\big) = u(x), \quad \forall t.$$

For $t \to \infty$ it follows by the theorem on majorized convergence and using that $u \in C(\overline{D})$

$$u(x) = E_x u\big(B_{\tau_D}\big) = E_x h\big(B_{\tau_D}\big)$$

because on the boundary u and h coincide. Thus, the uniqueness of the solution follows.

(c) **Recurrence of Brownian motion**

The basic problem of this section is the question of recurrence of the Brownian motion in $\mathbb{R}^d$, i.e., for any starting point x any arbitrarily small sphere in $\mathbb{R}^d$ is reached at some time point by the Brownian motion. W.l.g., we consider a sphere with center in 0 (Fig. 4.1).

Let B be a d-dimensional Brownian motion, i.e., the components of B are independent Brownian motions. We consider a circular ring $D := B(0, R) -$

Fig. 4.1 Recurrence

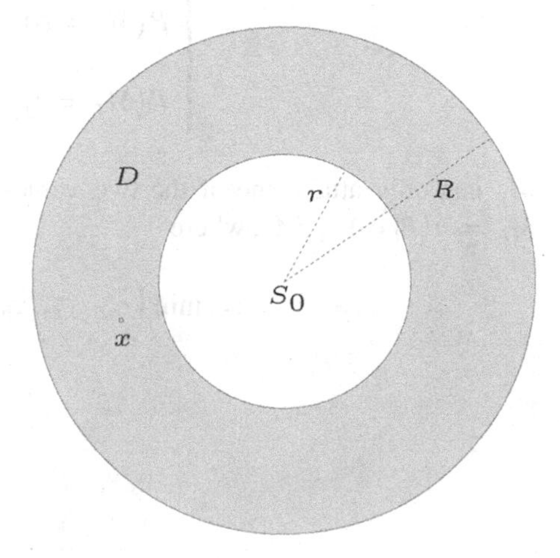

$B(0, r)$ around zero with $0 < r < R < \infty$. In the circular ring at point x starts the Brownian motion B. With what probability is the small circle $B(0, r)$ more likely to be earlier reached than the complement of the large circle $B(0, R)$?

This problem can be treated with a property of harmonic functions that follows from the Itô formula. We consider for dimension d the harmonic function u

$$d = 2, \quad u(x) := -\log|x| = u(|x|),$$
$$d \geq 3, \quad u(x) := |x|^{2-d} = u(|x|).$$

u is harmonic on D, i.e., Laplace's equation holds

$$\Delta u = 0 \; in \; D \quad \left(therefore\ use\ \frac{\partial|x|}{\partial x_i} = \frac{x_i}{|x|}\right).$$

We consider a Brownian motion starting at a point x of the circular ring D,

$$B_0 = x \in D.$$

Then, according to Proposition 4.15 for a stopping time $\widetilde{\tau} \leq \tau_R$ it holds

$$M_t := u(B_{t\wedge\widetilde{\tau}}) \in \mathcal{M}_c$$

because on a bounded domain u is bounded. We now use a well-known property of continuous real martingales.

Let $M \in \mathcal{M}_c$ be a continuous real martingale. We start M at a point $x \in [a, b]$ and wait until one of the two boundaries a, b is reached. We assume that this stopping time is finite, $\tau := \tau_{a,b} < \infty$. Then, according to the optional sampling theorem.

$$\begin{cases} P(M_\tau = a) = \dfrac{b-x}{b-a} & and \\[2ex] P(M_\tau = b) = \dfrac{x-a}{b-a} \end{cases} \tag{4.8}$$

M_τ can only attain one of the two values a or b (Fig. 4.2). Now let specifically $M_t := u(B_{t\wedge\widetilde{\tau}}) \in \mathcal{M}_c$, where

$$\widetilde{\tau} := \min\left\{\tau_{\partial B(0,r)}, \tau_{\partial B(0,R)}\right\}, \quad x \in D.$$

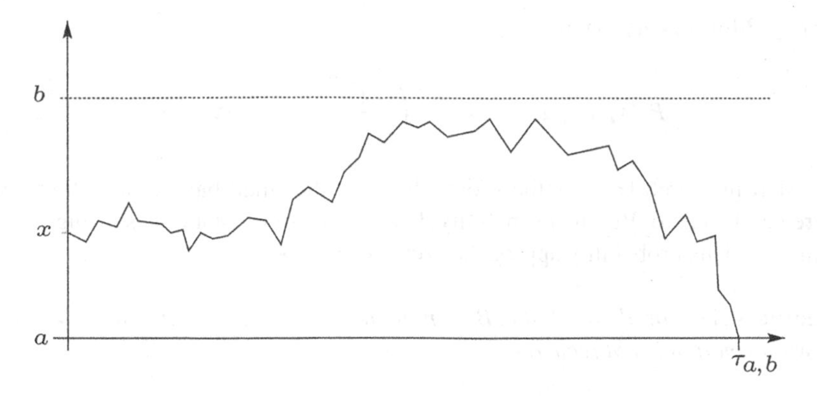

Fig. 4.2 Real continuous martingale M, stopping time $\tau_{a,b}$

$\widetilde{\tau}$ is the first time when the inner or the outer sphere surface is reached. From the behavior of the Brownian motion, we know that the outer sphere is reached in finite time because $\limsup |B_t| = \infty$. Therefore $\widetilde{\tau}$ is a finite stopping time. We now apply (4.8) to the above martingale. Let

$$a := u(R) \quad \text{and} \quad b := u(r).$$

Then $u(r) > u(R)$, because u is antitone, $u \downarrow$. Let $\tau := \tau_{a,b}$ then it follows by (4.8)

$$\begin{aligned} P_x\big(\tau_{\partial B(0,r)}(B) < \tau_{\partial B(0,R)}(B)\big) &= P_{u(x)}\big(M_\tau = u(r)\big) \\ &= \begin{cases} \dfrac{\log R - \log(|x|)}{\log R - \log r}, & d = 2, \\ \dfrac{|x|^{2-d} - R^{2-d}}{r^{2-d} - R^{2-d}}, & d \geq 3. \end{cases} \end{aligned}$$

The problem for Brownian motion is transformed to a stopping problem for the transformed Brownian motion. The probability that the smaller sphere is reached before the larger sphere is reached is equal to the probability that $M_\tau = u(r)$, i.e. $u(B_t)$ reaches $u(r)$ first. $u(r) = b$ gives the upper bound and $u(R) = a$ gives the lower bound after the transformation. The transformation is strictly antitonic, i.e., the order is preserved.

For $R \to \infty$ it follows for $d = 2$:

$$\begin{aligned} P_x\big(\exists t > 0; \quad B_t \in B(0,r)\big) &= P_x\big(\tau_{B(0,r)}(B) < \infty\big) \\ &= \lim_{R\to\infty} P_x\big(\tau_{B(0,r)}(B) < \tau_{B(0,R)}(B)\big) = 1. \end{aligned}$$

For $d \geq 3$ follows analogously

$$P_x\left(\tau_{B(0,r)}(B) < \infty\right) = \left(\frac{|x|}{r}\right)^{2-d}, \qquad |x| > r.$$

At a start in x with $|x| > r$ the probability that the small ball is hit in finite time is greater than zero. But the probability depends on how distant the starting point is from zero. This probability approaches zero for $|x| \to \infty$.

Theorem 4.17 *For $d = 2$ the Brownian motion is recurrent, for $d \geq 3$ the Brownian motion is not recurrent.*

Interpretation: The result is analogous to the case of a random walk on the d-dimensional lattice. For dimensions 1 and 2 the random walk is recurrent, but for higher dimensions it is not recurrent. If there is too much space, the small sphere in $d \geq 3$ will not be found with positive probability by the Brownian motion.

Another important application of the Itô formula is the characterization of the Brownian motion by Lévy. A Brownian motion is a process with continuous paths defined by two properties:

(1) The increments of X are independent $\Longleftrightarrow$ $X_t - X_s, \mathcal{A}_s$ are independent for $s \leq t$, i.e., for all $s < t$ the increments are independent of the past in $\mathcal{A}_s$.
(2) The increments of X are normally distributed, $X_t - X_s \sim N(0, t-s)$.

According to Proposition 4.15 it follows that harmonic functions applied to the Brownian motion are continuous local martingales. The question arises: Does this martingale property characterize Brownian motion? The surprising result is: just two martingale functionals are sufficient to characterize Brownian motion, namely.

$$B_t \quad \text{and} \quad B_t^2 - t.$$

These are just the first two Hermitian polynomials applied to the Brownian motion. Hermitian polynomials applied to Brownian motion generally yield martingales.

Theorem 4.18 (Characterization of the Brownian Motion according to Lévy) *Let $X = (X_t, \mathcal{A}_t)$ be a continuous d-dimensional process with $X_0 = 0$ then the following statements are equivalent:*

(1) X is a Brownian motion
(2) $X \in \mathcal{M}_{\text{loc},c}$ and $\langle X^i, X^j \rangle_t = \delta_{i,j} t \quad \forall i, j, t$

(3) $X \in \mathcal{M}_{\mathrm{loc},c}$ *and* $\forall\, f_k \in \mathcal{L}^2(\mathbb{R}_+, \lambda_+)$, $1 \le c \le d$

$$\mathcal{E}_t^{if} := \exp\left\{ i \sum_{k=1}^{d} \int_0^t f_k(s)\, \mathrm{d}X_s^k + \frac{1}{2} \sum_{k=1}^{d} \int_0^t f_k^2(s)\, \mathrm{d}s \right\}$$

is a complex martingale.

Remark 4.19 Thus, a Brownian motion is the only continuous local martingale for which the quadratic covariation is equal to $\delta_{ij}t$.

Proof

(1) ⇒ (2) according to Lemma 4.13 in the version for complex martingales.
(2) ⇒ (3) By Theorem 4.7 with $\lambda = i$ and with $M_t := \int_0^t f_k(s)\, \mathrm{d}X_s^k$ it holds $\mathcal{E}^{it} \in \mathcal{M}_{\mathrm{loc},c}$.
M_t is bounded; therefore, by Corollary 4.10 (b) it holds that $\mathcal{E}^{it} \in \mathcal{M}_c$ is a martingale.
(3) ⇒ (1) Let f be specifically chosen as $f := \xi \cdot \mathbb{1}_{[0,T]}$, $\xi \in \mathbb{R}^d$, $T > 0$, i.e. f is constant in each component ξ_i on $[0, T]$. Then the stochastic exponential is given by

$$\mathcal{E}_t^{if} = \left\{ \exp\left(i \langle \xi, X_{t \wedge T} \rangle \right) + \frac{1}{2} |\xi|^2 (t \wedge T) \right\} \in \mathcal{M}_c.$$

It follows with the martingale property from (3).

$$E\left(e^{i\langle \xi, X_t\rangle + \frac{1}{2}|\xi|^2 t} \mid \mathcal{A}_s\right) = e^{i\langle \xi, X_s\rangle + \frac{1}{2}|\xi|^2 \cdot s}, \quad 0 \le s \le t \le T. \tag{4.9}$$

This implies

$$E\left(e^{i\langle \xi, X_t - X_s\rangle} \mid \mathcal{A}_s\right) = e^{-\frac{1}{2}|\xi|^2 (t-s)}.$$

So the conditional expectation is independent of $\mathcal{A}_s$ and, therefore, coincides with the expected value.

$$\begin{aligned} E\left(e^{i\langle \xi, X_t - X_s\rangle} \mid \mathcal{A}_s\right) &= e^{-\frac{1}{2}|\xi|^2 (t-s)} \\ &= E\, E\left(e^{i\langle \xi, X_t - X_s\rangle} \mid \mathcal{A}_s\right) \\ &= E\left(e^{i\langle \xi, X_t - X_s\rangle}\right) \\ &= \varphi_{X_t - X_s}(\xi). \end{aligned}$$

But this is the characteristic function of the normal distribution. So the increments are $X_t - X_s$ are independent of $\mathcal{A}_s$ and the increments of X are independent and normally distributed. But these are exactly the defining properties of Brownian motion: $X_t - X_s$ is independent of $\mathcal{A}_s$ and $X_t - X_s \stackrel{d}{=} N(0, t - s)$. Thus X is a Brownian motion.

□

Because of the special importance we formulate the theorem for $d = 1$ separately.

Corollary 4.20 *Let $d = 1$ and $B = (B_t)$ be a continuous process with $B_0 = 0$, then it follows:*

$$\begin{aligned}
&B \text{ is a Brownian motion} \\
&\iff B \in \mathcal{M}_{\text{loc},c} \text{ and } \langle B \rangle_t = t, \quad \forall t \\
&\iff (B_t), \quad (B_t^2 - t) \in \mathcal{M}_{\text{loc},c}.
\end{aligned}$$

Thus a Brownian motion is the unique continuous local martingale such that $\langle B \rangle_t = t$. For the class of jump processes, this property (local martingale and $\langle N \rangle_t = t$) characterizes the compensated Poisson processes. Thus the Poisson process and Brownian motion occupy a prominent position as prototypes of these classes of processes.

A fundamental problem concerning Brownian motion B and more general classes of processes is the description of more general L^2-functionals $Y = F(B) \in L^2$ or equivalently of functionals F on the infinite-dimensional Wiener space, (C, W), W the Wiener measure. In financial mathematics, associated completeness theorems are fundamental. One way to describe such functionals is based on the construction of a basis using Hermitian polynomials, another way of description is based on stochastic integrals. Here we first treat the method of Hermitian polynomials.

The **Hermitian polynomials** h_n are defined as normalized solutions of the equations

$$\frac{d^n}{dx^n} \exp\Big(-\frac{x^2}{2}\Big) = (-1)^n h_n(x) e^{-\frac{x^2}{2}}, \quad h_n(0) = 1.$$

h_n is called a Hermitian polynomial of degree n. The Hermitian polynomials are orthogonal polynomials to the weight function $e^{-\frac{x^2}{2}}$, i.e., they are closely coupled to the normal distribution density:

$$\int h_n(x) h_m(x) e^{-\frac{x^2}{2}} \,\mathrm{d}x = 0, \quad n \neq m.$$

The functions (h_n) form an orthogonal basis in the space $L^2(\exp(-x^2/2)\,\mathrm{d}x)$, i.e. in the L^2-space of the standard normal distribution with the weight function $\exp(-x^2/2)$.

One can also introduce the Hermitian polynomials in a related way.

$$\sum_{n\geq 0} \frac{u^n}{n!} h_n(x) = \exp\left(ux - \frac{u^2}{2}\right). \tag{4.10}$$

Expanding and rearranging the exponential function yields the above representation. The $h_n(x)$ then occur as weights of u^n, i.e., (4.10) provides a representation of the generating function of the polynomial sequence (h_n). Coded in the exponential function are the $h_n(x)$ as coefficients in x and on the other hand all $h_n(x)$ in the expansion by the second variable u. One can normalize this representation in the following way with another parameter a:

$$\begin{aligned}\exp\left(ux - \frac{au^2}{2}\right) &= \exp\left(u\sqrt{a}\left(\frac{x}{\sqrt{a}}\right)^2 - \frac{(u\sqrt{a})^2}{2}\right)\\ &\overset{(4.10)}{=} \sum_{n\geq 0} \frac{u^n}{n!} \cdot \underbrace{a^{n/2} \cdot h_n\left(\frac{x}{\sqrt{a}}\right)}_{=:H_n(x,a)}, \quad H_n(x,0) := x^n.\end{aligned} \tag{4.11}$$

The H_n retain the orthogonality properties of the Hermitian polynomials. With respect to the variables x they are polynomials of degree n in x. The first standardized Hermitian polynomials $H_n = H_n(x, a)$ are

n	0	1	2	3	4	5
H_n	1	x	$x^2 - a$	$x^3 - 3ax$	$x^4 - 6ax^2 + 3a^2$	$x^5 - 10ax^3 + 15a^2x$

The space of square-integrable functions of the Brownian motion can be described in terms of the standardized Hermitian polynomials. These form a basis in the L^2-space when one inserts the Brownian motion and its quadratic variation. Moreover, they allow the construction of a large class of continuous local martingales as functionals of given local martingales M and their predictable quadratic variations $\langle M \rangle$.

Proposition 4.21 *Let $M \in \mathcal{M}_{\mathrm{loc},c}$, $M_0 = 0$, then it follows*

$$L_t^{(n)} := H_n\big(M_t, \langle M \rangle_t\big) \in \mathcal{M}_{\mathrm{loc},c}$$

$L_t^{(n)}$ *has a representation as a multiple stochastic integral*

$$L_t^{(n)} = n! \int_0^t \mathrm{d}M_{s_1} \Big(\int_0^{s_1} \mathrm{d}M_{s_2} \cdots \Big(\int_0^{s_{n-1}} \cdots \mathrm{d}M_{s_n} \Big)\Big).$$

The stochastic exponential of M has the expansion

$$\mathcal{E}^\lambda(M)_t = \sum_{n=0}^{\infty} \frac{\lambda^n}{n!} L_t^{(n)}.$$

Proof The exponential martingale $\mathcal{E}^\lambda(M)$ is a continuous local martingale and it is a solution of the stochastic differential equation (cf. Corollary 4.8):

$$\mathcal{E}^\lambda(M)_t = 1 + \lambda \int_0^t \mathcal{E}^\lambda(M)_s \,\mathrm{d}M_s.$$

With $\mathcal{E}^\lambda(M)_s = \exp\big(\lambda M_s - \frac{1}{2}\lambda^2 \langle M\rangle_s\big)$ it follows by (4.11) with $u = \lambda$, $a = \langle M\rangle_s$, $x = M_s$.

$$\mathcal{E}^\lambda(M)_t = 1 + \lambda \int_0^t \sum_{n=0}^{\infty} \frac{\lambda^n}{n!} L_s^{(n)} \,\mathrm{d}M_s.$$

Now let τ be a stopping time such that M^τ, $\langle M\rangle^\tau$ are bounded. Then it follows by the theorem of dominated convergence that the integral and the sum can be interchanged and thus it holds

$$\mathcal{E}^\lambda(M)_t = 1 + \sum_{n=0}^{\infty} \frac{\lambda^{n+1}}{n!} \int_0^t L_s^{(n)} \,\mathrm{d}M_s = \sum_{n=0}^{\infty} \frac{\lambda^n}{n!} L_s^{(n)}.$$

By comparison of coefficients we get

$$L^{(0)} \equiv 1, \quad L_t^{(n+1)} = (n+1) \int_0^t L_s^{(n)} \,\mathrm{d}M_s.$$

So $L^{(n)} \in \mathcal{M}_{\mathrm{loc},c}$. Thus the normalized Hermitian polynomials define a system of local martingales. □

4.2 Martingale Representation

Let B be a Brownian motion with Brownian filtration $(\mathcal{A}_t) = (\mathcal{A}_t^B)$, i.e., the completed right continuous filtration generated by the Brownian motion. The goal of this section is to represent martingales with respect to this filtration as stochastic integrals

$$M_t = M_0 + \int_0^t H_s \,\mathrm{d}B_s.$$

It will be shown below that any martingale admits such a representation. In financial mathematics, the integrand H from such a representation can be interpreted as a trading strategy.

The proof of the above integral representation is based on the special class of integrands,

$$\Phi := \left\{ f = \sum_{j=1}^{n} \lambda_j \mathbb{1}_{(t_{j-1},t_j]} \right\},$$

of the set of staircase functions with a bounded support. Let

$$M_t^f := (f \cdot B)_t = \int_0^t f(s) \,\mathrm{d}B_s \in \mathcal{M}$$

be the corresponding stochastic integral with respect to the Brownian motion. The integrands $f \in \Phi$ are deterministic and predictable. The predictable variation of M^f is

$$\langle M^f \rangle_t = \int_0^t f^2(s) \,\mathrm{d}s.$$

Let $\mathcal{E}^f$ be the associated stochastic exponential generated by M^f

$$\mathcal{E}^f := \exp\left(M^f - \frac{1}{2} \langle M^f \rangle \right).$$

A preliminary remark from Hilbert space theory follows:

A subset A of a Hilbert space H is called **total** in H,

$$\Longleftrightarrow \forall h \in H : \qquad h \perp A \Longrightarrow h = 0.$$

Total subsets are suitable for approximating elements of a Hilbert space. It holds

$$A \subset H \text{ is total} \iff \text{the linear hull of } \mathcal{A}, \ \text{lin}A, \ \text{is dense in } H.$$

This equivalence follows from an application of Hahn–Banach's theorem.

The following lemma shows that the set of stochastic exponentials of staircase functions is total in $L^2(\mathcal{A}_\infty, P)$.

Lemma 4.22 *Let $\mathcal{K} := \left\{\mathcal{E}_\infty^f;\ f \in \Phi\right\}$ be the set of stochastic exponentials of staircase functions. Then*

$$\mathcal{K} \ is\ total\ in\ L^2(\mathcal{A}_\infty, P),$$

equivalently, lin$\mathcal{K}$ *is dense in $L^2(\mathcal{A}_\infty, P)$.*

Proof Let $Y \in L^2(\mathcal{A}_\infty, P)$ and let $Y \perp \mathcal{K}$, then it is to show that: $Y \equiv 0$.

The assertion is equivalent to

$$Y \cdot P|_{\mathcal{A}_\infty} = 0,$$

i.e., the measure with density Y with respect to P restricted to $\mathcal{A}_\infty$ (a signed measure) is the zero measure. It suffices to show that the measure is the zero measure on a generator of $\mathcal{A}_\infty$.

We define for all n-tuples of complex numbers $z = (z_1, \dots, z_n) \in \mathbb{C}^n$

$$\varphi(z) := E\left[\exp\left\{\sum_{j=1}^n z_j\left(B_{t_j} - B_{t_{j-1}}\right)\right\}Y\right].$$

φ is analytic, that is, complex differentiable in each variable. One can easily specify a majorant here and swap integral and expectation value (similar to exponential families). For real arguments λ_j holds

$$\varphi(\lambda_1, \dots, \lambda_n) = E\exp\left\{\sum_{j=1}^n \lambda_j\left(B_{t_j} - B_{t_{j-1}}\right)\right\}Y = 0,$$

since $Y \perp \mathcal{K}$ and with $\lambda_j = 0$ it follows $EY = 0$. On the real axis, i.e., in $\mathbb{R}^n$ the function $\varphi = 0$. Then, using the uniqueness theorem for analytic functions, it follows that $\varphi = 0$ on $\mathbb{C}^n$.

Specifically, it follows from this

$$0 = \varphi(i\lambda_1, \dots, i\lambda_n) = E\exp\left\{i\sum \lambda_j\left(B_{t_j} - B_{t_{j-1}}\right)\right\}Y.$$

This can be written as the integral of the e-function with respect to the measure with density Y with respect to P. Then, by the uniqueness theorem for characteristic functions, it follows that the image measure of the random variable under the measure $Y \cdot P$ is the zero measure,

$$(Y \cdot P)^{(B_{t_1}, \ldots, B_{t_n} - B_{t_{n-1}})} = 0.$$

The successive differences of B_{t_i} produce the same σ algebra as $B_{t_1}, \ldots, B_{t_n}$. So it follows

$$Y \cdot P|_{\sigma(B_{t_1}, \ldots, B_{t_n})} = 0$$

and thus $Y \cdot P|_{\mathcal{A}_\infty} = 0$. □

Lemma 4.22 is the key to the following completeness theorem.

Theorem 4.23 (Completeness Theorem) *Let $F \in L^2(\mathcal{A}_\infty^B, P)$, be a square-integrable Brownian motion functional. Then there exists exactly one integrand $H \in \mathcal{L}^2(\mu) = \overline{\mathcal{L}^2}(B, \mu)$, μ the Doléans measure on $\mathcal{P}$, such that F can be represented as a stochastic integral*

$$F = EF + \int_0^\infty H_s \, \mathrm{d}B_s.$$

Any L^2-functional can be represented uniquely as a stochastic integral.

Proof Let $\mathcal{F}$ be the set of functionals for which such a representation exists,

$$\mathcal{F} := \left\{ F \in L^2(\mathcal{A}_\infty, P);\ \exists H \in L^2(\mu), F = EF + \int_0^\infty H_s \, \mathrm{d}B_s \right\} \subset L^2(\mathcal{A}_\infty, P).$$

$\mathcal{F}$ is a linear subspace of $L^2(\mathcal{A}_\infty, P)$

(1) $\mathcal{F}$ is closed.
To the proof let $F \in \mathcal{F}$. Then it holds with $E \int_0^\infty H_s \, \mathrm{d}B_s = 0$ and

$$\begin{aligned} E\left(\int_0^\infty H_s \, \mathrm{d}B_s\right)^2 &= E \int_0^\infty H_s^2 \, \mathrm{d}s \quad \text{(isometry property)} \\ EF^2 &= (EF)^2 + E \int_0^\infty H_s^2 \, \mathrm{d}s = (EF)^2 + \int H^2 \mathrm{d}\mu. \end{aligned} \tag{4.12}$$

Let $\{F^n = EF^n + \int_0^\infty H_s^n \,\mathrm{d}B_s\}$ be a Cauchy sequence in $\mathcal{F}$ with respect to L^2-convergence $L^2(\mathcal{A}_\infty, P)$. Then it follows in particular: (EF^n) is also a Cauchy sequence, because

$$|EF^n - EF^m| \leq \|F^n - F^m\|_2,$$

i.e., also the first moments form a Cauchy sequence and $(\int_0^\infty H^n \,\mathrm{d}B) = (H \cdot B)$ is a Cauchy sequence in $L^2(\mathcal{A}_\infty, P)$. The mappings

$$\begin{array}{ccccc} L^2(\mu) & \longrightarrow \mathcal{H}^2 & \longrightarrow L^2(\mathcal{A}_\infty, P) \\ H & \longrightarrow H \cdot B & \longrightarrow (H \cdot B)_\infty \end{array} \qquad \text{are isometries.}$$

So $(H^n) \subset L^2(\mu)$ a Cauchy sequence in the complete space $L^2(\mu)$. Therefore, there exists a limit $H \in L^2(\mu)$ such that $H^n \longrightarrow H \in L^2(\mu)$.

Because of the isometry property, convergence follows in $L^2(\mathcal{A}_\infty, P)$

$$F^n \longrightarrow (\lim EF^n) + \int_0^\infty H \,\mathrm{d}B.$$

Thus every Cauchy sequence in $\mathcal{F}$ has a limit in $\mathcal{F}$, i.e. $\mathcal{F}$ is closed.

(2) $\mathcal{K} = \left\{\mathcal{E}_\infty^f;\ f \in \Phi\right\} \subset \mathcal{F}$.

By the Itô formula it holds for $f \in \Phi$ and $M^f := f \cdot B$

$$\begin{aligned} \mathcal{E}_t^f &= \mathcal{E}(M^f)_t = 1 + \int_0^t \mathcal{E}_s^f \,\mathrm{d}M_s^f \\ &= 1 + \int_0^t \mathcal{E}_s^f f(s) \,\mathrm{d}B_s = 1 + \int_0^t H_s \,\mathrm{d}B_s \end{aligned}$$

with $H_s := \mathcal{E}_s^f f(s)$. $H \in L^2(\mu)$, because f is a bounded staircase function. Therefore, $\mathcal{E}^f$ has a representation as a stochastic integral, i.e. $\mathcal{E}_\infty^f \in \mathcal{F}$.

As a corollary to (2), we now get

(3) $\mathcal{F} \supset \operatorname{lin} \mathcal{K}$ and $\mathcal{F}$ is closed.

According to Lemma 4.22 it thus follows

$$\mathcal{F} = L^2(\mathcal{A}_\infty, P).$$

(4) Uniqueness
Let $H, G \in \mathcal{L}^2(\mu)$ and $F = EF + \int_0^\infty H_s \, dB_s = EF + \int_0^\infty G_s \, dB_s$ be two representations of F. Then it follows $0 = \int_0^\infty (H_s - G_s) \, dB_s$. So it also follows $0 = E \int_0^\infty (H_s - G_s) \, dB_s$ and thus

$$0 = E\left(\int_0^\infty (H_s - G_s) \, dB_s\right)^2 = E \int_0^\infty (H_s - G_s)^2 \, ds = \|H - G\|_2$$

the norm with respect to the Doléans measure.
From this follows $H = G \; [\mu]$. □

Remark 4.24 For $F \in L^2(\mu_\infty, P)$ thus there exists an integral representation with an integrand $H \in L^2(\mu, \mathcal{P}) = \overline{\mathcal{L}^2(B)} := \{f \in L(\mathcal{P});\; E \int_0^\infty f_s^2 \, ds < \infty\}$. Analogously, for square-integrable functionals of $(B_s; s \leq T)$ representations with integrands $H \in L^2(\mu_T, \mathcal{P})$ are obtained.

As a consequence, we now have the following representation theorem for martingales.

Theorem 4.25 (Martingale Representation Theorem) *Let $M \in \mathcal{H}^2 = \mathcal{H}^2(\mathcal{A}^B, P)$ (or $M \in \mathcal{M}^2$).*
Then there exists exactly one $H \in L^2(\mu)$ (or $H \in \mathcal{L}^2(B)$), such that

$$M_t = M_0 + \int_0^t H_s \, dB_s, \qquad \forall t \leq \infty \quad (\forall t < \infty).$$

Remark 4.26 In particular, every martingale in $\mathcal{H}^2$ (resp. $\mathcal{M}^2$) has a continuous version. The integral representation is such a continuous version.

Proof Let $M \in \mathcal{H}^2$, then $M_\infty \in L^2(\mathcal{A}_\infty^B, P)$. According to Theorem 4.23 there exists exactly one $H \in L^2(\mu)$ such that $M_\infty = M_0 + \int_0^\infty H_s \, dB_s$ and $M_0 = EM_0$ is constant. But it follows from this

$$\begin{aligned} M_t &= E(M_\infty \mid \mathcal{A}_t) \\ &= M_0 + E\left(\int_0^\infty H_s \, dB_s \mid \mathcal{A}_t\right) \\ &= M_0 + \int_0^t H_s \, dB_s + E\left(\int_t^\infty H_s \, dB_s \mid \mathcal{A}_t\right). \end{aligned}$$

$E\left(\int_t^\infty H_s \, dB_s \mid \mathcal{A}_t\right) = 0$ since the Brownian motion has independent increments and $H \in \mathcal{L}^2(B)$ is predictable. To prove this, first consider the class $\mathcal{E}$ and then apply the monotone class theorem. For $M \in \mathcal{M}^2$ we obtain the above integral representation on $[0, T_n]$, $T_n \uparrow \infty$. This then can be consistently extended to $[0, \infty)$. □

The stochastic integral representation can now be extended to the class $\mathcal{M}_{\text{loc}}$.

Theorem 4.27 *Let $M \in \mathcal{M}_{\text{loc}} = \mathcal{M}_{\text{loc}}(\mathcal{A}^B)$, then there exists a process $H \in \mathcal{L}^2_{\text{loc}}(B)$ such that*

$$M_t = M_0 + \int_0^t H_s \, dB_s. \tag{4.13}$$

In particular $\mathcal{M}_{\text{loc}}(\mathcal{A}^B) = \mathcal{M}_{\text{loc},c}(\mathcal{A}^B)$, i.e., any local martingale with respect to the Brownian filtration is continuous.

Proof

(1) For $M \in \mathcal{H}^2$ the above representation follows by Theorem 4.25.
(2) If M is uniformly integrable, then $M_\infty \in L^1$.
$L^2(\mathcal{A}_\infty) \subset L^1(\mathcal{A}_\infty)$ is dense in $L^1(\mathcal{A}_\infty)$. So there exists a sequence $(M^n) \subset \mathcal{H}^2$ such that $E|M_\infty - M^n_\infty| \longrightarrow 0$.
From this, using Doob's maximal inequality, it follows

$$P\left(\sup_t |M_t - M_t^n| > \lambda\right) \le \frac{1}{\lambda} E|M_\infty - M^n_\infty| \longrightarrow 0.$$

Therefore, according to Borel–Cantelli, there exists a P-almost surely convergent subsequence (n_k) such that

$$M^{n_k} \longrightarrow M \text{ almost surely, uniformly in } t.$$

It follows that M has a continuous version.
(3) $M \in \mathcal{M}_{\text{loc}}$. Then there exists a continuous version of M (with suitable stopping after (2).
Therefore, there further exists a sequence of stopping times $\tau_n \uparrow \infty$ such that M^{τ_n} is bounded and L^2 is bounded. M^{τ_n} is in $\mathcal{H}^2$ and, therefore, has a stochastic integral representation

$$M_t^{\tau_n} = c_n + \int_0^t H_s^n \, dB_s.$$

But it follows

$$c_n = c, \text{ and } H_t^{n+1} = H_t^n \text{ on } [0, \tau_n].$$

We define $H_t := H_t^n$ at $[0, \tau_n]$; then

$$H_s \in \mathcal{L}^2_{\text{loc}}(B) \quad \text{and} \quad M = c + \int_0^t H_s \, \mathrm{d}B_s.$$

□

The stochastic integral representation is also valid for the d-dimensional Brownian motion.

Theorem 4.28 *Let B be a d-dimensional Brownian motion and let $M \in \mathcal{M}_{\text{loc}}(\mathcal{A}^B)$. Then it follows:*

$$\exists\, H^i \in \mathcal{L}^2_{\text{loc}}(B),\, 1 \le i \le d,\ \exists\, c \in \mathbb{R}^1 \textit{ such that } M_t = c + \sum_{i=1}^d \int_0^t H_s^i \, \mathrm{d}B_s^i.$$

Proof The proof uses an analogous argument as in $d = 1$ with the multivariate stochastic exponential. □

Remark 4.29 How is the integrand of the stochastic integral representation obtained?

(1) Assuming that the predictable covariation is Lebesgue continuous, i.e., let $\langle M, B^i \rangle_t = \int_0^t H_s^i \, \mathrm{d}s$. Because of $\langle B^i, B^j \rangle_s = \delta_{ij} \cdot s$ it follows according to the Radon–Nikodým theorem

$$H_t^i = \frac{\mathrm{d}\langle M, B^i \rangle_t}{\mathrm{d}^1}$$

is equal to the Radon–Nikodým derivation. The Clark–Ocone formula (cf. Sect. 4.3.2) provides such a representation formula under more general conditions.

(2) An example of an explicit form for the integrand H:

If f is harmonic, i.e. $\Delta f = 0$, then $M_t := f(B_t)$ is a martingale. From the Itô formula it follows the explicit integral representation formula

$$M_t = f(B_t) = M_0 + \int_0^t \nabla f(B_s) \, \mathrm{d}B_s.$$

(3) Non-uniqueness of representation for $M \in \mathcal{M}_{\text{loc}}$:
For $M \in \mathcal{M}_{\text{loc}}$ as in Theorem 4.28, the integral representation is in general not unique.

Example Let $0 < a < T$ and $\tau := \inf\{t \geq 0; \int_a^t \frac{1}{T-u}\, \mathrm{d}B_u = -B_a\}$.

Then it follows from the strong oscillation of $\int_a^t \frac{1}{T-u}\, \mathrm{d}B_u$ in the vicinity of T: $P(\tau < T) = 1$. Let

$$\psi(\omega, s) := \begin{cases} 1 & 0 \leq s < a, \\ \frac{1}{T-s} & a \leq s < \tau, \\ 0 & \tau \leq s \leq T. \end{cases}$$

Then $\psi \in \mathcal{L}^2_{\text{loc}}$ and

$$\int_0^a \psi(\omega, s)\, \mathrm{d}B_s = B_a = -\int_a^T \Psi(\omega, s)\, \mathrm{d}B_s.$$

Thus $\psi \in \mathcal{L}^2_{\text{loc}}(B)$ and $\int_0^T \psi(\omega, s)\, \mathrm{d}B_s = 0$ but $\psi \not\equiv 0$ and $M = \psi \cdot B \in \mathcal{M}_{\text{loc}}$ is a local martingale.

Alternative Representation of $L^2(\mathcal{A}^B_\infty)$ by Multiple Stochastic Integrals
Let dimension $d = 1$. Let $\Delta_n = \{(s_1, \ldots, s_n) \in \mathbb{R}^n_+; s_1 > s_2 > \cdots > s_n\}$ and let $L^2(\Delta_n) = L^2(\Delta_n, \lambda^n_+ \,|_{\Delta_n})$ be the corresponding L^2-space. We consider the subspace E_n of $L^2(\Delta_n)$ defined by

$$E_n := \left\{ f \in L^2(\Delta_n);\ f(s) = \prod_{i=1}^n f_i(s_i),\ s = (s_1, \ldots, s_n),\ f_i \in L^2(\mathbb{R}_+) \right\}.$$

$E_n \subset L^2(\Delta_n)$ is total in $L^2(\Delta_n)$.
For $f = \prod_i f_i \in E_n$ define the multiple stochastic integral

$$I_n f := \int_0^\infty f_1(s_1)\, \mathrm{d}B_{s_1} \int_0^{s_1} f_2(s_2)\, \mathrm{d}B_{s_2} \cdots \underbrace{\int_0^{s_{n-1}} f_n(s_n)\, \mathrm{d}B_{s_n}}_{\in \mathcal{A}_{s_{n-1}}} \in \mathcal{L}^2(\mathcal{A}_\infty)$$

(cf.: Hermitian functionals in Proposition 4.21 for indicator functions f). Contrary to the notation used so far, here the integrands are to the right of the integrator. Assertion: I_n is norm-preserving,

$$\|I_n f\|_{L^2(\mathcal{A}_\infty)} = \|f\|_{L^2(\Delta_n)}, \qquad f \in E_n.$$

Proof Let $n = 2$, $f = \prod_{i=1}^{2} f_i \in E_2$, then

$$\|I_2 f\|_{L^2(\mathcal{A}_\infty)} = E\Bigg(\int_0^\infty \underbrace{\Big(f_1(s_1)\int_0^{s_1} f_2(s_2)\,\mathrm{d}B_{s_2}\Big)}_{=:H_{s_1}}\,\mathrm{d}B_{s_1}\Bigg)^2$$

$$= E\int_0^\infty H_{s_1}^2\,\mathrm{d}s_1$$

$$= \int_0^\infty \mathrm{d}s_1\big(f_1(s_1)^2\int_0^{s_1} f_2^2(s_2)\,\mathrm{d}s_2\big) = \|f\|^2_{L^2(\Delta_2)}.$$

Analogously for $n > 2$ with induction. □

The closure of the linear hull of multiple stochastic integrals

$$C_n := c\ell\text{lin}\{I_n(E_n)\} \subset L^2(\mathcal{A}_\infty)$$

is called ***n*-th Wiener chaos.**

(1) The mapping $I_n : E_n \to L^2(\mathcal{A}_\infty, P)$ is linear and continuous. There exists a unique continuous and linear extension $I_n : L^2(\Delta_n) \to C_n$. The extension $I_n : L^2(\Delta_n) \to C_n$ is an isometry.

In particular $C_n = I_n(L^2(\Delta_n))$; therefore, it holds $C_n \cong L^2(\Delta_n)$. Alternatively, for $f \in L^2(\Delta_n)$ we can directly define the multiple integral

$$I_n(f) = \int_0^\infty \cdots \int_0^{s_{n-1}} f(s_1, \dots, s_n)\,\mathrm{d}B_{s_1}\dots\mathrm{d}B_{s_n}.$$

(2) $C_n \perp C_m$, $m \neq n$. We first show: $C_1 \perp C_2$. For this consider

$$E\Bigg(\underbrace{\Big(\int_0^\infty f_1(s_1)\,\mathrm{d}B_{s_1}\Big)}_{\in C_1}\underbrace{\Big(\int_0^\infty g_1(s_1)\,\mathrm{d}B_{s_1}\Big)\Big(\int_0^{s_1} g_2(s_2)\,\mathrm{d}B_{s_2}\Big)}_{\in C_2}\Bigg)$$

$$= E\int_0^\infty \Big(f_1(s_1)g_1(s_1)\int_0^{s_1} g_2(s_2)\,\mathrm{d}B_{s_2}\Big)\,\mathrm{d}B_{s_1}$$

$$= \int_0^\infty f_1(s_1)g_1(s_1)\underbrace{E\Big(\int_0^{s_1} g_2(s_2)\,\mathrm{d}B_{s_2}\Big)}_{=0}\,\mathrm{d}B_{s_1} = 0.$$

The general case follows by induction with a similar argument.

Remark 4.30 The elements of $C_1 = c\ell \operatorname{lin}\{B_t;\ 0 \le t\} \subset \mathcal{L}^2(\mathcal{A}_\infty)$ are Gaussian random variables.
C_2 is the orthogonal complement of C_1 in the set of quadratic functions.

Theorem 4.31 (Wiener Chaos Representation) *The space of L^2-functionals of the Brownian motion has a representation as a direct topological sum.*

$$\mathcal{L}^2(\mathcal{A}_\infty^B) = \bigoplus_{n=0}^\infty C_n, \qquad C_0 \simeq \mathbb{R}^1,$$

i.e. $\forall\, Y \in \mathcal{L}^2(\mathcal{A}_\infty^B)$ *there exists a sequence* $f^n \in \mathcal{L}^2(\Delta_n)$, $n \in \mathbb{N}$, *with*

$$Y = \sum_{n=0}^\infty I_n(f^n) \qquad \text{in } \mathcal{L}^2(P). \tag{4.14}$$

Proof According to Proposition 4.21 it holds

$$\mathcal{E}_\infty^f = \mathcal{E}_\infty \underbrace{(M^f)}_{=f\cdot B} = \sum_{n=0}^\infty I_n(f^n),$$

with $f^n(s_1, \dots, s_n) = \prod_{i=1}^n f(s_i)$, $f = \sum_i \alpha_i \mathbb{1}_{(s_{i-1}, s_i)}$.
In general, for f with bounded compact support and $M^f = f \cdot B$:

$$\mathcal{E}(M^f)_\infty = \sum_{n=0}^\infty \underbrace{\frac{1}{n!} L_t^{(n)}}_{=I_n(f^n)} = \int_0^t f(s_1)\,\mathrm{d}B_{s_1} \int_0^{s_1} f(s_2)\,\mathrm{d}B_{s_2} \cdots \int_0^{s_{n-1}} f(s_n)\,\mathrm{d}B_{s_n}.$$

It follows:
The representation in (4.14) holds for $\operatorname{lin}(\{\mathcal{E}_\infty^f;\ f$ has a bounded, compact support$\})$.

According to Lemma 4.22 it follows that $\operatorname{lin}(\{\mathcal{E}_\infty^f;\ f$ bounded, compact support$\}) \subset L^2(\mathcal{A}_\infty^B)$ is dense. So there exists to $Y \in \mathcal{L}^2(\mathcal{A}_\infty, P)$ a sequence $(Y_n) \subset \operatorname{lin}(\{\mathcal{E}_\infty^f\})$ with

$$Y_n \to Y \text{ in } L^2, \quad \text{and} \quad Y_n = \sum_{m=0}^\infty \underbrace{I_m(f_m^n)}_{\in C_m} \xrightarrow{L^2} Y.$$

Y_m is for every m an orthogonal series. It follows that $(I_m(f_m^n))_n$ is for each m a Cauchy sequence, and hence has a limit:

$$I_m(f_m^n) \to G_m \in C_m.$$

Since I_m is an isometry, it follows that also the sequence $((f_m^n)_n)$ converges,

$$f_m^n \to f_m \text{ in } L^2(\Delta_m).$$

It follows $G_m = I_m(f_m)$, and thus the representation of Y:

$$Y = \sum_{m=0}^{\infty} I_m(f_m).$$

□

The basic martingale representation theorem states that the Brownian motion has the representation property for local martingales, i.e., one can represent martingales with respect to Brownian filtration $\mathcal{A}^B$ as an integral with respect to the Brownian motion. In general, this representation property does not hold for martingales $X \in \mathcal{M}^2$ or $\mathcal{H}^2$ with respect to the filtration $\mathcal{A}^X$ generated by those martingales.

Let for $X \in \mathcal{H}^2$

$$\overline{\mathcal{L}^2}(X) := \left\{ f \in L(\mathcal{P});\, E \int_0^\infty f_s^2 \mathrm{d}\langle X\rangle_s < \infty \right\} = L^2(\mu)$$

be the set of L^2-integrands on $[0, \infty)$. The Kunita–Watanabe decomposition identifies for a general functional $F \in L^2(\mathcal{A}_\infty, P)$ a uniquely representable part.

Theorem 4.32 (Kunita–Watanabe Decomposition) *Let be $X = (X_t) \in \mathcal{H}^2$, $F \in L^2(\mathcal{A}_\infty, P)$, $\mathcal{A}_\infty = \mathcal{A}_\infty^X$, then it follows: There exist unique processes $(H_s) \in L^2(\mu)$ and $L \in L^2(\mathcal{A}_\infty, P)$ such that*

(1) $F = EF + \int_0^\infty H_s \,\mathrm{d}X_s + L$

(2) $E\left(L \int_0^\infty f_s \,\mathrm{d}X_s\right) = 0, \quad \forall f \in L^2(\mu)$ *or equivalently* $\langle L, X\rangle = 0$, *for the martingale* $(L_t) = (E(L \mid \mathcal{A}_t))$ *generated by* L.

Proof Let $L_0^2 = \{\widetilde{F} \in L^2(\mathcal{A}_\infty, P);\, E\widetilde{F} = 0\}$ and be H the set of functionals that can be represented as a stochastic integral

$$H = \left\{ \widetilde{F} \in L_0^2; \quad \exists f \in \mathcal{L}^2(\mu) : \widetilde{F} = \int_0^\infty f_s \,\mathrm{d}X_s \right\}.$$

(1) H is stable under stopping, i.e., for each element $F \in H$ and each stopping time τ holds $F_\tau \in H$, where $F_t = E(F \mid \mathcal{A}_t)$ is the martingale generated by F.

For the proof let $F = \int_0^\infty f_s \,\mathrm{d}X_s \in H$. Then, according to the optional sampling theorem, it follows

$$F_\tau = E(F \mid \mathcal{A}_\tau) = \int_0^\infty f \mathbb{1}_{[0,\tau]} \,\mathrm{d}X_s$$

and

$$\begin{aligned} EF_\tau^2 &= E \int_0^\infty (f \mathbb{1}_{[0,\tau]})^2 \mathrm{d}\langle X \rangle \\ &\leq E \int_0^\infty f^2 \mathrm{d}\langle X \rangle < \infty. \end{aligned}$$

(2) The orthogonal space $H^\perp$ is stable, i.e. with $N \in H^\perp$ and with $N_t = E(N \mid \mathcal{A}_t)$, the martingale generated by N, satisfies $N_\tau \in H^\perp$ for each stopping time τ. It needs to be shown that for $F \in H$ it holds: $EFN_\tau = 0$.
However, this follows using the stability of H as follows:

$$\begin{aligned} EFN_\tau &= EE(FN_\tau \mid \mathcal{A}_\tau) = EE(F \mid \mathcal{A}_\tau) N_\tau = EN_\tau F_\tau \\ &= EF_\tau E(N \mid \mathcal{A}_\tau) = EE(F_\tau N \mid \mathcal{A}_\tau) \\ &= EF_\tau N = 0, \end{aligned}$$

as $N \in H^\perp$ and $F_\tau \in H$. So $N_\tau \in H^\perp$ and, therefore, $H^\perp$ is stable.

(3) Let $M \in \mathcal{H}^2$ and be w.l.g. $M_0 = 0$ (otherwise transition to $M - M_0$), then $M_\infty \in L_0^2$. We define

$$\begin{aligned} &Y_\infty := \widehat{E}(M_\infty \mid H) \text{ the orthogonal projection in } L^2(\mathcal{A}_\infty, P), \\ &L_\infty := M_\infty - Y_\infty, \\ &Y_t := E(Y_\infty \mid \mathcal{A}_t) \quad \text{and} \quad L_t := E(L_\infty \mid \mathcal{A}_t). \end{aligned}$$

Then it holds: $M_t = Y_t + L_t$, $Y_t = \int_0^t f_s \,\mathrm{d}X_s$ since $Y_\infty \in H$ and $(L_t, \mathcal{A}_t) \in \mathcal{H}^2$ with $L_\infty \in H^\perp$.

$H, H^\perp$ are stable. From this it follows $\forall F \in H$ and $\forall$ stopping times τ: $L_\tau \in H^\perp$, $F_\tau \in H$ and, therefore, $EL_\tau F_\tau = 0$.

From this, by the characterization of the martingale property by stopping times, it follows: $(L_t F_t) \in \mathcal{M}$.

This implies: $\langle F, L \rangle = 0, \forall F = f \cdot X \in H$, since $\langle F, L \rangle$ is the unambiguous predictable process A such that $FL - A \in \mathcal{M}$.

As consequence we obtain $\langle f \cdot X, L \rangle = f \cdot \langle X, L \rangle = 0, \forall f \in \mathcal{L}^2(\mu)$, or, equivalently: $\langle X, L \rangle = 0$.

(4) Uniqueness

Suppose there are two different representations of F

$$F = EF + H^1 \cdot X + L^1$$
$$= EF + H^2 \cdot X + L^2$$

Then it follows: $L^1 - L^2 = (H^2 - H^1) \cdot X$.

Since $L^1 - L^2 \in H^\perp$ it follows $\langle L^1 - L^2, (H^2 - H^1) \cdot X\rangle = 0$. This implies $(H^2 - H^1)^2 \cdot \langle X\rangle = 0$. In consequence it follows $H^1 = H^2$ in $L^2(\mu)$ and $L^1 = L^2$.

So F has a unique decomposition. □

Remark 4.33

(a) **Decomposition theorem for martingales** A direct corollary of Kunita–Watanabe's theorem is a decomposition theorem for martingales. Theorem 4.32 gives the decomposition of L^2-functionals.

Let $X = (X_t, \mathcal{A}_t) \in \mathcal{H}^2$, $\mathcal{A}_t = \mathcal{A}_t^X$ and $M \in \mathcal{H}^2 = \mathcal{H}^2(X)$ be a martingale with respect to this filtration. Then there exists a unique decomposition

$$M_t = M_0 + \int_0^t \varphi_s \, \mathrm{d}X_s + L_t$$

with $L \in \mathcal{H}^2$, $\varphi \in L^2(\mu)$ and $\langle L, X\rangle = 0$, i.e. L is perpendicular to X. L is the non-hedgeable part of M.

(b) There is also a **vector-valued version** of the decomposition theorem.

Let $X = (X^1, \ldots, X^d)$, $M \in \mathcal{H}^2(\mathcal{A}^X)$, then M has a decomposition

$$M_t = M_0 + \sum_{i=1}^d \int_0^t \varphi_s^i \, \mathrm{d}X_s^i + Z_t$$

with $Z \in \mathcal{H}^2$, and $\langle Z, X\rangle = 0$ and it holds $E \sum_{i=1}^d \int (\varphi_s^i)^2 \mathrm{d}\langle X^i\rangle_s < \infty$. This decomposition is unique.

(c) **Locally square-integrable martingales** One can get away from the assumption of square integrability by localization.

Let $X \in \mathcal{M}_{\mathrm{loc}}^2$, $N \in \mathcal{M}_{\mathrm{loc}}(\mathcal{A}^X)$ then it follows: There is exactly one decomposition

$$N_t = N_0 + H_t + L_t, \quad t < \infty$$

where H has a (possibly non-unique) representation of the form $H_t = \int_o^t \varphi_s \, \mathrm{d}X_s$ with $\varphi \in \mathcal{L}_{\mathrm{loc}}^2(X)$, $L \in \mathcal{M}_{0,\mathrm{loc}}$ and $\langle L, X\rangle = 0$.

Application: Quadratic Hedging in the Martingale Case
The theorem of Kunita–Watanabe theorem is significant in financial mathematics for the *mean variance hedging problem.* Let $X \in \mathcal{H}^2$ be a price process (with respect to a martingale measure Q), let $F \in L^2$ be a claim, i.e., a function of this process. What is sought is the best possible approximation of the claim F by a hedge, i.e., sought are $\vartheta_0 \in \mathbb{R}$ and a *trading strategy* $\varphi \in \mathcal{L}^2(\mu)$ such that

$$E\big(F - (\vartheta_0 + (\varphi \cdot X)_T)\big)^2 = \text{min!}.$$

By using the strategy φ and the initial value ϑ_0 a best approximation can, therefore, be found for the value F of the claim. The goal is to minimize the (quadratic) hedging error, i.e., what is sought is the L^2-projection of F on the space of hedgeable claims

$$\mathbb{R} + H = \{\vartheta_0 + (\varphi \cdot X)_T; \vartheta_0 \in \mathbb{R}, \varphi \in L^2(\mu)\}.$$

W.l.g. let $T = \infty$ otherwise one can pass from $X \longrightarrow X^T$. According to the Kunita–Watanabe decomposition theorem, Theorem 4.32, there exists a unique decomposition of F into a hedgeable part and a part L that is perpendicular to the hedgeable part, i.e.,

$$F = EF + \varphi \cdot X + L. \tag{4.15}$$

From this follows: $EF + \varphi \cdot X$ is the projection of F on the set of hedgeable claims H and it holds that the hedge error is given by

$$E(F - (EF + \varphi \cdot X))^2 = EL_T^2 = E\langle L\rangle_T. \tag{4.16}$$

The fundamental question is: how to determine the optimal hedging strategy φ?

To this end, consider the following.

From the Kunita–Watanabe decomposition of F in (4.15), it follows.

$$\langle X, F\rangle = \langle X, EF + \varphi \cdot X + L\rangle = \langle X, \varphi \cdot X\rangle + \underbrace{\langle X, L\rangle}_{=0} = \varphi \cdot \langle X\rangle,$$

i.e. φ is obtained as a predictable projection of the Radon–Nikodým derivative process.

$$\varphi = \frac{\mathrm{d}\langle X, F\rangle}{\mathrm{d}\langle X\rangle}. \tag{4.17}$$

For the case $X = B$ is the Brownian motion, the hedging error for each claim is zero. Each claim F can be perfectly hedged; the model of Brownian motion is complete. For the case of optimal hedging with respect to the underlying statistical measure P, some additional considerations are necessary. In general, L can only be orthogonal to the martingale component of S. This problem can be reduced to the case of a special martingale measure (minimal martingale measure) and is solved by the Föllmer–Schweizer decomposition (cf. Chap. 7) a generalization of the Kunita–Watanabe decomposition.

Extension of the Representation Problem

An extension of the representation problem is the following variant, which includes the construction of a suitable Brownian motion.

Given a filtration $(\mathcal{A}_t)$. For which (local) martingale $(M_t, \mathcal{A}_t)$ exists a Brownian motion B such that M has an integral representation, i.e.

$$M_t = M_0 + \int_0^t H_s \,\mathrm{d}B_s.$$

Proposition 4.34 *Let $M \in \mathcal{M}_{\mathrm{loc},c}$ and let $\langle M \rangle \sim \lambda_+^1$ a.s. in ω. Then there exists $f \in \mathcal{L}^0(\mathcal{A}^M)$, $f > 0[\lambda_+ \otimes P]$ and there exists a Brownian motion B with respect to $\mathcal{A}^M$ such that*

$$\frac{\mathrm{d}\langle M \rangle_t}{\mathrm{d}\lambda_t} = f_t P \text{ a.s. and } M_t = M_0 + \int_0^t f_s^{\frac{1}{2}} \,\mathrm{d}B_s$$

Proof According to Lebesgue's differentiation theorem there exists

$$f_t = \lim n(\langle M \rangle_t - \langle M \rangle_{t-\frac{1}{n}}) \quad \lambda \otimes P \text{ a.s.}$$

and

$$f_t = \frac{\mathrm{d}\langle M \rangle_t}{\mathrm{d}\lambda_t} \in \mathcal{L}(\mathcal{P}).$$

$f^{-\frac{1}{2}} \in \mathcal{L}^2_{\mathrm{loc}}(M)$, because

$$\int_0^t (f_s^{-\frac{1}{2}})^2 \,\mathrm{d}\langle M \rangle M_s = \int_0^t f_s^{-1} f_s \,\mathrm{d}\lambda(s) = t < \infty.$$

Now define $B_t := \int_0^t f_s^{-\frac{1}{2}} \,\mathrm{d}M_s$. Then $B \in \mathcal{M}_{\mathrm{loc},c}$ and it holds

$$\langle B \rangle_t = \int_0^t f_s^{-1} \mathrm{d}\langle M \rangle_s = t.$$

According to Lévy's theorem B is a Brownian motion and it is

$$\begin{aligned} M_t &= M_0 + \int_0^t f_s^{\frac{1}{2}} (f_s^{-\frac{1}{2}} \, \mathrm{d}M_s) \\ &= M_0 + \int_0^t f_s^{\frac{1}{2}} \, \mathrm{d}B_s. \end{aligned}$$

□

In the case where the variation process is only Lebesgue continuous, $\langle M \rangle \ll \lambda_+$ there is an analogous statement formulated with an extension of the basic space.

Proposition 4.35 *Let $M = (M^1, \dots, M^d) \in \mathcal{M}_{\mathrm{loc},c}$ and let $\langle M^i \rangle \ll \lambda_t$, $1 \le i \le d$, then there exists a d-dimensional Brownian motion B on an extension $(\widetilde{\Omega}, \widetilde{\mathcal{A}}, \widetilde{P}) \supset (\Omega, \mathcal{A}, P)$ of the basic space and there exists $f \in \mathcal{L}^2_{\mathrm{loc}}(B)$ with values in $\mathbb{R}^{d \times d}$ such that*

$$M_t = M_0 + \int_0^t f_s \cdot \mathrm{d}B_s$$

with $(f \cdot B)^i = \sum_j f^{i,j} \cdot B^j$.

4.3 Measure Change, Theorem of Girsanov

Topic of this section is the study of properties of processes when the measure changes. Let X be a semimartingale with respect to an underlying measure P, $(X, P) \in \mathcal{S}$ and let Q be a P-continuous probability measure, $Q \ll P$. Then a fundamental question is whether also (X, Q) is a semimartingale and how the *drift* and *martingale components* of X change by the transition from P to Q. We want to decompose the process (X, Q) into drift and martingale components. This is of great importance in financial mathematics, because it turns out that the prices of derivatives and the form of optimal trading strategies can be computed simply by introducing suitable new measures Q. To motivate the general form of Girsanov's theorem, we consider a simple example in discrete time.

Example 4.36 Let $Z_1, \dots, Z_n$ be independent, normally distributed, $Z_i \sim N(0, 1)$, $Z = (Z_1, \dots, Z_n)$ random variables on $(\Omega, \mathcal{A}, P)$. We define a new measure $\widetilde{P}$ using the Radon–Nikodým derivative.

$$\frac{\mathrm{d}\widetilde{P}}{\mathrm{d}P} := \exp\left(\sum_{i=1}^n \mu_i Z_i - \frac{1}{2} \sum_{i=1}^n \mu_i^2 \right).$$

This density has a similar shape to that in the exponential martingale. It holds:

$$Ee^{\mu \cdot Z} = e^{\frac{1}{2}\|\mu\|^2}$$

is the Laplace transform of the one dimensional normally distributed random variable $\mu \cdot Z$. It follows that the new measure $\widetilde{P}$ is a probability measure,

$$\widetilde{P} \in M^1(\Omega, \mathcal{A}).$$

How is the sequence (Z_i) distributed with respect to the new measure $\widetilde{P}$?

It holds:

$$\begin{aligned}
\frac{\mathrm{d}\tilde{P}^Z}{\mathrm{d}\lambda^n}(z) &= \frac{\mathrm{d}\tilde{P}^Z}{\mathrm{d}P^Z}(z) \cdot \frac{\mathrm{d}P^Z}{\mathrm{d}\lambda^n}(z) \\
&= \exp\left(\sum \mu_i z_i - \frac{1}{2}\mu_i^2\right) \cdot \frac{1}{(2\pi)^{n/2}} \cdot \exp\left(-\frac{1}{2}\sum z_i^2\right) \\
&= \frac{1}{(2\pi)^{n/2}} \cdot \exp\left(-\frac{1}{2}\left(\sum z_i - \mu_i\right)^2\right).
\end{aligned}$$

It follows: $\widetilde{Z}_i := Z_i - \mu_i$, $1 \le i \le n$, is an i.i.d. sequence with respect to $\widetilde{P}$. This implies that the sequence (Z_i) with respect to $\widetilde{P}$ has the following representation:

$$Z_i = \widetilde{Z}_i + \mu_i \textit{ with an i.i.d. sequence } (\widetilde{Z}_i) \sim N(0, 1) \textit{ and with shift } (\mu_i).$$

With respect to the new measure $\widetilde{P}$ the sequence (Z_i) is, therefore, an i.i.d. normally distributed sequence with shifts μ_i. Girsanov's theorem shows that a similar behavior holds in a more general form.

Denotation Let $P, Q \in M^1(\Omega, \mathcal{A})$ and be $(\mathcal{A}_t) \subset \mathcal{A}$ is a filtration, without restriction one can choose $\mathcal{A} = \mathcal{A}_\infty$. Then we define

- $Q \ll P$, Q is **absolutely continuous** with respect to $P \iff \mathcal{N}_P \subset \mathcal{N}_Q$
- $Q \sim P$, Q is **equivalent** to $P \iff \mathcal{N}_P = \mathcal{N}_Q$
- $Q \overset{\text{loc}}{\ll} P$, Q is **locally continuous** with respect to $P \iff Q_t \ll P_t, \ \forall t > 0$, where $Q_t := Q|_{\mathcal{A}_t}$ and $P_t := P|_{\mathcal{A}_t}$ are the measures restricted to $\mathcal{A}_t$.

  For stopping times τ be $P_\tau := P|_{\mathcal{A}_\tau}$, $P_{\tau_-} := P|_{\mathcal{A}_{\tau_-}}$. The statement $Q \overset{\text{loc}}{\ll} P$ is equivalent to the condition that a localizing sequence (τ_n) of stopping times exists with $Q_{\tau_n} \ll P_{\tau_n}$, $n \in \mathbb{N}$.
- $Q \overset{\text{loc}}{\sim} P$, Q is **locally equivalent** with respect to P

  $\iff Q \overset{\text{loc}}{\ll} P$ and $P \overset{\text{loc}}{\ll} Q$.

Proposition 4.37

(a) Let $Q \ll P$ and let $D_t := E\left(\frac{dQ}{dP} \mid \mathcal{A}_t\right)$ be the martingale generated by the density quotient $\frac{dQ}{dP}$. Then it follows

$$D_t = \frac{dQ_t}{dP_t} \, [P], \quad \forall t,$$

i.e., D_t coincides with the density quotient of Q to P on the restricted σ-algebra $\mathcal{A}_t$.

(b) Let $Q \overset{\text{loc}}{\ll} P$ be locally P-continuous, then the density process D is a P-martingale

$$D_t = \frac{dQ_t}{dP_t} \in \mathcal{M}(P) \text{ and it holds}$$

(D_t) is uniformly integrable $\Longleftrightarrow Q \ll P$ and then it holds

$$D_t := E\left(\frac{dQ}{dP} \mid \mathcal{A}_t\right). \tag{4.18}$$

Proof

(a) follows directly by verifying the Radon–Nikodým equation.

(b) According to (a) it follows that (D_t) is a P-martingale. The rest of the assertion follows by the closure theorem for martingales: (D_t) is uniformly integrable if and only if $D_t \longrightarrow D_\infty$ in L^1 and P a.s.

From the L^1-convergence it then follows:

$$Q = D_\infty P \quad \text{i.e.} \quad D_\infty = \frac{dQ}{dP}.$$

$$D_t = E\left(\frac{dQ}{dP} \mid \mathcal{A}_t\right).$$

□

Remark 4.38 Local continuity of measures $Q \overset{\text{loc}}{\ll} P$ does not in general imply continuity $Q \ll P$ (at infinity). There are a number of 0–1 laws (e.g., for Gaussian measures) which state that if there is local continuity in the limit, then the measures are either orthogonal or equivalent.

In preparation for Girsanov's theorem for the Brownian motion, we need a martingale characterization of Brownian motion that is related to one of the three equivalences from the Lévy characterization of Brownian motion: If the stochastic exponential

$$\mathcal{E}_t^{if} := \exp\left\{ i \sum_{k=1}^{d} \int_0^d f_k(s)\,\mathrm{d}X_s^k + \frac{1}{2} \sum_{k=1}^{d} \int_0^t f_k^2(s)\,\mathrm{d}s \right\}$$

is a complex martingale, then X is a Brownian motion. The following is a variant of this characterization:

Proposition 4.39 *Let $(X_t, \mathcal{A}_t)$ be a continuous process such that*

$$Z_t = Z_t^{(u)} := \exp\left\{ uX_t - \frac{1}{2}u^2 t \right\} \in \mathcal{M}_{\mathrm{loc}}, \quad u \in (-\varepsilon, \varepsilon),$$

then $(X_t, \mathcal{A}_t)$ is a Brownian motion. □

Proof There is a localizing sequence $\tau_n \uparrow \infty$ such that for all $0 \leq s < t$, and all $A \in \mathcal{A}_s$ it holds:

$$\int_A Z_{s\wedge\tau_n}^{(u)}\,\mathrm{d}P = \int_A Z_{t\wedge\tau_n}^{(u)}\,\mathrm{d}P, \tag{4.19}$$

i.e., the martingale property holds when localizing Z with a sequence τ_n. In the proof of the characterization theorem this has already been used. In an exponential family with a real parameter, it holds by the theorem of majorized convergence, that one can swap differentiation (derivative w.r.t. u) and integration:

$$\frac{\partial}{\partial u} \int \cdots = \int \frac{\partial}{\partial u} \cdots, \quad u \in U_\varepsilon(0).$$

This property is needed for a small interval $U_\varepsilon(0)$ around zero. Derivation of both sides in (4.19) results in

$$\int_A \left(X_{t\wedge\tau_n} - u(t \wedge \tau_n)\right) Z_{t\wedge\tau_n}^{(u)}\,\mathrm{d}P = \int_A \left(X_{s\wedge\tau_n} - u(s \wedge \tau_n)\right) Z_{s\wedge\tau_n}^{(u)}\,\mathrm{d}P, \quad A \in \mathcal{A}_s.$$

Differentiation again w.r.t. u provides

$$\int_A \Big(\left((X_{t\wedge\tau_n} - u(t \wedge \tau_n)\right)^2 - (t \wedge \tau_n) \Big) Z_{t\wedge\tau_n}^{(u)}\,\mathrm{d}P$$

$$= \int_A \Big(\left(X_{s\wedge\tau_n} - u(s \wedge \tau_n)\right)^2 - (s \wedge \tau_n) \Big) Z_{s\wedge\tau_n}^{(u)}\,\mathrm{d}P.$$

We now have two equations that hold in an neighborhood of zero. For $u = 0$ the first equation says X_t is a continuous local martingale, $X \in \mathcal{M}_{\mathrm{loc},c}$. The second equation says, $X_t^2 - t \in \mathcal{M}_{\text{loc},c}$. By Lévy's theorem it follows that X is a Brownian motion. □

The following theorem of Girsanov describes changes in measure for the Brownian motion.

Theorem 4.40 (Girsanov's Theorem for the Brownian Motion) *Let $(X_t, \mathcal{A}_t)$ be a Brownian motion with respect to P, let $\varphi \in \mathcal{L}^0(X)$ and let*

$$L_t := \exp\left(\int_0^t \varphi_s \mathrm{d}X_s - \frac{1}{2}\int_0^t \varphi_s^2 \,\mathrm{d}s\right)$$

be the exponential martingale of $\varphi \cdot X$. Then it holds:

(a) $(L_t, \mathcal{A}_t)$ is a local martingale and a supermartingale, and it holds:

$$L_t = 1 + \int_0^t L_s \varphi_s \,\mathrm{d}X_s$$

(b) If $EL_T = 1$, then $(L_t, \mathcal{A}_t)_{0 \le t \le T}$ is a martingale. Let

$$\frac{\mathrm{d}\widetilde{P}}{\mathrm{d}P} := L_T \text{ and } \widetilde{X}_t := X_t - \int_0^t \varphi_s \,\mathrm{d}s,$$

then $\widetilde{P}$ is a probability measure on $(\Omega, \mathcal{A})$ and $(\widetilde{X}_t, \mathcal{A}_t, \widetilde{P})$ is a Brownian motion. Concerning the new measure $\widetilde{P}$, therefore, X is a Brownian motion with a random drift.

Remark 4.41 We will deal with a reverse of Girsanov's theorem in the following. Continuous measure changes in Brownian motion correspond to the addition of a random drift and the densities of continuous measure changes are described by stochastic exponentials.

Proof

(a) Part (a) and the first part of (b) follow from Proposition 4.12 via exponential martingales.
(b) According to Proposition 4.39, it is to show that

$$Z_t^{(u)} := \exp\left\{u\left(X_t - \int_0^t \varphi_s \,\mathrm{d}s\right) - \frac{1}{2}u^2 t\right\} \in \mathcal{M}_{\text{loc}}(\widetilde{P}).$$

Let, therefore, D_t be the density quotient process.

$$D_t := \frac{\mathrm{d}\widetilde{P}_t}{\mathrm{d}P_t} = E_P\Big(\frac{\mathrm{d}\widetilde{P}}{\mathrm{d}P} \mid \mathcal{A}_t\Big) = E_P\big(L_T \mid \mathcal{A}_t\big) = L_t \text{ as } (L_t) \in \mathcal{M}.$$

Then it holds: $(Z_t^{(u)} D_t)$ is a local martingale w.r.t. P, since

$$\begin{aligned} Z_t^{(u)} D_t &= Z_t^{(u)} L_t \\ &= \exp\left\{ \int_0^t (\varphi_s + u)\,\mathrm{d}X_s - \frac{1}{2}\int_0^t (\varphi_s + u)^2\,\mathrm{d}s \right\} \in \mathcal{M}_{\mathrm{loc}}(P), \end{aligned}$$

using that (X, P) is a Brownian motion. Let τ_n be a localizing sequence such that $\big(Z^{(u)}D\big)^{\tau_n} \in \mathcal{M}$ and let σ is a bounded stopping time. According to the optional sampling theorem $D_\tau = E\left(\frac{\mathrm{d}\widetilde{P}}{\mathrm{d}P} \mid \mathcal{A}_\tau\right)$ for all bounded stopping times τ. Further, again according to the optional sampling theorem

$$\begin{aligned} E_{\widetilde{P}} Z^{(u)}_{\tau_n \wedge \sigma} &= E_P Z^{(u)}_{\tau_n \wedge \sigma} D_{\tau_n \wedge \sigma} \\ &= E_P Z_0^{(u)} D_0 \qquad \text{as } Z_{\tau_n \wedge t} D_{\tau_n \wedge t} \mathcal{M}(P) \\ &= E_{\widetilde{P}} Z_0^{(u)}, \qquad \forall \text{ bounded stopping times } \sigma. \end{aligned}$$

With the characterization of the martingale property by stopping times it follows $Z^{(u)} \in \mathcal{M}_{\mathrm{loc}}(\widetilde{P})$ and thus according to Proposition 4.37 the assertion.

□

Remark 4.42

(a) Sufficient for the requirement $E L_T = 1$ is the Novikov condition $E \exp\big(\frac{1}{2}\int_0^\infty \varphi_s^2\,\mathrm{d}s\big) < \infty$ as well as the Kazamaki condition.
(b) There is also an analogous multivariate version of Girsanov's theorem as well as a version on $[0, \infty)$. For $d \geq 1$ the density quotient process is of the form

$$L_t = \exp\left\{ (\varphi \cdot X)_t - \frac{1}{2}\sum_{i=1}^d \int_0^t (\varphi_s^i)^2\,\mathrm{d}s \right\}.$$

If X is a d-dimensional Brownian motion with respect to P, then

$$(\widetilde{X}_t^i) = \left(X_t^i - \int_0^t \varphi_s^i\,\mathrm{d}s \right)$$

is a Brownian motion with respect to the continuous measure $\widetilde{P}$ with density process L bzgl. P.

In the above version of Girsanov's theorem, we are concerned with the question how a Brownian motion behaves under a measure transformation. For the generalization to the case of general martingales or semimartingales we need the following statement about the density process.

Proposition 4.43 *Let $Q \overset{\text{loc}}{\ll} P$ then the density process is $D_t > 0$ Q-a.s. $\forall\, t$ and it holds,* $\inf\{t \in \mathbb{R}_+;\ D_t = 0$ *or* $D_{t-} = 0\} = \infty$ *Q-a.s. if $Q \ll P$.*

Proof It is sufficient to consider the case $Q \ll P$ otherwise one can reduce the problem to the case $[0, t]$, $t > 0$. Let $\tau := \inf\{t;\ D_t = 0 \text{ or } D_{t-} = 0\}$. (D_t) is a nonnegative local martingale and hence a nonnegative supermartingale.

According to Corollary 3.10 it follows $D = 0$ on $[\tau, \infty)$. So $D_\infty - \frac{\mathrm{d}Q}{\mathrm{d}P} = 0$ on $\{\tau < \infty\}$.

Because of $Q = D_\infty P$, therefore, follows $Q(\{\tau < \infty\}) = 0$, i.e. $\inf\{t \in \mathbb{R}_+;\ D_t = 0 \text{ or } D_{t-} = 0\} = \infty$. Thus the assertion follows. □

Proposition 4.44

(a) If $Q \overset{\text{loc}}{\ll} P$, then for all stopping times τ

$$Q = D_\tau P \qquad \text{on } \mathcal{A}_\tau \cap \{\tau < \infty\},$$

i.e. Q has the stopped density process as density with respect to P on $\mathcal{A}_\tau$ on $\tau < \infty$.

(b) If $Q \ll P$, then

$$Q = D_\tau P \text{ on } \mathcal{A}_\tau.$$

Proof First, we prove (b).

(b) According to Proposition 4.37, the density process in the continuous case is given by

$$D_t = E\Big(\frac{\mathrm{d}Q}{\mathrm{d}P} \mid \mathcal{A}_t\Big).$$

This is a uniformly integrable martingale, and $D_t \longrightarrow D_\infty = \frac{\mathrm{d}Q}{\mathrm{d}P}$ in L^1 and a.s.

By the optional sampling theorem, for all stopping times τ it follows :

$$D_\tau = E\Big(\frac{\mathrm{d}Q}{\mathrm{d}P} \mid \mathcal{A}_\tau\Big) = \frac{\mathrm{d}Q|_{\mathcal{A}_\tau}}{\mathrm{d}P|_{\mathcal{A}_\tau}}.$$

(a) For $Q \overset{\text{loc}}{\ll} P$ D is uniformly integrable on finite intervals $[0, t]$.
Let $A \in \mathcal{A}_\tau$, then it follows for all t:

$$A \cap \{\tau \leq t\} \in \mathcal{A}_{\tau \wedge t}.$$

From this follows with the optional sampling theorem:

$$\begin{aligned} Q\big(A \cap \{\tau \leq t\}\big) &= \int_{A \cap \{\tau \leq t\}} E\big(D_t \mid \mathcal{A}_{\tau \wedge t}\big)\,\mathrm{d}P \\ &= \int_{A \cap \{\tau \leq t\}} D_{\tau \wedge t}\,\mathrm{d}P = \int_{A \cap \{\tau \leq t\}} D_\tau\,\mathrm{d}P; \end{aligned}$$

so the assertion holds on $[0, t]$.
For $t \uparrow \infty$ it follows:

$$Q\big(A \cap \{\tau < \infty\}\big) = \int_{A \cap \{\tau < \infty\}} D_\tau\,\mathrm{d}P.$$

□

The martingale property of a process M under a measure Q, $Q \overset{\text{loc}}{\ll} P$, one can infer from the martingale property of MD with respect to P.

Proposition 4.45 *Let $Q \overset{\text{loc}}{\ll} P$ with density process D and let M be an adapted càdlàg process, then it holds:*

(1) $M \in \mathcal{M}(Q) \iff MD \in \mathcal{M}(P)$
(2) If $Q \overset{\text{loc}}{\sim} P$*, then* $M \in \mathcal{M}_{\text{loc}}(Q) \iff MD \in \mathcal{M}_{\text{loc}}(P)$.
For the backward direction "$\Longleftarrow$" local continuity, $Q \overset{\text{loc}}{\ll} P$ *is sufficient.*

Proof

(1) We first show that the integrability conditions are the same:

$$\begin{aligned} M_t \in L^1(Q) &\iff E_P D_t |M_t| = E_Q |M_t| < \infty \\ &\iff (DM)_t \in L^1(P). \end{aligned}$$

Thus it holds:

$$\begin{aligned} M \in \mathcal{M}(Q) &\Longleftrightarrow M_t \in L^1(Q) \text{ and } E_Q M_t \mathbb{1}_A \\ &\quad = E_Q M_s \mathbb{1}_A \ \forall s \le t, \forall A \in \mathcal{A}_s \\ &\Longleftrightarrow (MD)_t \in L^1(P) \text{ and } E_P(D_t M_t)\mathbb{1}_A \\ &\quad = E_P(D_s M_s)\mathbb{1}_A \quad \forall s \le t, \forall A \in \mathcal{A}_s \\ &\Longleftrightarrow MD \in \mathcal{M}(P). \end{aligned}$$

(2) follows from (1) by localization. □

As an application, we obtain a Bayesian formula for conditional expectations.

Corollary 4.46 (Bayesian Formula for Conditional Expectations) *Let $Q \overset{\text{loc}}{\ll} P$ with density process D and let $s \le t$, and $X \in \mathcal{L}_+(\mathcal{A}_t) \cup \mathcal{L}^1(\mathcal{A}_t, Q)$ then*

$$E_Q(X \mid \mathcal{A}_s) = \frac{E_P(XD_t \mid \mathcal{A}_s)}{D_s}[P]$$

Proof Note that according to Proposition 4.43 the density $D_s > 0 \quad Q$ a.s.

Let $X \in L^1(\mathfrak{A}_t, Q)$, and be $M_u := E_Q(X \mid \mathfrak{A}_u)$. Then it follows $M \in \mathcal{M}(Q)$, and, therefore, according to Proposition 4.45 $DM \in \mathcal{M}(P)$.

From this it follows

$$D_s E_Q(X \mid \mathfrak{A}_s) = D_s M_s = E_P(D_t M_t \mid \mathfrak{A}_s) = E_P(XD_t \mid \mathfrak{A}_s),$$

therefore, the proposition.

Case $X \in \mathcal{L}_+(\mathfrak{A}_t)$ is analogous. □

The following theorem extends the statement of Girsanov's theorem to continuous local martingales under the assumption of a continuous density process.

Theorem 4.47 (Girsanov's Theorem for Continuous Local Martingales) *Let $Q \overset{\text{loc}}{\ll} P$ and let the density process D be continuous, then it holds:*

(a) If X is a continuous semimartingale with respect to P, then X is also a continuous semimartingale with respect to Q: $X \in \mathcal{S}_c(P) \Longrightarrow X \in \mathcal{S}_c(Q)$.
(b) If $M \in \mathcal{M}_{\text{loc},c}(P)$, then it holds:

$$\widetilde{M} := M - D^{-1} \cdot \langle M, D\rangle \in \mathcal{M}_{\text{loc},c}(Q) \quad \textit{and} \quad \langle \widetilde{M}\rangle^Q = \langle M\rangle^P$$

The mapping $\widetilde{\cdot} : \mathcal{M}_{\text{loc},c}(P) \to \mathcal{M}_{\text{loc},c}(Q), M \mapsto \widetilde{M}$ is called ***Girsanov transform***.

(c) If $D > 0[P]$ (i.e. $P \overset{\text{loc}}{\sim} Q$), then it follows:
(1) $\exists!\ L \in \mathcal{M}_{\text{loc},c}(P): \quad D = \mathcal{E}(L) = \exp\left(L - \frac{1}{2}\langle L\rangle\right)$, D is the stochastic exponential of L.
(2) $L = \log D_0 + \int_0^{\cdot} D_s^{-1} \mathrm{d}D_s = \mathcal{L}(D)$, L is the stochastic logarithm of D.
(3) $P = \mathcal{E}(-\widetilde{L})Q$

Proof

(a) follows from (b) since $X \in \mathcal{S}(P)$ has a decomposition

$$X = M + A \text{ with } M \in \mathcal{M}_{\text{loc},c} \text{ and } A \in \mathcal{V}_c.$$

Then by (b) it follows $\widetilde{M} = M - D^{-1} \cdot \langle M, D\rangle \in \mathcal{M}_{\text{loc},c}(Q)$.
From this then follows the decomposition

$$X = \widetilde{M} + \underbrace{(A + D^{-1} \cdot \langle M, D\rangle)}_{\in \mathcal{V}_c} \in \mathcal{S}_c(Q);$$

Thus X is a continuous Q-semimartingale.

(b) If $XD \in \mathcal{M}_{\text{loc}}(P)$ then it follows by Proposition 4.45 $X \in \mathcal{M}_{\text{loc}}(Q)$. So to show is: $\widetilde{M}D \in \mathcal{M}_{\text{loc},c}(P)$.
Using the partial integration formula for continuous semimartingales it holds:

$$\begin{aligned}\widetilde{M}_t D_t &= M_0 D_0 + \int_0^t \widetilde{M}_s \mathrm{d}D_s + \int_0^t D_s \mathrm{d}\widetilde{M}_s + \langle \widetilde{M}, D\rangle_t \\ &= M_0 D_0 + \int_0^t \widetilde{M}_s \mathrm{d}D_s + \int_0^t D_s\, \mathrm{d}M_s - \langle M, D\rangle_t \\ &\qquad + \langle M, D\rangle_t \in \mathcal{M}_{\text{loc},c}(P).\end{aligned}$$

To this purpose, we use the following relation:

$$\begin{aligned}D \cdot \widetilde{M} + \langle \widetilde{M}, D\rangle &= \underbrace{D \cdot M}_{\in \mathcal{M}_{\text{loc},c}} - \underbrace{D \cdot (D^{-1} \cdot \langle M, D\rangle)}_{=\langle M, D\rangle} + \langle M, D\rangle \\ &\qquad - D^{-1} \cdot \langle \langle M, D\rangle, D\rangle.\end{aligned}$$

The last term is zero, since $\langle M, D\rangle$ is predictable and of finite variation.

Finally, it holds that $\langle \widetilde{M} \rangle^Q = \langle M \rangle^Q = [M]^Q = [M]^P = \langle M \rangle^P$

(c) According to the Itô formula

$$\log D_t = \underbrace{\log D_0 + \int_0^t D_s^{-1} \mathrm{d}D_s}_{=L_t} - \frac{1}{2} \int_0^t D_s^{-2} \mathrm{d}\langle D \rangle_s = L_t - \frac{1}{2} \langle L \rangle_t.$$

But it follows that D is the stochastic exponential of L is, $D_t = \mathcal{E}(L)_t$ and, therefore, also $L = \mathfrak{L}(D)$, L is the stochastic logarithm of D.

To prove (c): By definition and the first part of (b).

$$\begin{aligned}\widetilde{M} &= M - D^{-1} \cdot \langle M, D \rangle \\ &= M - \langle M, D^{-1} \cdot D \rangle \\ &= M - \langle M, L \rangle \in \mathcal{M}_{\mathrm{loc},c}(Q)\end{aligned}$$

is a locally continuous Q-martingale. Further : $\widetilde{L} = L - \langle L \rangle \in \mathcal{M}_{\mathrm{loc},c}(Q)$.

It follows that:

$$\mathcal{E}(-\widetilde{L}) = \exp(-L + \langle L \rangle - \frac{1}{2} \langle L \rangle) = \mathcal{E}(L)^{-1}.$$

Thus it follows that

$$P = \mathcal{E}(L)^{-1} Q = \mathcal{E}(-\widetilde{L}) Q.$$

□

Thus, continuous locally non-negative martingales, in particular continuous density quotients of equivalent measures, are characterized by having a representation as a stochastic exponential of a continuous local martingale.

Remark 4.48 In particular, for the case of the Brownian motion $M = B$ it holds for the Girsanov transform $\widetilde{M}$:

$$\widetilde{M} \in \mathcal{M}_{\mathrm{loc},c}(Q) \quad \text{and} \quad \langle \widetilde{M} \rangle_t^Q = \langle M \rangle_t^P = t.$$

By Lévy's theorem it follows: $(\widetilde{M}, Q)$ is a Brownian motion. The density process D has the form

$$D = \mathcal{E}(L) = \exp\left(L - \frac{1}{2} \langle L \rangle\right) \text{ with } L \in \mathcal{M}_{\mathrm{loc},c}(P).$$

According to the martingale representation theorem L is of the form $L = L_0 + \varphi \cdot B$ with $\varphi \in L^0(B)$. The Girsanov transform

$$\widetilde{B}_t = B_t - \langle B, L\rangle_t = B_t - \int_0^t \varphi_s \, ds$$

thus yields a stochastic shift and we obtain Girsanov's theorem on Brownian motion as a special case of the general theorem.

There is also a version of Girsanov's theorem for non-continuous local martingales. We use in the proof some statements from Jacod and Shiryaev (1987, I, Sections 3.5 and 4.49).

Theorem 4.49 (General Measure Transformation) *Let $Q \overset{\text{loc}}{\ll} P$ with density process D and let $M \in \mathcal{M}_{\text{loc}}(P)$ with $M_0 = 0$ and $[M, D] \in \mathcal{A}_{\text{loc}}(P)$. Then it holds:*

(a) $\widetilde{M} := M - \frac{1}{D_-} \cdot \langle M, D\rangle = M - \langle M, \mathfrak{L}(D)\rangle \in \mathcal{M}_{\text{loc}}(Q)$, $\mathfrak{L}(D)$ the stochastic logarithm
(b) $\langle M^c\rangle^P = \langle \widetilde{M}^c\rangle^Q$; the predictable quadratic variations of the continuous martingale components of M, $\widetilde{M}$ with respect to P and Q are identical.

Proof

(a) $\frac{1}{D_-}$ is locally bounded according to Proposition 4.43 such that $\widetilde{M}$ is well-defined.
$A := \frac{1}{D_-} \cdot \langle M, D\rangle \in \mathcal{V} \cap \mathcal{P}$ is predictable, since in $\mathcal{A}_{\text{loc}}(P)$ (cf. Jacod and Shiryaev (1987, I, Sect. 3.5)). It follows by the partial integration formula for semimartingales

$$AD = A \cdot D + D_- \cdot A = A \cdot D + \langle M, D\rangle$$

(cf. Jacod and Shiryaev (1987, I, Sect. 4.4)).
Again with partial integration we get

$$\begin{aligned}\widetilde{M}D = MD - AD &= \underbrace{M_- \cdot D}_{\in \mathcal{M}_{\text{loc}}(P)} + \underbrace{D_- \cdot M}_{\in \mathcal{M}_{\text{loc}}(P)} + [M, D] - \underbrace{A \cdot D}_{\in \mathcal{M}_{\text{loc}}(P)} - \langle M, D\rangle \\ &= \underbrace{Y}_{\in \mathcal{M}_{\text{loc}}(P)} + \underbrace{[M, D] - \langle M, D\rangle}_{\in \mathcal{M}_{\text{loc}}(P)} \in \mathcal{M}_{\text{loc}}(P)\end{aligned}$$

With $Y := M_- \cdot D + D_- \cdot M - A \cdot D \in \mathcal{M}_{\text{loc}}(P)$ this martingale property of $\widetilde{M}D$ with respect to P implies according to Proposition 4.45

$$\widetilde{M} \in \mathcal{M}_{\text{loc}}(Q).$$

(b) $M = \underbrace{\widetilde{M}}_{\in \mathcal{M}_{\text{loc}}(Q)} + \underbrace{A}_{\in \mathcal{V}_P} \in \mathcal{S}^Q$ is a special semimartingale with respect to Q.
Equality of the quadratic variation of M under P and Q hods. For because of $P \overset{\text{loc}}{\ll} Q$ the P_t stochastic limit implies the same Q_t stochastic limit (transition to a.s. convergent subsequences), i.e. $[M]^Q = [M]^P$.

For $A \in \mathcal{V}$ the continuous martingale part of A equals zero, $A^c = 0$ since $A^c \in \mathcal{V} \cap \mathcal{M}_{\text{loc},c}$.

Thus it follows: $\widetilde{M}^c = M^c$.

Therefore, as a result we get:

$$\begin{aligned}\langle \widetilde{M}^c \rangle^Q = \langle M^c \rangle^Q &= [M]^Q - \sum_{s \le \cdot} (\Delta M_s)^2 \\ &= [M]^P - \sum_{s \le \cdot} (\Delta M_s)^2 = \langle \widetilde{M}^c \rangle^P.\end{aligned}$$

□

Similarly, a generalization of Proposition 4.49 to a version of Girsanov's theorem for general semimartingales holds.

Proposition 4.50 *Let $Q \overset{\text{loc}}{\ll} P$, $X \in \mathcal{S}^P$, $\varphi \in \mathcal{L}_P^0(X)$, then it follows:*

$$X \in \mathcal{S}^Q, \varphi \in \mathcal{L}_Q^0(X)$$

and it holds:

$$[X]^Q = [X]^P, \quad (\varphi \cdot X)^Q = (\varphi \cdot X)^P \quad \text{and} \quad \langle X^c \rangle^Q = \langle X^c \rangle^P.$$

Proof For the proof, we refer the reader to Jacod and Shiryaev (1987, III, Sect. 3.13). □

Remark 4.51

(a) Let $\varphi \in \mathcal{L}_{\text{loc}}^0(M)$, then it follows $\varphi \in \mathcal{L}_{\text{loc}}^0(\widetilde{M})$.
For the Girsanov transformation of $\varphi \cdot M$ it holds:

$$\widetilde{\varphi \cdot M} = \varphi \cdot \widetilde{M}.$$

Proof Since $\langle M \rangle = \langle \widetilde{M} \rangle$ it holds $\mathcal{L}_{\text{loc}}^0(M) = \mathcal{L}_{\text{loc}}^0(\widetilde{M})$.
φ is locally bounded and, therefore, it holds

$$\varphi \cdot \widetilde{M} = \varphi \cdot M - (\varphi D^{-1}) \cdot \langle M, D \rangle = \varphi \cdot M - D^{-1} \cdot \langle \varphi \cdot M, D \rangle = \widetilde{\varphi \cdot M}.$$ □

(b) (P, Q) is called **Girsanov pair**, if $P \sim Q$ on $\mathcal{A}_\infty$ and D is continuous. Then it follows:

$$\mathcal{S}(P) = \mathcal{S}(Q)$$

and the Girsanov transformation $G_P^Q : \mathcal{S}_c(P) \longrightarrow \mathcal{S}_c(Q)$, $M \mapsto \widetilde{M}$ is bijective.

4.3.1 Applications of the Girsanov Theorem

The Girsanov transform is one of the important operations that allow to perform analysis on infinite-dimensional spaces. We outline two applications of Girsanov's theorem.

4.3.1.1 Theorem of Cameron–Martin

Let $(C_d[0, T], \|\cdot\|_\infty, W)$ be the Wiener space, that is W is the Wiener measure on $C_d[0, T]$ the distribution of a d-dimensional Brownian motion on $[0, T]$. The standard construction of the Brownian motion B on $\Omega = C_d[0, T]$ then is given by the evaluation mappings β_t:

$$B_t(\omega) = \beta_t(\omega) = \omega(t), \quad w \in C_d[0, T].$$

For each continuous function $h \in C_d[0, T]$, $h : [0, T] \longrightarrow \mathbb{R}^d$, let $\tau_h(\omega)$ be the h-shift (the translation by h), i.e.

$$(\tau_h\omega)(t) = \beta_t\big(\tau_h(\omega)\big) = \beta_t(\omega) + h(t)$$

We consider the image measure of the Wiener measure under the h-shift

$$W_h := W^{\tau_h} = P^{B+h}.$$

W_h is the distribution of the Brownian motion shifted by h (Fig. 4.3).

A fundamental question is: for which functions $h \in C_d[0, T]$ does it hold, that $W_h \approx W$, i.e. the translated measure W_h is equivalent to W and how to determine in this case the density $\frac{\mathrm{d}W_h}{\mathrm{d}W}$.

This is a basic problem for the statistical analysis of stochastic processes resp. measures on the infinite-dimensional space of continuous functions. For two Gaussian measures on $C_d[0, T]$ a 0–1 law holds. Either the measures are orthogonal, that is, they can be perfectly distinguished statistically, or they are equivalent. If B^σ is a Brownian motion with variance $\mathrm{var}(B_t^\sigma) = \sigma^2 t$ then B^{σ_1} and B^{σ_2} for $\sigma_1 \neq \sigma_2$are orthogonal. Since the quadratic variation of B^σ is $[B^\sigma]_t = \sigma^2 t$ it follows that B^{σ_1} and B^{σ_2} are concentrated on disjoint path sets implying orthogonality of the measures.

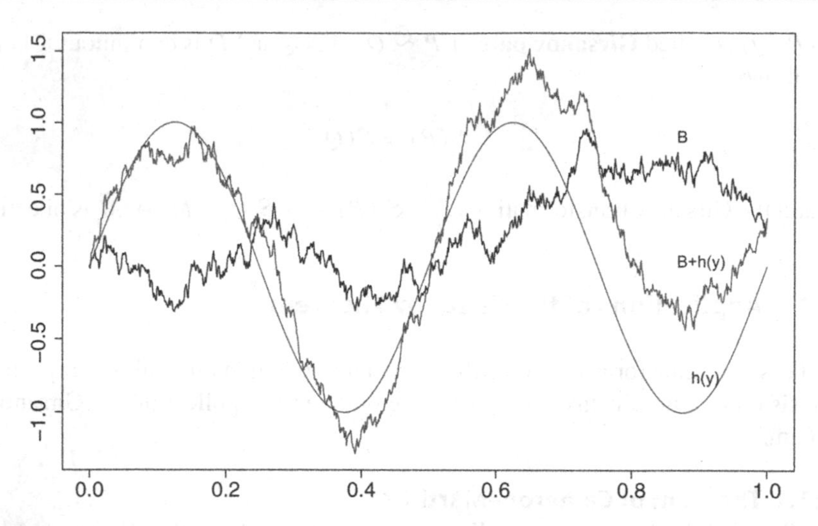

Fig. 4.3 Brownian motion, normal and shifted, B = Brownian motion; $h(y) = \cos(4\pi y)$, $B + h$ = addition of B and h

Cameron-Martin's theorem answers the question for which continuous functions $h \in C_d[0, T]$ $W_h \approx W$ resp. $W_h \perp W$ holds and gives a formula for the density $\mathrm{d}W_h / \mathrm{d}W$. Obviously, for the equivalence of W_h and W necessary that $h_i(0) = 0$ for $h = (h_1, \dots, h_d)$.

Definition 4.52 Let

$$H := \Big\{ h \in C_d[0, T];\ h = (h_1, \dots h_d),\ h_i \text{ absolutely continuous},\ h_i(0) = 0,$$

$$\text{and} \int_0^T |h_i'(s)|^2 \,\mathrm{d}s < \infty \Big\}.$$

The subspace $H \subset C_d[0, T]$ is called **Cameron–Martin space.**

Remark 4.53 The Cameron–Martin space H is a Hilbert space with scalar product $\langle h, g \rangle := \sum_i \int_0^T h_i'(s) g_i'(s)\,\mathrm{d}s$ and $\|h\|_H := \langle h, h \rangle^{1/2}$.

$H \subset C_{d,0}[0, T] = \{h \in C_d[0, T];\ h(0) = 0\}$ and H is dense with respect to the supremum norm $\|\cdot\|_\infty$. For $h \in H$, $h' \circ \beta = \int_0^\cdot h' \cdot \mathrm{d}\beta = \sum_{i=1}^d \int_0^\cdot h_i' \mathrm{d}\beta^i$ satisfies the Novikov condition. Therefore, the stochastic exponential $\mathcal{E}(h' \circ \beta)$ is a martingale.

The following theorem was proved by Cameron–Martin in 1949 by functional analytic methods. The proof given here is based on an inventive combination of Girsanov's theorem with the martingale representation theorem.

Theorem 4.54 (Theorem of Cameron–Martin) *The shifts W_h of the Wiener measure are either equivalent or orthogonal to W and it holds:*

(a) $W_h \approx W \Longleftrightarrow h \in H$
(b) For $h \in H$ the Cameron–Martin formula holds

$$\frac{\mathrm{d}W_h}{\mathrm{d}W} = \mathcal{E}\big((h' \circ \beta)_T\big).$$

Proof

(a) "$\Longrightarrow$": Let $W_h \approx W$, then according to Girsanov's theorem (in multivariate version) it follows:
There is a continuous martingale L with respect to the Wiener measure W such that

$$\frac{\mathrm{d}W_h}{\mathrm{d}W} = \mathcal{E}(L)_T \quad \text{and} \quad \widetilde{B} := \beta - \langle \beta, L \rangle,$$

(i.e. $\widetilde{B}^i := \beta^i - \langle \beta^i, L \rangle$) is a Brownian motion with respect to W_h.
By definition of W_h with respect to W_h also the representation

$$\beta = B + h,$$

with a Brownian motion B with respect to W_h holds.
So it holds with respect to W_h:

$$\begin{aligned} h &= \beta - B \\ &= (\widetilde{B} - B) + \langle \beta, L \rangle_t \in \mathcal{S}_c(W_h), \end{aligned}$$

h is a deterministic continuous semimartingale with respect to W_h.
It follows that h has finite variation, $V^h(T) < \infty$.
This gives us two decompositions of β with regard to W_h,

$$\beta = B + h = \widetilde{B} + \langle \beta, L \rangle,$$

with Brownian motions B, $\widetilde{B}$. Since h, $\langle \beta, L \rangle$ are predictable, it follows that β is a special semimartingale. Because of the uniqueness property of the decomposition of special semimartingales, it thus follows that

$$h = \langle \beta, L \rangle \text{ a.s.}$$

L is a martingale in the filtration generated by β. By the martingale representation theorem there exists a predictable process $\Phi \in \mathcal{L}^0(B)$,

$\int_0^T |\Phi_s^i|^2\, ds < \infty$, $1 \le i \le d$, such that with respect to the Wiener measure W

$$L = L_0 + \Phi \cdot \beta.$$

From this it follows: $h_i(t) = \langle \beta^i, L\rangle_t$
$= (\Phi^i \cdot \langle \beta^i, \beta^i\rangle)_t$
$= \int_0^t \Phi_s^i\, ds$, for all $t \in [0, T]$.

Therefore, by Lebesgue's differentiation theorem h_i is absolutely continuous and it holds

$$\Phi_s^i = h_i'(s) \text{ almost surely.}$$

So (independently of $\omega \in \Omega$) $h \in H$ since $\Phi \in \mathcal{L}^0(B)$ and it holds

$$\begin{aligned}\frac{dW_h}{dW} &= \mathcal{E}(L)_T \\ &= \mathcal{E}(\Phi \cdot \beta)_T \\ &= \mathcal{E}(h' \cdot \beta)_T.\end{aligned}$$

The Radon–Nikodým density of W_h is thus given by the stochastic exponential of $h' \cdot \beta$.

"$\Longleftarrow$": For $h \in H$ let Q be the probability measure $Q := \mathcal{E}((h' \cdot \beta))_T \cdot W$

According to Theorem 4.40 of Girsanov it holds: $Q \sim W$ and with respect to Q it holds:

$$\begin{aligned}\beta = (\beta^i) &= \left(B^i + \int_0^{\cdot} h_i'(s)\, ds\right) \\ &= (B^i + h_i) = B + h, \quad \text{because } h_i(0) = 0\end{aligned}$$

with a Brownian motion (B, Q). β is, therefore, a Brownian motion with shift and, therefore, it holds: $Q = W_h$.

Let $h \in C_d[0, T]$, then β is with respect to W_h a Gaussian process. The 0–1 law for Gaussian measures holds:

$$W, W_h \text{ are equivalent } W_h \sim W \quad \text{or} \quad W, W_h \text{ are orthogonal } W_h \perp W.$$

Either W_h and W can be distinguished on the basis of one observation; this is exactly the case if $W \perp W_h \Longleftrightarrow h \notin H$.

Or W, W_h are equivalent and $W \sim W_h \Longleftrightarrow h \in H$. □

Denote by supp μ the support of a measure μ, i.e. the smallest closed set A such that $\mu(A^c) = 0$. As a consequence of the above consideration one obtains also a simple argument for the determination of the support of the Wiener measure W.

Corollary 4.55

$$\operatorname{supp} W = C_{d,0}\big([0, T]\big) = \{f \in C_d([0, T]);\ f(0) = 0\}.$$

Proof There exists an element $x \in \operatorname{supp} W \neq \emptyset$. Then it follows: $x + H \subset \operatorname{supp} W$, for $W_h \approx W$ for $h \in H$. H is a dense subset of $C_{d,0}[0, T]$. Thus the assertion follows. □

4.3.2 The Clark Formula

Another application of Girsanov's theorem is the derivation of a formula for the integrand of the martingale representation theorem for the case of the Brownian motion: the Clark formula.

Let $(\mathcal{A}_t) = (\mathcal{A}_t^\beta)$ be the Brownian filtration in $C_d[0, T]$ and let β be the standard construction of the Brownian motion. Let X be an L^2-functional of the Brownian motion

$$X := F(\beta) \in \mathcal{L}^2(\mathcal{A}_T).$$

According to the martingale representation theorem X has a representation of the form

$$X = EX + \int_0^T \Phi_s \cdot \mathrm{d}\beta_s.$$

How can one determine Φ as a functional of F. For the special case of harmonic functions F we had obtained a solution to this problem using the Itô formula. We make the following assumption.

Assumption F is Lipschitz, i.e.

1. $\exists K: \quad |F(\beta + \Psi) - F(\beta)| \leq K\|\Psi\|$.
2. There is a kernel F' from $C_d[0, T]$ to $[0, T]$ such that $\forall\, \Psi \in H$

$$\lim_{\varepsilon \to 0} \frac{1}{\varepsilon}\big(F(\beta + \varepsilon\Psi) - F(\beta)\big) = \int_0^T \Psi(t) F'(\beta, \mathrm{d}t) \text{ a.s.}$$

The directional derivative of F is given by a measure $F'(\beta, dt)$.

Remark 4.56

(a) If F is Fréchet differentiable with bounded derivative, then conditions (1) and (2) hold. The continuous linear forms on $C_d[0, T]$ are the bounded measures on [0,T].
(b) In the following theorem we need the **predictable projection.**

Let $X \in L(\mathcal{A} \otimes \mathcal{B}_+)$ be a measurable process, then there exists exactly one predictable process ${}^pX \in L(\mathcal{P})$ such that for all predictable stopping times τ

$$^pX_\tau = E(X_\tau \mid \mathcal{A}_{\tau-}) \text{ on } \{\tau < \infty\}.$$

${}^pX$ is the **predictable projection** of X, ${}^pX := \widehat{\pi}_{\mathcal{P}}(X)$.

${}^pX$ is characterized by the equations

$$E \int_0^T X_s \mathrm{d}A_s = E \int_0^T {}^pX_s \mathrm{d}A_s \qquad \forall A_s \uparrow \in \mathcal{L}(\mathcal{P}). \tag{4.20}$$

In many cases, the predictable projection ${}^pX$ of X is given by the left-hand limit ${}^pX_t = X_{t-}$ or also by $E_P(X_t \mid \mathcal{A}_{t-})$.

The following theorem determines the integrand Φ of the martingale representation of $F(\beta)$ as the predictable projection of the derivative measure $F'(\beta, (t, T])$.

Theorem 4.57 (Clark Formula) *Let be* $X = F(\beta) \in L^2(\mathcal{A}_T, W)$ *and let the assumptions* (1), (2) *hold. Then the integrand* Φ *in the martingale representation* $F(\beta) = EX + \int_0^T \Phi_s \mathrm{d}B_s$ *of* X *is given by:*

$$\Phi = \widehat{\pi}_{\mathcal{P}}\big(F'(\beta, (t, T])\big).$$

Φ *is the predictable projection of* $F'(\beta, (t, F])$ *in* $L^2(\mathcal{P}, \mu)$.

Proof Let $u \in B(\mathcal{P})$ be bounded, $\mathcal{P}$-measurable and be

$$\psi_t := \int_0^t u_s \,\mathrm{d}s, \qquad \widetilde{\psi}_t := \int_0^t u_s \mathrm{d}\beta_s \quad \text{and} \quad Q := \mathcal{E}(\varepsilon\widetilde{\psi})_T W.$$

By Girsanov's theorem, it follows:

$$\beta - \varepsilon\psi \text{ is a Brownian motion with respect to } Q.$$

Thus the expected value of $F(\beta)$ can also be calculated by a change of measure:

$$\begin{aligned} E_W F(\beta) &= E_Q F(\beta - \varepsilon\psi) \\ &= E_W F(\beta - \varepsilon\psi)\mathcal{E}(\varepsilon\widetilde{\psi})_T. \end{aligned}$$

Equivalent to this is the following equation

$$E_W\big[(F(\beta - \varepsilon\psi) - F(\beta))(\mathcal{E}(\varepsilon\widetilde{\psi})_T - 1) + E_W(F(\beta - \varepsilon\psi) - F(\beta))$$
$$+ E_W F(\beta)(\mathcal{E}(\varepsilon\widetilde{\psi})_T - 1)\big] = 0.$$

In this form, the differentiability assumptions on F can now be applied. Multiplication by $\frac{1}{\varepsilon}$ leads to:

$$\frac{1}{\varepsilon}(\ldots) = 0 = I_1(\varepsilon) + I_2(\varepsilon) + I_3(\varepsilon), \quad \forall\, \varepsilon > 0, \tag{4.21}$$

where $I_j(\varepsilon)$ denote the three resulting terms.

For $\varepsilon \to 0$ converges $I_1(\varepsilon) \to 0$ because F satisfies the Lipschitz condition. $\frac{1}{\varepsilon}|F(\beta - \varepsilon\psi) - F(\beta)| \leq \|\psi\|$ and $\mathcal{E}(\varepsilon\widetilde{\psi})_T \to 1$.

To: $I_2(\varepsilon) + I_3(\varepsilon) \to 0$:

Using the differential equation for the stochastic exponential, the following holds.

$$\varepsilon^{-1}\big(\mathcal{E}(\varepsilon\widetilde{\psi})_T - 1\big) = \int_0^T \mathcal{E}(\varepsilon\widetilde{\psi})_s u_s \mathrm{d}\beta_s.$$

The integrand is majorized and converges pointwise to u. According to the theorem on majorized convergence for stochastic integrals, it follows

$$\varepsilon^{-1}(\mathcal{E}(\varepsilon\widetilde{\psi})_T - 1) \longrightarrow \int_0^T u_s \mathrm{d}\beta_s \quad \text{in } \mathcal{L}^2(P, \mathcal{A}_T).$$

Thus it holds: $I_2(\varepsilon) + I_3(\varepsilon) \longrightarrow 0$ and, therefore, by (4.21) and Assumption 2:

$$E_W F(\beta) \int_0^T u_s \mathrm{d}\beta_s = -\lim_{\varepsilon \downarrow 0} I_2(\varepsilon) = E_W F(\beta) \int_0^T \psi_t F'(\beta, \mathrm{d}t). \tag{4.22}$$

With the properties of the stochastic integral, it follows from (4.22)

$$\begin{aligned}
E_W \int_0^T \Phi_s u_s \,\mathrm{d}s &= E_W \int_0^T \Phi_s \mathrm{d}\beta_s \int_0^T u_s \mathrm{d}\beta_s \\
&= E_W F(\beta) \int_0^T u_s \mathrm{d}\beta_s \\
&= E_W \int_0^T \psi_t F'(\beta, \mathrm{d}t) \qquad \text{with } \psi_t = \int_0^t u_s \,\mathrm{d}s
\end{aligned}$$

$$= E_W \int_0^T u_t F'(\beta, (t, T])\,\mathrm{d}t \qquad \text{by partial integration}$$

$$= E_W \int_0^T u_t \widehat{\pi}_{\mathcal{P}}\big(F'(\beta, (t, T])\big)\,\mathrm{d}t, \qquad \text{as } u \in \mathcal{L}(\mathcal{P}).$$

The last equality follows from the projection equations for the predictable projection in (4.20). However, this equation implies that the predictable projection $\widehat{\pi}_{\mathcal{P}}(F'(\beta, (t, T)))$ of $F'(\beta, (t, T])$ is equal to the integrand Φ,

$$\Phi = \widehat{\pi}_{\mathcal{P}}\big(F'(\beta, (t, T])\big) \text{ in } \mathcal{L}^2(\mathcal{P}, \mu).$$

□

4.4 Stochastic Differential Equations

Stochastic differential equations are an important tool for modeling stochastic time-dependent processes. The local behavior of the processes can often be well described and then leads to stochastic differential equations as suitable models. Diffusion processes governed by local drift and diffusion coefficients are an important class of examples. Two examples of stochastic differential equations have already occurred so far.

Remark 4.58

(1) For a continuous local martingale $M \in \mathcal{M}_{\mathrm{loc},c}$ the **stochastic exponential**

$$\mathcal{E}(M)_t = \exp\big(M_t - \langle M\rangle_t\big)$$

is the unique solution of the stochastic integral equation

$$X_t = 1 + \int_0^t X_s\,\mathrm{d}M_s.$$

Equivalent to this integral equation is the stochastic differential equation

$$\begin{cases} dX_t &= X_t\,\mathrm{d}M_t, \\ X_0 &= 1. \end{cases}$$

(2) The geometric Brownian motion is described by the stochastic differential equation

$$\mathrm{d}X_t = \mu X_t\,\mathrm{d}t + \sigma X_t\,\mathrm{d}B_t, \qquad X_0 = x_0.$$

Here we have a local drift and a local variation term. The drift term is proportional to X_t and has the value μX_t. The variation term σX_t is also proportional to X_t. The solution to this stochastic differential equation is the **geometric Brownian motion**

$$X_t = x_0 \exp\left(\left(\mu - \frac{1}{2}\sigma^2\right)t + \sigma B_t\right).$$

The geometric Brownian motion is a standard model in financial mathematics. One can see some properties of X directly from the explicit solution, e.g., the evolution of the expected value:

$$EX_t = x_0 e^{\mu t}.$$

This follows from the fact that the stochastic exponential $\mathcal{E}(\sigma B)$ is a martingale,

$$\exp\left(\sigma B_t - \frac{1}{2}\sigma^2 t\right) \in \mathcal{M}.$$

For $t \to \infty$ it holds $B_t/t \to 0$. Therefore, it follows: $X_t \longrightarrow 0$ if $\frac{\sigma^2}{2} > \mu > 0$.
If $\mu > \sigma^2/2$ then $X_t \sim e^{\left(\mu - \frac{\sigma^2}{2}\right)t}$.

We now want to study more generally the following type of stochastic differential equations: Given a Brownian motion B on $(\Omega, \mathcal{A}, P)$ with natural filtration $\mathcal{A} = \mathcal{A}^B$ and let X be a solution of the stochastic differential equation

$$\begin{cases} \mathrm{d}X_t &= b(t, X_t)\,\mathrm{d}t + \sigma(t, X_t)\mathrm{d}B_t \\ X_0 &= Z, \end{cases} \tag{4.23}$$

i.e. $X_t = Z + \int_0^t b(s, X_s)\,\mathrm{d}s + \int_0^t \sigma(s, X_s)\mathrm{d}B_s$. b is a local drift term and σ is a local diffusion term. One can also consider the above stochastic differential equation on an interval $[s, t]$. The initial value Z should in this case lie in $\mathcal{L}(\mathcal{A}_s)$. Solutions X of (4.23) are called **generalized diffusion processes**, since they are generated by the diffusion of a Brownian motion. More precisely, we speak of **strong solutions.** One can introduce an additional degree of freedom, namely the construction of a suitable Brownian motion, and then obtain an extended class of solutions—the **weak solutions.**

The stochastic differential equations in (4.23) describe the Markovian case of diffusion equations, where $b = b(t, X_t)$, $\sigma = \sigma(t, X_t)$ depends only on the current state X_t of the process. **Non-Markovian diffusion equations** are a generalization of (4.23) with general predictable drift and diffusion coefficients b_t, σ_t.

Remark 4.59 The differential equation (4.23) can be understood as a generalization of the deterministic differential equations in which the stochastic diffusion term is absent, or equivalently in which $\sigma = 0$,

$$X_t = X_0 + \int_0^t b(s, X_s)\, ds.$$

For such deterministic differential equations, the solution behavior is well studied. Under a local Lipschitz condition with respect to x for the coefficient $b = b(x, t)$ in x, boundedness of b on compacta in $[0, \infty) \times \mathbb{R}^d$, i.e., in space and time, one can apply the iteration procedure of Picard Lindelöf and obtain: The iteration

$$X_t^{(0)} = x,$$

$$X_t^{(n+1)} = x + \int_0^t b(s, X_s^{(n)})\, ds, \quad n \geq 0,$$

converges for sufficiently small times t to a solution in $[0, t]$. The solution is unique.

However, one cannot expect a global solution. For example, consider the differential equation

$$X_t = 1 + \int_0^t X_s^2\, ds,$$

then for the coefficient $b(s, x) = x^2$ the local Lipschitz condition holds. The unique local solution of the differential equation with $x = 1$ is

$$X_t = \frac{1}{1-t}.$$

This solution explodes for $t \uparrow 1$. There is no global solution.

A fundamental question for stochastic differential equations, as in the case of deterministic equations, is the existence and uniqueness of solutions. There are examples of coefficients where a deterministic solution does not exist, or exists only on compact intervals, but where there is a global solution with the diffusion term. The first general existence and uniqueness statement is an analogue of Picard Lindelöf's theorem for deterministic differential equations.

Theorem 4.60 *Let b, σ be two measurable coefficient functions which are locally Lipschitz in x, i.e. $\forall\, T > 0, \exists K = K_T < \infty$ with*

$$|b(t, x) - b(t, y)| + |\sigma(t, x) - \sigma(t, y)| \leq K|x - y|, \quad 0 \leq t \leq T.$$

Further, the coefficients b, σ *have an at most linear growth in* x, *i.e.* $|b(t,x)| + |\sigma(t,x)| \le K(1+|x|), \quad 0 \le t \le T, \quad \forall x$ *and the initial value has a finite second moment,* $EZ^2 < \infty$, *then it holds:*
There exists a unique solution to the stochastic differential equation (4.23) in $[0,T]$ *i.e. if* X *and* Y *are two solutions, then it holds:* $X_t = Y_t \quad \forall t \in [0,T]\,[P]$.
The solution has bounded second moments: $E \sup_{s \le T} |X_s|^2 < \infty$.

Proof The proof relies on an iteration and fixed point argument similar to the deterministic case. Let $\mathcal{E}$ be the class of all (A_t)-adapted continuous processes $X = (X_t)_{0 \le t \le T}$ so that $\|X\|^2 := E \sup_{0 \le t \le T} |X_t|^2 < \infty$. This ground space $(\mathcal{E}, \|\cdot\|)$ is a Banach space, that is, a complete, separable, normed vector space. Now we make use of a fixed point argument: For a process $X \in \mathcal{E}$ we define a new process

$$\Phi(X)_t := Z + \int_0^t b(s, X_s)\,\mathrm{d}s + \int_0^t \sigma(s, X_s)\mathrm{d}B_s.$$

The above assumptions ensure the well-definedness of these integrals; Φ is a well-defined mapping $\Phi : \mathcal{E} \to \mathcal{E}$. We are looking for a fixed point of Φ. To apply Banach's fixed point theorem, we show that Φ is a contraction with respect to the introduced norm. With the inequality $(a+b)^2 \le 2(a^2+b^2)$ it holds:

$$\begin{aligned}|\Phi(X)_t - \Phi(Y)_t|^2 \le 2\Bigg[&\sup_{s \le T}\left(\int_0^t b(s, X_s) - b(s, Y_s)\right)\mathrm{d}s\Bigg]^2\\ &+ \sup_{s \le T}\left(\int_0^t (\sigma(s, X_s) - \sigma(s, Y_s)\mathrm{d}B_s)\right)^2.\end{aligned}$$

According to the Doob inequality and using the Lipschitz condition, it follows

$$\begin{aligned}E \sup_{t \le T}\left|\int_0^t \Big(\sigma(s, X_s) - \sigma(s, Y_s)\Big)\mathrm{d}B_s\right|^2 &\le 4E\left(\int_0^T (\sigma(s, X_s) - \sigma(s, Y_s))\mathrm{d}B_s\right)^2\\ &= 4E\int_0^T (\sigma(s, X_s) - \sigma(s, Y_s))^2\,\mathrm{d}s\\ &\le 4K^2 E \sup_{s \le T} |X_s - Y_s|^2 T.\end{aligned}$$

As a consequence, one obtains:

$$\|\Phi(X) - \Phi(Y)\| \le (4K^2T^2 + 4K^2T)^{1/2}\|X - Y\|.$$

Furthermore, using the inequality $(a+b+c)^2 \leq 3a^2+3b^2+3c^2$ one obtains for the initial value $\Phi(0)$ the estimate

$$|\Phi(0)_t|^2 \leq 3\left(Z^2 + \sup_{t\leq T}\left|\int_0^t b(s,0)\,\mathrm{d}s\right|^2 + \sup_{t\leq T}\left|\int_0^t \sigma(s,0)\,\mathrm{d}s\right|^2\right)$$

and it follows with the triangle inequality as above

$$E\sup_{t\leq T}|\Phi(0)_t|^2 \leq 3\big(EZ^2 + K^2T^2 + 4K^2T\big) < \infty.$$

As a result we obtain: $\Phi : \mathcal{E} \to \mathcal{E}$ is a Lipschitz mapping with the Lipschitz constant.

$$k(T) := \big(2(K^2T^2 + 4K^2T)\big)^{1/2}.$$

If one chooses T sufficiently small ($T \leq T_0$), then $k(T) < 1$, i.e. the mapping is a contraction on $\mathcal{E}$. By Banach's fixed point theorem it follows:

Φ has exactly one fixed point $X \in \mathcal{E}$ for $T \leq T_0$, i.e. the differential equation (4.23) restricted to $[0, T_0]$ has a solution in $\mathcal{E}$, and in $\mathcal{E}$ the solution is unique.

It remains to show: Every solution of (4.23) lies in $\mathcal{E}$.

Let X be a solution of (4.23) and let

$$\tau_n := \inf\{s \geq 0;\ |X_s| \geq n\} \text{ and let } f^{(n)}(t) := E\sup_{s\leq t\wedge\tau_n}|X_s|^2.$$

Then it holds as above with the Lipschitz and *linear growth* condition

$$\begin{aligned}
E\sup_{s\leq t\wedge\tau_n}|X_s|^2 &\leq 3\Big(EZ^2 + E\int_0^{t\wedge\tau_n} K^2(1+|X_s|^2)\,\mathrm{d}s \\
&\qquad + 4E\int_0^{t\wedge\tau_n} K^2(1+|X_s|)^2\,\mathrm{d}s\Big) \\
&\leq 3\left(EZ^2 + 10K^2\int_0^t \left(1 + E\sup_{u\leq s\wedge\tau_n}|X_u|^2\right)\mathrm{d}s\right) \\
&\leq 3(EZ^2 + 10K^2T) + 30K^2\int_0^t f^{(n)}(s)\,\mathrm{d}s.
\end{aligned}$$

So with suitable constants a, b:

$$f^{(n)}(t) \leq a + b\int_0^t f^{(n)}(s)\,\mathrm{d}s, \qquad 0 \leq t \leq T.$$

Continuous functions f satisfying such a recursive inequality can grow exponentially at most. This is the content of the following lemma of Gronwall from the theory of differential equations.

Lemma 4.61 (Gronwall's Lemma) *Let f be continuous on $[0, T]$ and for $t \leq T$ let*

$$f(t) \leq a + b \int_0^t f(s)\,\mathrm{d}s. \tag{4.24}$$

Then it follows

$$f(T) \leq a\big(1 + e^{bT}\big).$$

Proof Let $u(t) := e^{-bt} \int_0^t f(s)\,\mathrm{d}s$. Then $u(0) = 0$ and by assumption it holds:

$$u'(t) = e^{-bt}\left(f(t) - b\int_0^t f(s)\,\mathrm{d}s\right) \leq ae^{-bt}.$$

It follows:

$$u(T) \leq a \int_0^T e^{-bt}\,\mathrm{d}t = \frac{a}{b}\big(- e^{-bT} + 1\big) \leq \frac{a}{b}.$$

Thus we obtain as a result the desired bound

$$f(T) \leq a + b\int_0^T f(s)\,\mathrm{d}s \leq a + bu(T)e^{bT} \leq a\big(1 + e^{bT}\big).$$

□

Thus it follows in the proof of Theorem 4.60: $f^{(n)}(T) < K < \infty$, $K = K_T$ independent of n. By Fatou's lemma, it follows:

$$E \sup_{s \leq T} |X_s|^2 = E \varliminf_{n} \sup_{s \leq T \wedge \tau_n} |X_s|^2 \leq K < \infty.$$

Thus X satisfies the integrability condition and it follows: $X \in \mathcal{E}$. This implies the existence and uniqueness of solutions for $t \leq T_0$.

For arbitrary times T we consider the intervals $[0, \frac{T}{n}], [\frac{T}{n}, \frac{2T}{n}], \dots, [\frac{n-1}{n}T, T]$. For $n \geq n_0$ sufficiently large so that for $\frac{T}{n} \leq T_0$ it follows the existence of the solution with initial value $X_0^{(1)} = Z$ in the first interval $[0, \frac{T}{n}]$. The bound given in the proof is independent of the initial value. In the second interval we consider as new initial value $X_0^{(2)} = X_{\frac{T}{n}}$ from the solution on the first interval. This procedure can be iterated to go through all intervals. In each case, we use the value on the right edge as the new initial value. The solution can be combined consistently, since the bound only depends on EZ^2. For EZ^2 we had found a universal upper bound. Therefore, the solution intervals do not decrease. So as a result we get a unique solution on $[0, T]$. □

Two examples of the determination of solutions of stochastic differential equations follow.

Example 4.62 **(Brownian Bridge)** A continuous Gaussian process with normally distributed, finite-dimensional marginal distributions $B^0 = (B_t^0)_{0 \le t \le 1}$, on the time interval $[0, 1]$ is called Brownian bridge if

$$EB_t^0 = 0, \quad 0 \le t \le 1 \quad \textit{and} \quad \text{Cov}(B_s^0, B_t^0) = s(1-t), \quad 0 \le s \le t \le 1.$$

Compared to the Brownian bridge, the Brownian motion has the covariance function $\min(s, t)$. In particular, (Fig. 4.4)

$$\text{Var} B_s^0 = s(1-s)$$

and $B_0^0 = B_1^0 = 0$, i.e., a Brownian bridge starts and ends at zero. The special importance of the Brownian bridge B^0 results from the fact that the empirical process V_n has the Brownian bridge as a limit process,

$$V_n(s) = \sqrt{n}\big(F_n(s) - s\big), \quad s \in [0, 1]$$

$$V_n \xrightarrow{\mathcal{D}} B^0.$$

Therefore, as with Donsker's invariance principle, all continuous functionals of V_n converge to the corresponding functional of the Brownian bridge. In statistics, many

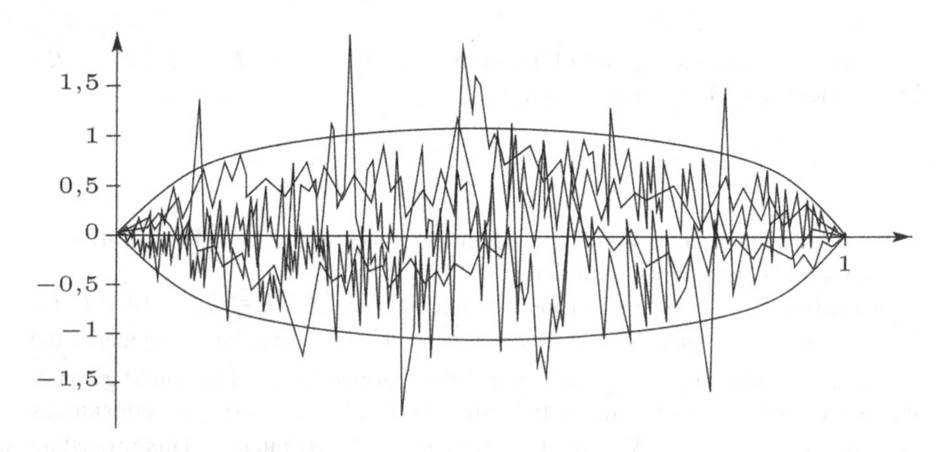

Fig. 4.4 Brownian bridge

empirical quantities of interest are given or approximated by functionals of the empirical process. Thus, one gets the possibility to approximate the distribution of statistical functionals by the distribution of functionals of Brownian bridge.

One way to construct a Brownian bridge is as follows: Take a Brownian motion under the condition that it has value zero at the point one (i.e., at the end):

$$B^0 \stackrel{d}{=} B | B_1 = 0.$$

So one considers the paths of Brownian motion that end up in the 'neighborhood of zero' at time 1. A second way to construct a Brownian bridge is as follows: Define

$$B_t^0 := B_t - t B_1.$$

B^0 is a Brownian bridge, because for this process $E B_t^0 = 0$ and

$$\begin{aligned} E B_s^0 B_t^0 &= E(B_s - sB_1)(B_t - tB_1) \\ &= (s \wedge t) - ts - st + st = s(1-t), \qquad \text{for } s \le t. \end{aligned}$$

B^0 is a Gaussian process with the covariance of Brownian bridge. Thus B^0 is a Brownian bridge.

The Brownian bridge constructed above B^0 is not a process adapted to the filtration of Brownian motion. Indeed, for the definition one uses the value of the Brownian motion at the point $t = 1$. The question is: Can one also generate a Brownian bridge on the Brownian filtration from the Brownian motion itself?

The idea is to construct a Brownian bridge using a differential equation where a drift in the zero direction is introduced. Let the drift $b(t, X_t)$ be given at time t by

$$b(t, x) := -\frac{x}{1-t}.$$

This drift drives the process in the residual time $1 - t$ in the direction of zero. The corresponding stochastic differential equation is given by

$$\mathrm{d}X_t = -\frac{X_t}{1-t}\,\mathrm{d}t + \mathrm{d}B_t. \tag{4.25}$$

To solve this differential equation, we use the 'variation of coefficients' method familiar from the theory of differential equations. We consider the solution ansatz

$$X_t = a(t)\left(x_0 + \int_0^t b(s)\,\mathrm{d}B_s\right), \tag{4.26}$$

with a, b differentiable. With partial integration it follows

$$\mathrm{d}X_t = a'(t)\Big(x_0 + \int_0^t b(s)\mathrm{d}B_s\Big)\,\mathrm{d}t + a(t)b(t)\mathrm{d}B_t.$$

If $a(0) = 1$, $a(t) > 0$, $\forall t$ then X solves the following stochastic differential equation

$$\begin{cases} \mathrm{d}X_t &= \frac{a'(t)}{a(t)} X_t\,\mathrm{d}t + a(t)b(t)\mathrm{d}B_t \\ X_0 &= x_0 \end{cases} \tag{4.27}$$

To solve the differential equation (4.25), we obtain by comparing coefficients

$$\frac{a'(t)}{a(t)} = -\frac{1}{1-t}, \qquad a(t)b(t) = 1.$$

These equations have the solution $a(t) = 1 - t$ and $b(t) = \frac{1}{1-t}$. Using $x_0 = 0$ it follows:

$$X_t = (1-t)\int_0^t \frac{1}{1-s}\mathrm{d}B_s, \qquad 0 \le t \le 1,$$

is a solution of the differential equation (4.25).

Assertion: X is a Brownian bridge. By construction X is a continuous Gaussian process and $EX_t = 0$. For $s \le t$ it follows with the independence of the increments of the Brownian motion

$$\begin{aligned} EX_sX_t &= (1-s)(1-t)E\Big(\int_0^s \frac{1}{1-u}\mathrm{d}B_u\Big)\Big(\int_0^t \frac{1}{1-v}\mathrm{d}B_v\Big) \\ &= (1-s)(1-t)E\Big(\int_0^s \frac{1}{1-u}\mathrm{d}B_u\Big)^2 \\ &= (1-s)(1-t)\int_0^s \frac{1}{(1-u)^2}\,du = s(1-t) = \mathrm{Cov}(B_s^0, B_t^0) \end{aligned}$$

The solution X of the stochastic differential equation (4.25) thus has the covariance of a Brownian bridge

$$\mathrm{Cov}(X_s, X_t) = s(1-t) = \mathrm{Cov}(B_s^0, B_t^0).$$

Thus X is a Brownian bridge and X is adapted to the filtration of Brownian motion B.

Example 4.63 **(Ornstein–Uhlenbeck Process)** The Ornstein–Uhlenbeck model dates back to Langevin in 1908 and Ornstein and Uhlenbeck in 1930. The Ornstein–Uhlenbeck process is a model for the velocity of molecular particles and is defined as the solution of the stochastic differential equation

$$\begin{cases} \mathrm{d}X_t &= -\alpha X_t\,\mathrm{d}t + \sigma \mathrm{d}B_t, \\ X_0 &= x_0, \end{cases}$$

σ and α are positive constants.

The Ornstein–Uhlenbeck model is a modification of the Brownian motion model. Since the paths of Brownian motion are nowhere differentiable, one cannot define a velocity for Brownian motion. The Ornstein–Uhlenbeck process is 'mean reverting'. For $X_t > 0$ there is a negative, for $X_t < 0$ a positive drift proportional to X_t which drives the process towards zero. This 'mean reverting' property causes a damping of the paths of the Brownian motion and leads to a stationary limit distribution. However, the paths remain non-differentiable.

To solve the differential equation, we use the method for variation of the coefficients in (4.27) with

$$\frac{a'(t)}{a(t)} = -\alpha, \quad a(t)b(t) = \sigma, \quad a(0) = 1.$$

It then follows:

$$a(t) = e^{-\alpha t}, \qquad b(t) = \sigma e^{\alpha t}$$

and we obtain

$$\begin{aligned} X_t &= e^{-\alpha t}\left(x_0 + \sigma \int_0^t e^{\alpha s}\mathrm{d}B_s\right) \\ &= x_0 e^{-\alpha t} + \sigma \int_0^t e^{-\alpha(t-s)}\mathrm{d}B_s. \end{aligned}$$

X is a Gaussian process with

$$\begin{aligned} EX_t &= x_0 e^{-\alpha t} \longrightarrow 0 \qquad (t \to \infty) \\ \mathrm{Var} X_t &= \sigma^2 \int_0^t e^{-2\alpha(t-s)}\,\mathrm{d}s \\ &= \frac{\sigma^2}{2\alpha}\left(1 - e^{-2\alpha t}\right) \longrightarrow \frac{\sigma^2}{2\alpha} \end{aligned}$$

The variance does not vanish but converges in the limit towards $\frac{\sigma^2}{2\alpha}$. The stronger the drift α is, the smaller becomes the variance. The process X_t converges in distribution to the normal distribution

$$X_t \xrightarrow{\mathcal{D}} N\Big(0, \frac{\sigma^2}{2\alpha}\Big),$$

the stationary distribution of this process. If one starts the Ornstein–Uhlenbeck process in the stationary distribution, then one obtains a stationary process. By analogy with Markov chains in discrete time, the Ornstein–Uhlenbeck process is a Markov process in continuous time and converges for $t \to \infty$ to the stationary distribution.

The existence and uniqueness theorems can be dealt with in the same way for the multidimensional case. Let $B = (B^1, \dots, B^r)$ be an r-dimensional Brownian motion, i.e., a vector of r independent Brownian motions which drives the differential equation. We call the generated filtration $(\mathcal{A}_t)$. Let

$$\begin{aligned} b : \mathbb{R}^+ \times \mathbb{R}^d &\longrightarrow \mathbb{R}^d \\ (s, x) &\longrightarrow b(s, x) \end{aligned}$$

be a drift vector, σ be a matrix of diffusion coefficients given by a $d \times r$-matrix

$$\begin{aligned} \sigma : \mathbb{R}^+ \times \mathbb{R}^d &\longrightarrow \mathbb{R}^{d\times r} \\ (s, x) &\longrightarrow \big(\sigma_{ij}(s, x)\big) \end{aligned}$$

and $Z \in L_d(\mathfrak{A}_0)$ an initial vector. The diffusion process X is then defined by the following stochastic differential equation:

$$\begin{cases} X_t = Z + \displaystyle\int_0^t b(s, X_s)\,\mathrm{d}s + \int_0^t \sigma(s, X_s)\,\mathrm{d}B_s \\ X_t^i = Z^i + \displaystyle\int_0^t b^i(s, X_s)\,\mathrm{d}s + \sum_{j=1}^r \int_0^t \sigma_{ij}(s, X_s)\,\mathrm{d}B_s^j. \end{cases} \tag{4.28}$$

Theorem 4.64 *Let the Lipschitz condition:*

$$|b(t, x) - b(t, y)| + |\sigma(t, x) - \sigma(t, y)| \le K|x - y|$$

and the linear growth condition:

$$|b(t, x)| + |\sigma(t, x)| \le K\big(1 + |x|\big), \quad E|Z|^2 < \infty, \quad 0 \le t \le T \textit{ hold.}$$

Then there exists exactly one solution to the stochastic differential equation (4.28), such that

$$E \sup_{s\leq T} |X_s|^2 < \infty.$$

Example 4.65 (Linear Differential Equations) For the special class of linear stochastic differential equations of the form

$$\underbrace{\mathrm{d}X_t}_{d\times 1} = \big(\underbrace{A(t)}_{d\times d}\ \underbrace{X_t}_{d\times 1} + \underbrace{a(t)}_{d\times 1} \big)\,\mathrm{d}t + \underbrace{\sigma(t)}_{d\times r}\ \underbrace{\mathrm{d}B_t}_{r\times 1}, \quad X_0 = \xi, \tag{4.29}$$

with locally bounded, deterministic coefficients A, a, σ, solutions can be given in explicit form. Such linear differential equations arise, for example, in economics, in models with many factors (multi-factor models). To find the solution of (4.29), we consider the corresponding inhomogeneous deterministic linear differential equation where the stochastic term is missing

$$\dot{\zeta}(t) = A(t)\zeta(t) + a(t), \qquad \zeta(0) = \xi \in \mathbb{R}^d. \tag{4.30}$$

By standard existence and uniqueness statements, there exists a unique solution to (4.30). To determine this solution, we consider the associated homogeneous matrix differential equation:

$$\dot{\Phi}(t) = A(t)\,\underbrace{\Phi(t)}_{d\times d}, \qquad \Phi(0) = I.$$

$\Phi(t)$ is a $d \times d$-matrix. The matrix system has a unique solution Φ the fundamental solution of the associated homogeneous differential equation

$$\dot{\zeta}(t) = A(t)\zeta(t).$$

$\Phi(t)$ is nonsingular for all t. Because otherwise there exists a $\lambda \in \mathbb{R}^d$, $\lambda \neq 0$, and t_0 such that $\Phi(t_0)\lambda = 0$. But $\Phi(t)\lambda$ is unique solution of the homogeneous equation with $\zeta(t_0) = 0$. It follows: $\Phi(t)\lambda = 0, \forall t$ in contradiction to $\Phi(0) = I$.

The solution of the deterministic equation (4.30) is now available via the fundamental solution

$$\zeta(t) = \Phi(t)\left[\zeta(0) + \int_0^t \Phi^{-1}(s)\cdot a(s)\,\mathrm{d}s\right]. \tag{4.31}$$

For the stochastic differential equation we get a solution ansatz in a similar way

$$X_t = \Phi(t)\left[X_0 + \int_0^t \Phi^{-1}(s)\cdot a(s)\,\mathrm{d}s + \int_0^t \Phi^{-1}(s)\cdot \sigma(s)\mathrm{d}B_s\right]. \tag{4.32}$$

By means of partial integration (product formula) we can verify with the Itô formula that X is a solution of the linear stochastic differential equation (4.29). Uniqueness follows according to the existence and uniqueness theorems.

Remark 4.66 Solutions of the SDE (4.23) are Markov processes, i.e. it holds:

$$E(f(X_t) \mid \mathfrak{A}_s) = E(f(X_t)|X_s) = \psi(X_s), \quad \textit{for } s < t,$$

with $\psi(x) := E(f(X_t) \mid X_s = x)$, as in the case of Brownian motion. The Markov property is a consequence of the **flow property.**

Let $(X_s^{t,x})_{s \geq t}$ be a solution of (4.23), starting in x at time t,

$$X_s^{t,x} = x + \int_t^s b(u, X_u^{t,x})\,\mathrm{d}u + \int_t^s \sigma(u, X_u^{t,x})\mathrm{d}B_u, \quad t \leq s.$$

Then the flow property holds:

$$X_s^{0,x} = X_s^{t, X_t^{0,x}}, \quad s > t. \tag{4.33}$$

(4.33) is a corollary of the uniqueness of solutions and the continuity of solutions in (s, t, x).

In $d = 1$ there is an existence and uniqueness statement under weakened assumptions on the diffusion coefficient, cf. Karatzas and Shreve (1991, Prop. 5.2.13).

Theorem 4.67 (Yamada, Watanabe) $d = 1$: *Let there exist* $K < \infty$ *such that*

$$|b(t, x) - b(t, y)| \leq K|x - y| \textit{ and } |\sigma(t, x) - \sigma(t, y)| \leq h(|x - y|),$$

for a function $h \uparrow$, $h(0) = 0$ *and* $\int_0^\varepsilon h^{-2}(u)\,\mathrm{d}u = \infty$, $\forall\, \varepsilon > 0$ *(e.g.* $\frac{1}{2} \leq \alpha \leq 1$ *and* $h(u) = u^\alpha$*).*

Then there exists exactly one solution of the SDE (4.23).

Example 4.68 The solution X of the stochastic differential equation

$$\mathrm{d}X_t = a(b - X_t^+)\,\mathrm{d}t + \sqrt{\gamma X_t^+}\mathrm{d}B_t \quad \textit{with } X_0 = x \geq 0, \gamma > 0, a, b \geq 0$$

is called Feller's branching diffusion with immigration, or in financial mathematics **Cox–Ingersoll–Ross model** for the time evolution of interest rates. With $h(x) = Cx^\alpha$, $\alpha \in [\frac{1}{2}, 1]$ the condition of rate Theorem 4.67 is satisfied with $\alpha = \frac{1}{2}$, $K = \sqrt{\gamma} + \alpha$. It holds $X_t \geq 0, \forall t$ and $X_t > 0, \forall t > 0$ if $\frac{2ab}{\gamma} \geq 1$, $X_t = 0$ infinitely often with probability 1 if $\frac{2ab}{\gamma} < 1$.

In the special case $a = b = 0$ (no drift) holds with the help of the Itô formula

$$e^{-\lambda X_t} - e^{-\lambda x} - \gamma\frac{\lambda^2}{2}\int_0^t e^{-\lambda X_s} X_s \,\mathrm{d}s = \lambda\int_0^t e^{-\lambda X_s}\sqrt{\gamma X_s}\mathrm{d}B_s \in \mathcal{M}.$$

The Laplace transform $\varphi(t,\lambda,x) = E_x e^{-\lambda X_t}$ of X_t solves the differential equation

$$\frac{\mathrm{d}}{\mathrm{d}t}\varphi(t,\lambda,x) = \gamma\frac{\lambda^2}{2}EX_t e^{-\lambda X_t} = -\frac{\gamma\lambda^2}{2}\frac{\mathrm{d}}{\mathrm{d}\lambda}\varphi(t,\lambda,x), \quad \varphi(0,\lambda,x) = e^{-\lambda x}.$$

The unique solution of this equation is:

$$\varphi(t,\lambda,x) = \exp\left(-\frac{\lambda}{\gamma/2t+1}x\right).$$

For $\gamma = 2$ this is identical to the limit of rescaled Galton–Watson branching processes.

4.4.1 Strong Solution—Weak Solution of Stochastic Differential Equations

The problem of *strong solutions* of stochastic differential equations concerns the question:

Given a probability space $(\Omega, \mathcal{A}, P)$, a Brownian motion B with filtration $\mathfrak{A}^B$ on $(\Omega, \mathcal{A}, \mathcal{P})$, we are looking for a solution of (4.23), adapted to $\mathfrak{A}^B$.

By extension of the notion of strong solution, the **weak solution** has an additional degree of freedom:

We are looking for a probability space $(\Omega, \mathcal{A}, P)$, a filtration and a Brownian motion. B with respect to this filtration and a solution X of (4.23) with respect to this Brownian motion B, i.e., (X, B) is a solution of

$$X_t = X_0 + \int_0^t \sigma(s, X_s)\,\mathrm{d}B_s + \int_0^t b(s, X_s)\,\mathrm{d}s \tag{4.34}$$

$$X_0 \stackrel{d}{=} \mu, \quad \text{the initial distribution of } X_0 \text{ is given and equal to } \mu.$$

This additional degree of freedom allows the construction of solutions even in some cases where no strong solution exists. An example of such a situation is the following.

Example 4.69 We consider the differential equation

$$\mathrm{d}X_t = \operatorname{sgn}(X_t)\mathrm{d}B_t, \quad X_0 = 0, \tag{4.35}$$

where $\operatorname{sgn} x := \mathbb{1}_{(0,\infty)}(x) - \mathbb{1}_{(-\infty,0)}(x)$ denotes the sign function. (4.35) is equivalent to

$$X_t = X_0 + \int_0^t \operatorname{sgn}(X_s)\,\mathrm{d}B_s \tag{4.36}$$

$$\Longleftrightarrow B_t = \int_0^t \mathrm{d}B_s = \int_0^t \operatorname{sgn}(X_s)\mathrm{d}X_s, \quad \forall t \geq 0. \tag{4.37}$$

(4.37) follows from (4.35) by multiplication by $\operatorname{sgn} X_t$.

Assertion: There exists a weak solution of (4.36).

For let X be a Brownian motion on $(\Omega, \mathfrak{A}, P)$, $\mathfrak{A} = \mathfrak{A}^X$, then define B by (4.37). Then it follows:

B is a continuous martingale and $\langle B\rangle_t = \int_0^t (\operatorname{sgn}(X_s))^2\,\mathrm{d}s = t$. It follows that B is a Brownian motion. So (B, X) is a weak solution of the differential equation in (4.36). But there is no strong solution !

For let (X, B) be any weak solution. Then as above X is a continuous martingale with $\langle X\rangle_t = t$, that is, a Brownian motion. By an approximation argument (cf. Karatzas and Shreve (1991, p. 562)), one obtains

$$\sigma(B) \subset \sigma(|X_s|; s \leq t) \subset \sigma(X_s; s \leq t).$$

The second inclusion is proper. So X is not adapted to $\sigma(B)$.

An example of the extended possibilities of weak solutions is provided by the following theorem. In the case $\sigma \equiv 1$ only a boundedness, a linear growth and measurability assumption is posed on the drift term b but no Lipschitz condition.

Theorem 4.70 *The stochastic differential equation*

$$\mathrm{d}X_t = b(t, X_t)\,\mathrm{d}t + \mathrm{d}B_t, \quad 0 \leq t \leq T, \tag{4.38}$$

$b(t, x)$ bounded and measurable, and $|b(t, x)| \leq K\big(1 + |x|\big)$, has a weak solution with initial distribution

$$P^{X_0} = \mu, \quad \mu \in M^1(\mathbb{R}^d, \mathcal{B}^d).$$

Proof Let $X = (X_t, \mathfrak{A}_t, (P_x)_x, (\Omega, \mathfrak{A}))$ be a d-dimensional Brownian motion with start in x with respect to P_x. Then

$$\begin{aligned} Z_t :&= \exp\left(\sum_{j=1}^d \int_0^t b_j(s, X_s)\mathrm{d}X_s^j - \frac{1}{2}\int_0^t |b(s, X_s)|^2\,\mathrm{d}s\right) \\ &= \exp\left((b \cdot X)_t - \frac{1}{2}\langle b \cdot X\rangle_t\right) \end{aligned}$$

is the exponential martingale. Z is a martingale with respect to P_x, because by the linear growth condition and the boundedness of b the Novikov condition is satisfied. We define

$$\frac{\mathrm{d}Q_x}{\mathrm{d}P_x} := Z_T.$$

According to the Theorem of Girsanov it follows

$$B_t := X_t - X_0 - \int_0^t b(s, X_s)\,\mathrm{d}s, \quad 0 \le t \le T$$

is a Brownian motion with respect to Q_x, with $Q_x(B_0 = 0) = 1\ \forall x$.
Define $Q_\mu(A) := \int Q_x(A)\mathrm{d}\mu(x)$, then it follows

$$X_t = X_0 + \int_0^t b(s, X_s)\,\mathrm{d}s + B_t \text{ with respect to } Q_\mu.$$

With respect to Q_μ B is a Brownian motion and

$$(Q_\mu)^{X_0} = \int Q_x^{X_0}\mathrm{d}\mu(x) = \int \varepsilon_x \mathrm{d}\mu(x) = \mu$$

$$Q_x^{X_0}(A) = 1_A(x).$$

Thus it follows that $(X, B, (\Omega, \mathfrak{A}, Q_\mu), (\mathfrak{A}_t))$ is a weak solution of (4.38). □

Remark 4.71

(a) The statement of Theorem 4.70 can be generalized as follows. Let X be a solution of the stochastic differential equation

$$\mathrm{d}X_t = \mu_t\,\mathrm{d}t + \sigma_t \cdot \mathrm{d}B_t.$$

Let ν_t is another drift such that $\vartheta_t := \frac{\mu_t - \nu_t}{\sigma_t}$ is bounded, then it follows that

$$M_t = \exp\Big(- \vartheta_t \cdot B_t - \frac{1}{2}\int_0^t \|\vartheta_s\|^2\,\mathrm{d}s\Big) \in \mathcal{M}_c(P).$$

With $\frac{\mathrm{d}Q}{\mathrm{d}P} := M_T$ it follows by Girsanov's Theorem

$$\widetilde{B}_t = B_t + \int_0^t \vartheta_s\,\mathrm{d}s \textit{ is a Brownian motion with respect to } Q$$

and it holds

$$\mathrm{d}X_t = \nu_t \,\mathrm{d}t + \sigma_t \mathrm{d}\widetilde{B}_t \textit{ with respect to } Q. \tag{4.39}$$

Proof:

$$\begin{aligned} \mathrm{d}X_t &= \mu_t \,\mathrm{d}t + \sigma_t \cdot \mathrm{d}B_t \\ &= \nu_t \,\mathrm{d}t + \sigma_t \cdot \mathrm{d}B_t + (\mu_t - \nu_t)\,\mathrm{d}t \\ &= \nu_t \,\mathrm{d}t + \sigma_t (\underbrace{\mathrm{d}B_t + \nu_t \,\mathrm{d}t}_{\mathrm{d}\widetilde{B}_t}). \end{aligned}$$

The pair $(X, \widetilde{B})$ is thus on the new probability space with measure Q a weak solution of (4.39). □

(b) Girsanov can also be used to prove the **weak uniqueness of solutions,** i.e., the uniqueness of the distributions of weak solutions can be proved (cf. Karatzas and Shreve (1991, p. 304)).

Definition 4.72 (Pathwise Uniqueness) The solution of the stochastic differential equation (4.23) with initial distribution $\mu \in M^1(\mathbb{R}^d, \mathcal{B}^d)$ is called **pathwise unique** if for every two weak solutions (X, B_1), (Y, B_2) on $(\Omega, \mathfrak{A}, P)$ with respect to filtrations $(\mathfrak{A}_t)$ and $(\mathfrak{B}_t)$ and initial distribution μ it holds:

$$P(X_t = Y_t, \forall t \geq 0) = 1.$$

Remark 4.73 For the stochastic differential equation $\mathrm{d}X_t = \operatorname{sgn}(X_t)\,\mathrm{d}B_t$ from Example 4.69 with X also $-X$ is a solution, i.e., pathwise uniqueness does not hold. The solution is however **weakly unique**, i.e., for two solutions X, Y with the same initial distribution μ it holds $P^X = P^Y$.

Proposition 4.74 *The following statements are equivalent :*

(a) The stochastic differential equation (4.23) has a unique strong solution.
(b) There exists a weak solution and pathwise uniqueness holds (for the initial distribution $\mu = P^{X_0}$).

Under (a), (b) it holds: the solution is weakly unique (Karatzas and Shreve 1991).

Definition 4.75

(a) $X \in C^d$ is called the solution of the **local martingale problem** with drift b and diffusion a, initial distribution μ (term: $X \in LMP(a, b, \mu)$) if:

$$\begin{aligned} &P^{X_0} = \mu \textit{ and } M := X - \int_0^{\cdot} b(s, X_s)\,\mathrm{d}s \in \mathcal{M}_{\mathrm{loc},c} \\ &\textit{with } \langle M^i, M^j \rangle_t = \textstyle\int_0^t a_{ij}(s, X_s)\,\mathrm{d}s \quad \forall i, j, t \geq 0 \end{aligned} \tag{4.40}$$

(b) The solution of $LMP(a, b, \mu)$ is unique, if for two solutions X and Y it holds:

$$P^X = P^Y.$$

The local martingale problem is defined by the martingale property of M. There is an important connection between the notion of weak solution and the local martingale problem.

Theorem 4.76 (Weak Solution—Local Martingale Problem)

(a) X is exactly then a solution of the local martingale problem $LMP(\underbrace{\sigma\sigma^T}_{=:a}, b, \mu)$ if there exists a Brownian motion B, if necessary, on an extension $\widetilde{\Omega} \supset \Omega$ such that (X,B) is a weak solution of (4.23).

(b) There exists a unique weak solution of (4.23) with initial distribution μ $\iff$ The local martingale problem $LMP(\sigma\sigma^T, b, \mu)$ is uniquely solvable.

Proof We give the proof for $m = d = 1$; d is the dimension and m is the number of Brownian motions. We consider the stochastic differential equation

$$\underbrace{\mathrm{d}X_t}_{d\times 1} = \underbrace{\sigma(t, X_t)}_{d\times m}\underbrace{\mathrm{d}B_t}_{m\times 1} + \underbrace{b(t, X_t)}_{d\times 1}\,\mathrm{d}t.$$

"$\Leftarrow$" If (X, B) is a weak solution of the stochastic differential equation in (4.23), then it follows

$$X_t - \int_0^t b(s, X_s)\,\mathrm{d}s = \int_0^t \sigma(s, X_s)\mathrm{d}B_s \in \mathcal{M}_{\mathrm{loc},c}.$$

So X solves the local martingale problem $LMP(\sigma\sigma^T, b, \mu)$.
"$\Rightarrow$" Now be X be a solution of $LMP(\sigma^2, b, \mu)$ and $M_t := X_t - \int_0^t b(s, X_s)\,\mathrm{d}s \in \mathcal{M}_{\mathrm{loc},c}$. Then there exists an extension $\widetilde{\Omega} \supset \Omega$ and a Brownian motion $\widetilde{B}$ on $(\widetilde{\Omega}, \widetilde{\mathfrak{A}}, \widetilde{P})$, such that (cf. e.g. Klenke (2006, p. 542))

$$M_t = \int_0^t |\sigma(s, X_s)|\mathrm{d}\widetilde{B}_s.$$

With $B_t := \int_0^t \mathrm{sgn}(\sigma(s, X_s))\mathrm{d}\widetilde{B}_s$ then the following representation holds

$$M_t = \int_0^t \sigma(s, X_s)\mathrm{d}B_s.$$

So (X, B) is a weak solution of the stochastic differential equation (4.23). □

The connection between weak solutions and the local martingale problem can now be used to prove general existence theorems for weak solutions and for the local martingale problem, respectively. The formulation of the local martingale problem is particularly suitable for the application of approximation arguments and convergence in distribution.

Theorem 4.77 *Let b, σ_{ij} $\mathbb{R}_+ \times \mathbb{R}^d \to \mathbb{R}$ continuous and bounded. Then for all initial distributions $\mu \in M^1(\mathbb{R}^d, \mathcal{B}^d)$ with $\int \|x\|^{2m} \mathrm{d}\mu(x) < \infty$ for a $m > 1$ there exists a solution of the local martingale problem $LMP(a, b, \mu)$ with $a = \sigma\sigma^T$, and a weak solution of the stochastic differential equation in (4.23).*

Proof Sketch We first introduce a discretized version of the stochastic differential equation. Let $t_j^n := \frac{\delta}{2^n}$ and $\Psi_n(t) = t_j^n$ for $t \in (t_j^n, t_{j+1}^n]$ the dyadic rational approximation of g. Then for $y \in C_d([0, \infty))$

$$b^{(n)}(t, y) := b(y(\Psi_n(t))), \quad \sigma^{(n)}(t, y) := \sigma(y(\Psi_(t))))$$

is progressively measurable.

Let further be $(B_t, \mathcal{A}_t^B)$ be a d-dimensional Brownian motion on $(\Omega, \mathcal{A}, P)$ and ξ a random variable independent of B with $P^\xi = \mu$. We define the sequence $(X^{(n)}) = (X_t^{(n)}, \mathcal{A}_t)$ recursively by an Euler scheme:

$$X_0^{(n)} := \xi$$
$$X_t^{(n)} := X_{t_j}^{(n)} + b\left(X_{t_j^{(n)}}^{(n)}\right)(t - t_j^{(n)}) + \sigma\left(X_{t_j^{(n)}}^{(n)}\right)\left(B_t - B_{t_j^{(n)}}\right), \quad t_j^{(n)} < t \le t_{j+1}^{(n)}.$$

Then, by definition of $b^{(n)}, \sigma^{(n)}$:

$$X_t^{(n)} = \xi + \int_0^t b^{(n)}(s, X^{(n)})\, \mathrm{d}s + \int_0^t \sigma^{(n)}(s, X^{(n)}) \mathrm{d}B_s. \tag{4.41}$$

$X^{(n)}$ thus solves the modified stochastic differential equation with approximate coefficients.

Now, as in the proof of the general existence and uniqueness theorem (Theorem 4.60), the following estimate is also valid for moments of order $2m > 2$ (cf. Karatzas and Shreve (1991, Problem 3.15)). The proof uses the Burkholder-Davis inequality instead of Doob's inequality. If $\|b(t, y)\|^2 + \|\sigma(t, y)\|^2 \le K(1 + \max_{s \le t} \|y(s)\|^2)$ and if (X, B) is a weak solution of the diffusion equation (4.23) with $E\|X_0\|^{2m} < \infty$ then for all $T < \infty$

$$\begin{aligned} & E \max_{s \le t} \|X_s\|^{2m} \le C(1 + E\|X_0\|^{2m})e^{Ct}, \quad t \le T \\ \text{and} \quad & E\|X_t - X_s\|^{2m} \le C(1 + E\|X_0\|^{2m})(t - s)^m, \quad s < t \le T. \end{aligned} \tag{4.42}$$

For the solution $X^{(n)}$ of the discretized stochastic differential equation in (4.41) thus it holds:

$$\sup_n E|X_t^{(n)} - X_s^{(n)}|^{2m} \leq C(1 + E\|E\xi\|^{2m})(t-s)^m, \quad s \leq t < T.$$

Let $P^{(n)} := P^{X^{(n)}}$ be the associated probability measure on $C_d = C([0,\infty), \mathbb{R}^d)$. Then it follows that the sequence $(P^{(n)})$ is tight, because it satisfies the **stochastic Arzela–Ascoli criterion:**

$$\exists\, \nu, \alpha, \beta > 0 : \forall n \in \mathbb{N} : E\|X_0^{(n)}\|^\nu \leq M \text{ and}$$
$$E\|X_t^{(n)} - X_s^{(n)}\|^\alpha \leq C_T|t-s|^{1+\beta}, \quad \forall s, t \leq T, \forall T.$$

So there exists a convergent subsequence. W.l.g. it holds $P^{(n)} \xrightarrow{\mathcal{D}} P^*$, i.e. $\int f \,\mathrm{d}P^{(n)} \to \int f \,\mathrm{d}P^*$
for all continuous, bounded functions f on C_d.

We assert:

(1) $P^*(\{y \in C_d;\, y(0) \in \Gamma\}) = \mu(\Gamma)$,
i.e. P^* has the initial distribution μ. For $f \in C_b(\mathbb{R}^d)$ it holds

$$E^* f(y(0)) = \lim_n E^{(n)} f(y(0)) = \int f \,\mathrm{d}\mu,$$

since $P^{(n)} \xrightarrow{\mathcal{D}} P^*$ implies that $\mu = (P^{(n)})^{\pi_0} \xrightarrow{\mathcal{D}} (P^*)^{\pi_0}$. So P^* has the initial distribution μ.

(2) The standard construction on $C_d(\pi_t), P^*)$ is a solution of the associated martingale problem, i.e.

$$E^*\Big(f(y(t)) - f(y(s)) - \int_s^t \mathcal{A}_u f(y(u))\,du \mid \mathfrak{A}_s\Big) = 0[P^*] \tag{4.43}$$

for $0 \leq s < t < \infty$, $f \in C_K^2(\mathbb{R}^d)$ and with the associated differential operator

$$\mathcal{A}_t f(y) = b(t,y) \cdot \nabla f(y(t)) + \frac{1}{2}\sigma^T(t,y)\left(\frac{\partial^2 f(y(t))}{\partial x_i \partial x_j}\right)\sigma(t,y).$$

With Itô's formula, it follows that the formulation in (4.43) is equivalent to the formulation in the definition of the local martingale problem.

$X^{(n)}$ solves the stochastic differential equation with discretized coefficients. Therefore, by Theorem 4.76 $X^{(n)}$ is a solution of the associated local martingale problem, i.e., for $f \in C_K^2(\mathbb{R}^d)$ it holds:

$$\left(M_t^{f,n} := f(y(t)) - f(y(0)) - \int_0^t \mathcal{A}_u^{(n)} f(y)\, du, \mathfrak{A}_t \right) \in \mathcal{M}_c(P^{(n)}), \quad f \in C_K^2$$

with $\mathcal{A}_t^{(n)} f(y) := b^{(n)}(t, y) \cdot \nabla f(y(t)) + \frac{1}{2}(\sigma^{(n)})^T(t, y) D^2 f(y(t)) \sigma^{(n)}(t, y)$.

It follows for $\mathfrak{A}_s$-measurable $g \in C_b(C_d)$:

$$E^{(n)} \Big\{ \underbrace{f(y(t)) - f(y(s)) - \int_s^t \mathcal{A}_u^{(n)} f(y(u)) du}_{=F_n(y)=F_n^{s,t}(y)} \Big\} g(y) = 0.$$

Let $F(y) := f(y(t)) - f(y(s)) - \int_s^t \mathcal{A}_u f(y)(u)\, du$, then it holds: $F_n(y) - F(y) \to 0$ uniformly on compacta of $\mathbb{R}^d$ (cf. Karatzas and Shreve (1991, p. 325)). For $K \subset C_d$ compact it holds $M := \sup_{\substack{y \in K, \\ 0 \le s \le t}} \|y(s)\| < \infty$ and $\lim_n \sup_{y \in K} \omega^t(y, 2^{-n}) = 0$, $\forall t < \infty$ with the modulus of continuity ω^t on $[0, t]$. b, σ are uniformly continuous on $\{y;\ \|y\| \le M\}$. It follows that

$$\sup_{0 \le s \le t} \{ \|b^{(n)}(s, y) - b(y(s))\| + \|\sigma^{(n)}(s, y) - \sigma(y(s))\| \} \le \varepsilon, \quad \forall n \ge n_\varepsilon.$$

Thus it follows that $E^* F(y) g(y) = 0$, $\forall g$. Thus $((\pi_t), P^*)$ is a solution of the local martingale problem (4.43). By Theorem 4.76, it, therefore, follows that there exists a weak solution to the stochastic differential equation.

Remark 4.78

(a) An analogous existence statement for weak solutions holds also in the time-inhomogeneous case with bounded continuous coefficients $b = b(t, x)$, $\sigma = \sigma(t, x)$.
(b) From the solvability of the Cauchy problem $\frac{\partial u}{\partial t} = \mathcal{A} u$, $u(0, \cdot) = f$, u bounded $[0, T] \times \mathbb{R}^d$ weak uniqueness of solutions follows (Stroock and Varadhan 1979; Karatzas and Shreve 1991). Sufficient for the solvability of the Cauchy problem is that the coefficients are bounded and Hölder continuous and that the diffusion matrix a is uniformly positive definite.

4.5 Semigroup, PDE, and SDE Approach to Diffusion Processes

The aim of this section is to describe the construction of diffusion processes by means of semigroup theory, partial differential equations (PDE)—in particular the Kolmogorov equations—and by means of stochastic differential equations (SDE), and to discuss the interactions of these approaches. Let $X = (X_t)_{t\geq 0}$ be a Markov process in $\mathbb{R}^d$ with transition probabilities

$$P_{s,x}(t, A) := P(X_t \in A | X_s = x), \quad A \in \mathcal{B}^d, 0 \leq s < t \leq T, x \in \mathbb{R}^d.$$

Then the **Chapman–Kolmogorov equation** holds

$$P_{s,x}(t, A) = \int_{\mathbb{R}^d} P_{u,z}(t, A) P_{s,x}(u, \mathrm{d}z) \quad \forall s < u < t, x \in \mathbb{R}^d.$$

Examples are the Brownian motion B with transition kernel

$$P_{s,x}(t, A) = (2\pi(t-s))^{-\frac{d}{2}} \int_A e^{-|y-x|^2/2(t-s)} \,\mathrm{d}y$$

or the multidimensional Ornstein–Uhlenbeck process with transition kernel

$$P_{s,x}(t, A) = \left(\pi(1 - e^{-2(t-s)})\right)^{-\frac{d}{2}} \int_A \exp\left\{\frac{\left(y - e^{-(t-s)}x\right)^2}{1 - e^{-2(t-s)}}\right\} \mathrm{d}y.$$

Definition 4.79 An $\mathbb{R}^d$-valued Markov process $(X_t)_{0\leq t\leq T}$ is called **diffusion process,** if $\forall t \leq T$, $x \in \mathbb{R}^d$, $c > 0$ it holds:

(1) $\lim_{\varepsilon \searrow 0} \frac{1}{\varepsilon} \int_{\{|y-x|\geq c\}} P_{t,x}(t+\varepsilon, \mathrm{d}y) = 0$
(2) $\lim_{\varepsilon \searrow 0} \frac{1}{\varepsilon} \int_{\{|y-x|<c\}} (y_i - x_i) P_{t,x}(t+\varepsilon, \mathrm{d}y) =: \varrho_i(t, x)$ exists
(3) $\lim_{\varepsilon \searrow 0} \frac{1}{\varepsilon} \int_{\{|y-x|<c\}} (y_i - x_i)(y_j - x_j) P_{t,x}(t+\varepsilon, \mathrm{d}y) := a_{ij}(t, x)$ exists.

Remark 4.80

(a) From (1) it follows that X has no jumps.
Conditions (2) and (3) are because of (1) independent of $c > 0$.

$$\textit{So} \qquad \varrho_i(t, x) = \lim_{\varepsilon \searrow 0} \frac{1}{\varepsilon} \int (y_i - x_i) P_{t,x}(t+\varepsilon, \mathrm{d}y),$$

$$a_{ij}(t, x) = \lim_{\varepsilon \searrow 0} \frac{1}{\varepsilon} \int (y_i - x_i)(y_j - x_j) P_{t,x}(t+\varepsilon, \mathrm{d}y).$$

It further follows that:

$$\varrho(t,x) = \lim_{\varepsilon\searrow 0} \frac{1}{\varepsilon} E(X_{t+\varepsilon} - x \mid X_t = x)$$

$$a(t,x) = \lim_{\varepsilon\searrow 0} \frac{1}{\varepsilon} E\big((X_{t+\varepsilon} - x)(X_{t+\varepsilon} - x)^T \mid X_t = x\big).$$

ϱ describes the local drift and a describes the local diffusion.

$\varrho(t,x) = (\varrho_1(t,x), \dots, \varrho_d(t,x))$ is the **drift coefficient**, and the matrix $a(t,x) = (a_{ij}(t,x))_{i,j}$ is called **diffusion matrix** of the diffusion process X.

(b) The Brownian motion B is a diffusion process and it holds:

$$\frac{1}{\varepsilon} P(|B_{t+\varepsilon} - x| \geq c \mid B_t = x) = \frac{1}{\varepsilon} P(|B_{t+\varepsilon} - B_t| \geq c) \leq \frac{1}{\varepsilon c^4} \underbrace{E|B_\varepsilon|^4}_{=d(d+2)\varepsilon^2}$$

$$= \frac{1}{c^4} d(d+2)\varepsilon \to 0, \quad \varepsilon \searrow 0.$$

Thus $a(t,x) = I$, $\varrho(t,x) = 0$.

For the Ornstein–Uhlenbeck process, by an analogous calculation the following holds

$$a(t,x) = I, \varrho(t,x) = -x.$$

An important class of diffusion processes are the solutions of stochastic differential equations.

Theorem 4.81 *Let b, σ be continuous functions satisfying the conditions of Theorem 4.64, i.e., the linear growth and Lipschitz condition. Then it holds: the solution of the stochastic differential equation (4.23) $\mathrm{d}X_t = b(t, X_t)\,\mathrm{d}t + \sigma(t, X_t)\mathrm{d}B_t$, $X_0 = Z$, is a diffusion process with drift coefficient b and diffusion matrix $a(t,x) := \sigma(t,x)\sigma(t,x)^T$.*

Proof Sketch Using the Gronwall inequality, estimates of the moments are obtained as in the proof for the existence of solutions of stochastic differential equations:

$$\begin{cases} E\|X_t\|^4 & \leq (27EZ^4 + C_1T)e^{C_1t} \\ E\|X_t - Z\|^4 & \leq C_1(1 + 27EZ^4 + C_1T)t^2e^{C_1t}. \end{cases} \tag{4.44}$$

It follows:

$$\int_{\{\|y-x\|\geq c\}} P_{t,x}(t+\varepsilon, \mathrm{d}y) = P(\|X_{t+\varepsilon} - x\| \geq c \mid X_t = x)$$

$$= P\big(\|X_{t+\varepsilon}^{t,x} - x\| \geq c\big)$$

$$\leq \frac{E\|X_{t+\varepsilon}^{t,x} - x\|^4}{c^4} \leq \frac{1}{c^4} L\varepsilon^2 e^{C_1\varepsilon}$$

with a constant $L > 0$. From this follows condition 1 of the definition of a diffusion.

Further it is

$$X_{t+\varepsilon} = x + \int_t^{t+\varepsilon} \sigma(s, X_s)\,\mathrm{d}B_s + \int_t^{t+\varepsilon} b(s, X_s)\,\mathrm{d}s.$$

From this follows

$$\varrho(t,x) = \lim_{\varepsilon\searrow 0} E(X_{t+\varepsilon} - x \mid X_t = x)$$

$$= \lim_{\varepsilon\searrow 0} \frac{1}{\varepsilon} \int_t^{t+\varepsilon} E(b(s, X_s) \mid X_t = x)\,\mathrm{d}s = b(t,x).$$

Similarly, the formula for the diffusion coefficient is obtained. □

The general basic question in the theory of diffusion processes is: Given the diffusion matrix $a(t,x)$ and the drift coefficient $\varrho(t,x)$. Does there exist an associated diffusion process? What are its properties?

A necessary condition for existence is that $a(t,x)$ is positive semidefinite, because for $v \neq 0$ it holds:

$$v^T a(t,x) v = \lim_{\varepsilon\searrow 0} \frac{1}{\varepsilon} E(v^T (X_{t+\varepsilon} - x)(X_{t+\varepsilon} - x)^T v \mid X_t = x)$$

$$= \lim_{\varepsilon\searrow 0} \frac{1}{\varepsilon} E\Big(|(X_{t+\varepsilon} - x)^T v|^2 \mid X_t = x\Big) \geq 0.$$

There are three different approaches to the treatment of diffusion processes:

1. Semigroup theory (Hille–Yosida theory).
2. Partial differential equations (Kolmogorov equation)
3. Stochastic differential equations (Itô theory)

1. Semigroup Access

The study of processes with the help of semigroup theory is a suitable tool for general Markov processes, in particular, therefore, also for diffusion processes $(X_t)_{0\leq t\leq T}$. We restrict ourselves to the stationary case with diffusion matrix

$a(t, x) = a(x)$ and drift coefficient $\varrho(t, x) = \varrho(x)$, $\varrho(x) = \lim_{t \searrow 0} \frac{1}{t} E((X_t - x)|X_0 = x)$ and $a(x) = \lim_{t \searrow 0} \frac{1}{t} E((X_t - x)(X_t - x)^T |X_0 = x)$.

Definition 4.82

(a) A family $(T_t, t \geq 0)$ of linear operators on a Banach space $(\mathbb{B}, \|\cdot\|)$ is called a **contraction semigroup,** if
  a1) $\lim_{t \searrow 0} T_t h = h \quad \forall h \in \mathbb{B}$ (strong continuity)
  a2) $\|T_t h\| \leq \|h\|, \forall t \geq 0, h \in \mathbb{B}$ (contraction)
  a3) $T_0 = I$, and $T_{s+t} = T_s T_t, s, t \geq 0$ (semigroup property)

(b) If $(T_t, t \geq 0)$ is a contraction semigroup then $Ah = \lim_{t \searrow 0} \frac{1}{t}(T_t h - h)$ is called **infinitesimal generator,** $h \in \text{dom}\, A \subset \mathbb{B}$.

Remark 4.83 The infinitesimal generator A has, by a theorem of Yosida, an in $\mathbb{B}$ dense domain of definition, dom $A \subseteq \mathbb{B}$. In general A is not defined on the whole $\mathbb{B}$ and A is not in general a bounded operator.

Example 4.84 Let $\mathbb{B} := C_b[0, \infty)$ with supremum norm $\|\cdot\|_\infty$, $\|f\|_\infty := \sup_{x \in [0,\infty)} |f(x)|$. Let T_t be defined as a translation operator on $\mathbb{B} = C_b[0, \infty)$, $T_t f(x) := f(x + t)$, $f \in \mathbb{B}$, $x \in [0, \infty)$, $t \in [0, \infty)$. Then $\{T_t, t \geq 0\}$ is a contraction semigroup with infinitesimal generator $Af = f' = \frac{\mathrm{d}}{\mathrm{d}x} f$.

The domain of definition of A is dense in $C_b[0, \infty)$:

$$\text{dom} A = \{f \in C_b[0, \infty); f' \in C_b[0, \infty)\}\{\subset C_b[0, \infty)\ is\ dense.$$

A is not bounded. The image of the unit sphere is unbounded, i.e., there exists no constant K, so that for all f it holds $\|Af\| \leq K\|f\|$.

Now let $X = (X_t)$ be a stationary diffusion process, i.e. $\varrho(t, x) = \varrho(x)$ and $a(t, x) = a(x)$ in $\mathbb{R}^d$ with stationary transition kernel $P_t(x, A) = P(X_{t+s} \in A|X_s = x)$.

Transition Probability Assumptions

(A1) $\forall t > 0, c > 0 : \lim_{|x| \to \infty} P_t(x, \{y; |y| < c\}) = 0$
(A2) $\forall c > 0 : \lim_{t \searrow 0} P_t(x, \{y; |y - x| \geq c\}) = 0$ uniformly in $x \in \mathbb{R}^d$
(A3) $\forall f \in C_b, \forall t > 0 : T_t f(x) := \int f(y) P_t(x, \mathrm{d}y) \in C_b$ (Feller property)

Let $C_0 = C_0(\mathbb{R}^d) = \{f \in C(\mathbb{R}^d); \lim_{|x| \to \infty} f(x) = 0\}$. $(C_0, \|\cdot\|_\infty)$ is a Banach space.

Theorem 4.85 *Let $X = (X_t)$ be a stationary diffusion process and conditions* (A1)–(A3) *hold. Then $(T_t, t \geq 0)$ is a C_0-contraction semigroup with infinitesimal generator A, defined by:*

$$\begin{aligned} Af(x) &= \frac{1}{2} tr\ (a(x)D^2 f(x)) + \varrho(x) \cdot \nabla f(x) \\ &= \frac{1}{2} \sum_{i,j} a_{ij}(x) \frac{\partial^2 f}{\partial x_i \partial x_j}(x) + \sum_i \varrho_i(x) \frac{\partial f}{\partial x_i}. \end{aligned} \tag{4.45}$$

Proof We first show:

(1) $f \in C_0 \overset{!}{\Rightarrow} Tf \in C_0$

According to (A3) $T_t f$ is continuous. For $f \in C_0$ there exists a $c > 0$ such that for $|y| \geq c$ holds: $|f(y)| < \frac{\varepsilon}{2}$. It follows:

$$\begin{aligned} |T_t f(x)| &\leq \int_{\{|y| \geq c\}} |f(y)| P_t(x, \mathrm{d}y) + \int_{\{|y| \leq c\}} |f(y)| P_t(x, \mathrm{d}y) \\ &\leq \frac{\varepsilon}{2} + \|f\|_\infty P_t(x, \{|y| < c\}). \end{aligned}$$

Let w.l.g. : $\|f\|_\infty > 0$. According to (A1) there exists a constant $K > 0$ such that $\forall\, |x| > K$:

$$P_t(x, \{|y| < c\}) < \frac{\varepsilon}{2\|f\|_\infty}.$$

From this it follows $|T_t f(x)| \leq \frac{\varepsilon}{2} + \frac{\varepsilon}{2} = \varepsilon$ for $|x| > K$ and, therefore, it holds: $T_t f \in C_0$.

(2) $(T_t, t \geq 0)$, $T_0 := I$ is a C_0-contraction semigroup.

It holds according to the Chapman–Kolmogorov equation $T_{s+t} = T_s T_t$. From this follows condition (A3).

Contraction condition (A2):

$$\|T_t f\|_\infty = \sup_{x \in [0,\infty)} |T_t f(x)| \leq \sup_x \int |f(y)| P_t(x, \mathrm{d}y) \leq \|f\|_\infty;$$

so condition (A2) holds.

Condition (A1):

If $f \in C_0$ then f is uniformly continuous, i.e. $|x-y| < \delta \Rightarrow |f(x)-f(y)| < \frac{\varepsilon}{2}$. From this follows

$$\begin{aligned}|T_t f(x) - f(y)| &\le \int_{|y-x|<\delta} |f(x) - f(y)| P_t(x, \mathrm{d}y) \\ &+ \int_{|y-x|\ge\delta} |f(x) - f(y)| P_t(x, \mathrm{d}y) \\ &\le \frac{\varepsilon}{2} + 2\|f\|_\infty P_t(x, \{|y-x| \ge \delta\}) \\ &\le \frac{\varepsilon}{2} + 2\|f\|_\infty \frac{\varepsilon}{4\|f\|_\infty} = \varepsilon \qquad \text{after condition (A2).}\end{aligned}$$

From this follows condition (A1).

(3) Infinitesimal generator:

Let $f \in C_b^2 \cap C_0$; let f be 2-times continuously differentiable with bounded second derivatives. With Taylor expansion, it follows:

$f(y) = f(x) + (y-x)\nabla f(x) + \frac{1}{2} tr(y-x)(y-x)^T D^2 f(x) + R(y, x)$ with a remainder term R.

It follows:

$$\begin{aligned}T_t f(x) - f(x) &= E\big(f(X_t)|X_0 = x\big) - f(x) = E(X_t - x)\nabla f(x) \\ &\quad + \tfrac{1}{2} tr E(X_t - x)(X_t - x)^T D^2 f(x) + E R(X_t, x).\end{aligned}$$

By definition of ϱ, a it follows with usual estimation of remainder terms:

$$\lim_{t\downarrow 0} \frac{1}{t}\big(T_t f(x) - f(x)\big) = \varrho(x)\nabla f(x) + \frac{1}{2} tr\, a(x) D^2 f(x) = Af(x).$$

Still to be shown is the uniformity of convergence in x. This follows from the boundedness of f, $D^2 f$ and the uniform continuity.

□

Example 4.86 For a Brownian motion on $\mathbb{R}^d$ the infinitesimal generator is identical to the Laplace operator: $Af(x) = \frac{1}{2}\Delta f(x)$.

For the Ornstein–Uhlenbeck process on $\mathbb{R}^d$ the infinitesimal generator is

$$Af(x) = \frac{1}{2}\Delta f(x) - x\nabla f(x)$$

The basic question now is: For given operator A of the form (4.45) do there exist transition functions (T_t), (P_t), and a Markov process (X_t) such that A is an infinitesimal generator of X?

From the semigroup property it follows for $f \in \text{dom}A$:

$$\begin{aligned}\frac{\mathrm{d}}{\mathrm{d}t}T_t f &= \lim_{\varepsilon\downarrow 0}\frac{T_{t+\varepsilon}f - T_t f}{\varepsilon} = T_t \lim_{\varepsilon\downarrow 0}\frac{T_\varepsilon - I}{\varepsilon} f \\ &= T_t A f = A T_t f, \quad \forall t > 0.\end{aligned} \tag{4.46}$$

The usual ansatz to a solution: $T_t = e^{tA} = \sum_{n=0}^{\infty}\frac{t^n}{n!}A^n$ however, is well-defined only if A is a bounded operator, a rather atypical case.

An answer to the basic question posed above about the existence of a diffusion process X is essentially based on the following theorem of Hille and Yosida, cf. Yosida (1974) or Dynkin (1965).

Theorem 4.87 (Hille–Yosida Theorem) *Let A be a densely defined operator on a Banach space $\mathbb{B}$, i.e. A is well-defined on a dense subset of $\mathbb{B}$.*
Then A is an infinitesimal producer of a contraction semigroup if and only if for an $n \in \mathbb{N}$ $(I - \frac{1}{n}A)^{-1}$ exists, and $\|(I - \frac{1}{n}A)^{-1}\| \le 1$.

Remark 4.88 If $Af(x) = \frac{1}{2}tr\, a(x)D^2 f(x) + \varrho(x)\nabla f(x)$, $a(x)$ symmetric, positive semidefinite, then we can choose $\mathbb{B} = C_b^u(\mathbb{R}^d)$ as the set of uniformly continuous bounded functions or $\mathbb{B} = C_0(\mathbb{R}^d)$.

Procedure: The construction of diffusion processes using semigroup theory now consists of the following steps:

(1) To show: A is densely defined on $\mathbb{B}$, a Banach lattice of continuous functions on $\mathbb{R}^d$.
(2) To show: A satisfies the conditions from the Hille–Yosida theorem.
By Theorem 4.87, it then follows: $\exists$ contraction semigroup $(T_t, t \ge 0)$ on $\mathbb{B}$ with infinitesimal generator A.
(3) To show: $(T_t, t \ge 0)$ additionally satisfies the conditions
$T_t 1 = 1$, $T_t f \ge 0$ for $f \ge 0$ in the Banach lattice $\mathbb{B}$.
Then, by Riesz's theorem, it follows: $\exists$ Kernel $P_t(x, \cdot), t \ge 0, x \in \mathbb{R}^d$ such that $T_t f(x) = \int_{\mathbb{R}^d} f(y)P_t(x, \mathrm{d}y)$, $f \in \mathbb{B}$, $x \in \mathbb{R}^d$, $t \ge 0$
(4) Since (T_t) is a semigroup, it follows: The transition kernel (P_t) satisfies the Chapman–Kolmogorov equation.
So there exists a Markov process $X = (X_t)$ with transition probabilities (P_t). X is a diffusion process with diffusion coefficients ϱ and diffusion matrix $a = \sigma\sigma^T$.

2. PDE Access to Diffusion Processes, Kolmogorov Equations
Let $X = (X_t)_{0\le t\le T}$ be a diffusion process with drift and diffusion coefficients $\varrho(t, x), a(t, x)$. We first consider the case $d = 1$.

Let $P_{s,x}(t,\cdot)$ be the transition probability and $F_{s,x}(t,y) := P_{s,x}(t,(-\infty,y])$ the corresponding distribution function of $P^{X_t|X_s=x}$.
The Chapman–Kolmogorov equations provide

$$\begin{aligned} F_{s,x}(t,y) &= \int_{\mathbb{R}} F_{u,z}(t,y)\,\mathrm{d}F_{s,x}(u,z) \\ &= \int F_{s+\varepsilon,z}(t,y)\,\mathrm{d}F_{s,x}(s+\varepsilon,z) \quad \text{with } u = s+\varepsilon. \end{aligned}$$

With the approximation

$$\begin{aligned} F_{s+\varepsilon,z}(t,y) \approx F_{s+\varepsilon,x}(t,y) &+ \left(\frac{\partial}{\partial x}F_{s+\varepsilon,x}(t,y)\right)(z-x) \\ &+ \left(\frac{1}{2}\frac{\partial^2}{\partial x^2}F_{s+\varepsilon,x}(t,y)\right)(z-x)^2 \end{aligned}$$

then follows

$$\begin{aligned} F_{s,x}(t,y) \approx F_{s+\varepsilon,x}(t,y) &+ \frac{\partial}{\partial x}F_{s+\varepsilon,x}(t,y)\varepsilon\varrho(s,x) \\ &+ \frac{1}{2}\left(\frac{\partial^2}{\partial x^2}F_{s+\varepsilon,x}(t,y)\right)\varepsilon a(s,x). \end{aligned}$$

From this follows after forming differences, multiplication by $\frac{1}{\varepsilon}$ and with transition to the limit $\varepsilon \to 0$ the

Kolmogorov Backward Differential Equation

$$-\frac{\partial}{\partial s}F_{s,x}(t,y) = \varrho(s,x)\frac{\partial}{\partial x}F_{s,x}(t,y) + \frac{1}{2}a(s,x)\frac{\partial^2}{\partial x^2}F_{s,x}(t,y) \tag{4.47}$$

with boundary conditions $\lim_{s \nearrow t} F_{s,x}(t,y) = \begin{cases} 1, & x < y, \\ 0, & x > y. \end{cases}$

The above argument can be carried out exactly and yields the following existence statement.

Theorem 4.89 *Let ϱ, a be continuous and ϱ, a satisfy the Lipschitz condition and the linear growth condition in x. Further let there exist a $c > 0$ such that $a(t,x) \geq c, t \leq T, x \in \mathbb{R}$. Then it follows:*
The Kolmogorov backward equation (4.47) has a unique transition function F as a solution.
There exists a continuous Markov process X with $P_{s,x} \sim F_{s,x}$, and X is a diffusion process with coefficients $\varrho, \sigma = |a|^{\frac{1}{2}}$.

Remark 4.90

(a) An analogous result holds in dimension $d \geq 1$ under the condition: $\exists c > 0, \forall t$ and $\forall v \in \mathbb{R}^n \setminus \{0\}$ the 'uniform ellipticity inequality' holds $v^\top a(t,x)v \geq c|v|^2$.
(b) If X is stationary, then for all $s \geq 0$: $F_t(x,y) = F_{s,x}(s+t,y)$. Therefore, it follows:

$$\frac{\partial}{\partial s} F_{s,x}(s+t,y) = 0.$$

Thus, the stationarity equation holds

$$\frac{\partial}{\partial t} F_t(x,y) = -\frac{\partial}{\partial s} F_{s,x}(u,y)|_{u=s+t}.$$

From this follows the **Kolmogorov equation in the stationary case:**

$$\begin{cases} \frac{\partial}{\partial t} F_t(x,y) = \varrho(x)\frac{\partial}{\partial x} F_t(x,y) + \frac{1}{2}a(x)\frac{\partial^2}{\partial x^2} F_t(x,y) \\ \lim_{t \searrow 0} F_t(x,y) = \begin{cases} 1, & x < y, \\ 0, & x > y. \end{cases} \end{cases}$$

(c) Let $u(t,x) := \int f(y) P_t(x, \mathrm{d}y) = \int f(y) F_t(x, \mathrm{d}y)$ where $f \in C_b^{(2)}$. From the above equation it then follows

$$\begin{cases} \frac{\partial u}{\partial t} = \mathcal{A}u \\ u(0,x) = f(x), \ \mathcal{A} := \varrho(x)\,\mathrm{d}x + \frac{1}{2}a(x)\frac{\mathrm{d}^2}{\mathrm{d}x^2}, \end{cases} \tag{4.48}$$

i.e., u is a solution of the Cauchy problem to the differential operator $\mathcal{A}$. $\mathcal{A}$ is the infinitesimal generator of the Markov process X.
(d) **Kolmogorov forward equation**

Let $p_{s,x}(t, dy) = p_{s,x}(t,y)\,dy$ be the transition density of a Markov process X. Then, according to Chapman–Kolmogorov, it follows for $s \leq u \leq t$

$$p_{s,x}(t,y) = \int p_{u,z}(t,y) p_{s,x}(u,z)\,\mathrm{d}z$$

The Kolmogorov forward equation, also called Fokker–Planck equation, for p is obtained by variation at a time point in the future.

Theorem 4.91 (Kolmogorov Forward Equation) *a, ϱ satisfy the conditions of Theorem 4.89 and in addition let $\frac{\partial \varrho}{\partial x}, \frac{\partial a}{\partial x}, \frac{\partial^2 a}{\partial x^2}$ satisfy the Lipschitz and the linear growth condition.*

Then the ***Kolmogorov forward equation*** *for the transition density* p

$$\begin{aligned}\frac{\partial}{\partial t} p_{s,x}(t,y) &= -\frac{\partial}{\partial y}\big(\varrho(t,y)p_{s,x}(t,y)\big) + \frac{1}{2}\frac{\partial^2}{\partial y^2}\big(a(t,y)p_{s,x}(t,y)\big) \\ \lim_{t\searrow s} p_{s,x}(t,y) &= \delta_x(y)\end{aligned} \tag{4.49}$$

has a unique solution $p_{s,x}$. *There exists exactly one diffusion process* X *with transition density* p.

Proof (Idea) Let X be a diffusion process with transition density p to the above data. Let $\vartheta(t) := \int \xi(y)\, p_{s,x}(t,y)\,\mathrm{d}y$, $\xi \in C_b^2$, and let $\varepsilon > 0$.

The Chapman–Kolmogorov equation applied to $\vartheta(t+\varepsilon)$ yields:

$$\begin{aligned}\vartheta(t+\varepsilon) &= \int \xi(y)\Big(\int p_{t,z}(t+\varepsilon, y) p_{s,x}(t,z)\,\mathrm{d}z\Big)\,\mathrm{d}y \\ &= \int \Big[\int \xi(y) p_{t,z}(t+\varepsilon, y)\,\mathrm{d}y\Big] p_{s,x}(t,z)\,\mathrm{d}z \\ &\approx \int \Big[\int \big(\xi(z) + \xi'(z)(y-z) \\ &\qquad + \frac{1}{2}\xi''(z)(y-z)^2 p_{t,z}[t+\varepsilon, y]\,\mathrm{d}y\big)\Big] p_{s,x}(t,z)\,\mathrm{d}z \\ &= \int_{\mathbb{R}} \Big(\xi(z) + \xi'(z)\varepsilon\varrho(t,z) + \frac{1}{2}\xi''(z)\varepsilon a(t,z)\Big) p_{s,x}(t,z)\,\mathrm{d}z \\ &= \vartheta(t) + \varepsilon \int \Big(\xi'(z)\varrho(t,z) + \frac{1}{2}\xi''(z) a(t,z)\Big) p_{s,x}(t,z)\,\mathrm{d}z.\end{aligned}$$

It follows with partial integration

$$\begin{aligned}\vartheta'(t) &= \int \Big(\xi'(z)\varrho(t,z) + \frac{1}{2}\xi''(z) a(t,z)\Big) p_{s,x}(t,z)\,\mathrm{d}z \\ &= \int \xi(z)\Big(-\frac{\partial}{\partial z}(\varrho(t,z)p_{s,x}(t,z)) + \frac{1}{2}\frac{\partial^2}{\partial z^2}(a(t,z)p_{s,x}(t,z))\Big)\,\mathrm{d}z.\end{aligned}$$

On the other hand, by definition

$$\vartheta'(t) = \int \xi(z)\Big(\frac{\partial}{\partial t} p_{s,x}(t,z)\Big)\,\mathrm{d}z, \quad \forall \vartheta \in C_b^2.$$

From this, by comparing the integrands, the Kolmogorov forward equation follows in (4.48). □

In comparison to the backward differential equation in Theorem 4.89, the forward differential equation in Theorem 4.91 is stated for the transition densities p (instead of the transition distribution function F). This leads to the relatively stronger assumptions in Theorem 4.91. An analogous result to Theorem 4.91 also holds in the multivariate case $d \geq 1$.

3. Approach Through Stochastic Differential Equations (Itô Theory)
We assume that the coefficients ϱ, a satisfy the conditions of Theorem 4.89.
Let $\boldsymbol{d} = \mathbf{1}$ and let $\sigma(t, x) = \sqrt{a(t, x)}$.
Then

$$|\sigma(t, x) - \sigma(t, y)| = \frac{|a(t, x) - a(t, y)|}{\sqrt{a(t, x)} + \sqrt{a(t, y)}} \leq \frac{1}{2\sqrt{c}}|a(t, x) - a(t, y)|.$$

Thus σ satisfies the Lipschitz condition and the linear growth condition holds. Thus, by the existence and uniqueness theorem, there exists a unique solution of the stochastic differential equation (4.23).

Theorem 4.92 *Under the above assumptions, the unique solution X of the stochastic integral equation*

$$X_t = \xi + \int_0^t \sigma(s, X_s) \mathrm{d}B_s + \int_0^t \varrho(s, X_s) \,\mathrm{d}s$$

is a diffusion process with coefficients ϱ, a. The transition distribution function $F_{s,x}(t, y)$ from X_t for t, y is fixed, is the unique solution of the Kolmogorov backward equation.
Under the additional conditions in Theorem 4.91, there exists a transition density $p_{s,x}(t, y)$. p is the unique solution of the Kolmogorov forward equation.

Remark 4.93 If $d \geq 1$ then the solution $\sigma(t, x)$ of $\sigma(t, x)\sigma(t, y)^T = a(t, x)$ is not unique in general. If e.g. U is an orthogonal matrix, $UU^T = I$, then σU is also a solution of this equation and UB is also a Brownian motion. The transition distribution function of the solution X of the stochastic differential equation is independent of the choice of σ.

We compare the presented methods for constructing a diffusion process X by means of a special case:

Comparison of Methods
Let $d = 1$, $a(x) = 2$ and $\varrho(x) = -x$. Thus, the problem to solve is to construct a diffusion process in $d = 1$ with coefficients $a(x) = 2$ and $\varrho(x) = -x$ and the infinitesimal generator

$$Af(x) = f''(x) - xf'(x).$$

(1) Semigroup theory

To apply the semigroup theory, one has to check that $(I - \frac{1}{n}A)^{-1}$ exists, and whether $\|(I - \frac{1}{n}A)^{-1}\| \leq 1$ for a $n \in \mathbb{N}$; a task which is already not easy in this example; the spectrum of the operator has to be investigated.

(2) PDE method

We analyze the Kolmogorov forward equation

$$\begin{aligned} \frac{\partial}{\partial t} p_t(x,y) &= \frac{\partial}{\partial y}(y p_t(x,y)) + \frac{\partial^2}{\partial y^2} p_t(x,y) \\ \lim_{t \downarrow 0} p_t(x,y) &= \delta_x(y). \end{aligned} \tag{4.50}$$

We assume that the stationary limit $p(y) = \lim p_t(x,y)$ exists, and is independent of x.

For the density of the invariant measure for the diffusion X we obtain the **stationarity condition:** $\frac{\partial}{\partial t} p(y) = 0$.

Then it follows from (4.50) in the limit the equation for the stationary density:

$$p(y) + y p'(y) + p''(y) = 0.$$

This equation has as solution the density $p(y) = \frac{1}{\sqrt{2\pi}} e^{-y^2/2}$.

These preliminary considerations lead us to the approach with undetermined coefficients $\lambda(t)$, $\vartheta(t)$: $p_t(x,y) = \frac{1}{\sqrt{2\pi\vartheta(t)}} \cdot e^{\frac{-(y-\lambda(t)x)^2}{2\vartheta(t)}}$.

For this we get

$$\frac{\partial}{\partial t} p_t(x,y) = p_t(x,y) \cdot \left(\frac{\vartheta'(t)}{2\vartheta(t)} + \frac{\vartheta'(t)(y-\lambda(t)x)^2}{2\vartheta^2(t)} + \frac{\lambda'(t)x(y-\lambda(t)x)}{\vartheta(t)} \right).$$

Further it holds

$$\begin{aligned} &\frac{\partial}{\partial y}(y p_t(x,y)) + \frac{\partial^2}{\partial y^2} p_t(x,y) \\ &= p_t(x,y) \left(1 - \frac{y(y-\lambda(t)x)}{\vartheta(t)} + \frac{(y-\lambda(t)x)^2}{\vartheta^2(t)} - \frac{1}{\vartheta(t)} \right). \end{aligned}$$

Substituting in (4.50) results in

$$\begin{aligned} &-\frac{\vartheta'(t)}{2\vartheta(t)} + \vartheta'(t) \frac{(y-\lambda(t)x)^2}{2\vartheta^2(t)} + \frac{\lambda'(t)x(y-\lambda(t)x)}{\vartheta(t)} \\ &= 1 - \frac{y(y-\lambda(t)x)}{\vartheta(t)} + \frac{(y-\lambda(t)x)^2}{\vartheta(t)} - \frac{1}{\vartheta(t)}. \end{aligned}$$

Coefficient comparison yields the equivalence of this differential equation with:

$$\vartheta'(t) = -2\vartheta(t) + 2 \text{ and } \vartheta'(t)\lambda(t) + 2\vartheta(t)\lambda'(t) = -2\lambda(t).$$

From the initial conditions it follows: $\vartheta(0) = 0$, $\lambda(0) = 1$.

This implies the solution

$$\vartheta(t) = 1 - e^{-2t}, \quad \lambda(t) = e^{-t}$$

and thus the transition density

$$p_t(x, y) = \frac{1}{\sqrt{2\pi(1 - e^{-2t})}} \exp\left(-\frac{(y - e^{-t}x)^2}{2(1 - e^{-2t})}\right).$$

p is the transition density of the Ornstein–Uhlenbeck process.

(3) SDE Method

To determine the diffusion process by means of stochastic differential equations with. $a(x) = 2, \varrho(x) = -x$, and thus $\sigma(x) = \sqrt{2}$. The stochastic differential equation to be solved is :

$$\mathrm{d}X_t = \sqrt{2}\mathrm{d}B_t - X_t\,\mathrm{d}t \text{ where } X_0 = 0,$$

or the equivalent integral equation

$$X_t = x_0 + \int_0^t \sqrt{2}\mathrm{d}B_s - \int_0^t X_s\,\mathrm{d}s.$$

Its solution results directly from the Itô formula (cf. Example 4.63)

$$X_t = e^{-t}x_0 + \sqrt{2}\int_0^t e^{-(t-u)}\mathrm{d}B_u.$$

It follows that X_t is normally distributed:

$$X_t \sim N\left(e^{-t}x_0, 2\int_0^t e^{-(t-u)}\,\mathrm{d}u\right) = N\left(e^{-t}x_0, 1 - e^{-2t}\right)$$

and thus the transition density is given in explicit form. The SDE method is the easiest to carry out for this example.

PDEs, Infinitesimal Generator, and SDEs

In the following, we deal with some statements on the relation between PDEs, infinitesimal producer and SDEs. From the semigroup property in (4.48) or the martingale property of $f(X_t, t) - f(x, 0) - \int_0^t \mathcal{A} f(X_s, s)\,ds$, where

$$\mathcal{A} f = \frac{\partial f}{\partial s} + \mathcal{A}_s f, \ \mathcal{A}_s f(x) = \sum_i b_i(x) \frac{\partial f}{\partial x_i} + \sum_{i,k} a_{ik}(x) \frac{\partial^2 f}{\partial x_i \partial x_k},$$

it follows that $u(t, x) = E_x f(t, X_t)$ is the solution of a partial differential equation.

Corollary 4.94 *If X is a weak solution of the stochastic differential equation (4.28), $f \in C_K^{1,2}$ and if σ_{ij} is bounded on the support* supp f, *then it holds:*

$$E_x f(t, X_t) = f(0, x) + E_x \int_0^t \left(\frac{\partial f}{\partial s} + \mathcal{A}_s f \right) (s, X_s)\,ds. \tag{4.51}$$

In the homogeneous case let for $f \in C_K^2$:

$$A f(x) = \lim_{t \downarrow 0} \frac{E_x f(X_t) - f(x)}{t}$$

be the infinitesimal generator. Then it follows from (4.51) using the theorem of majorized convergence:

$$A f(x) = \lim_{t \downarrow 0} \frac{1}{t} E_x \int_0^t \mathcal{A} f(X_s)\,ds = \mathcal{A} f(x).$$

Corollary 4.95 *Let $f \in C_K^2$, let X be a solution of the stochastic differential equation*

$$dX_t = b(X_t)\,dt + \sigma(X_t) dB_t$$

and let $b, \sigma_{i,j}$ be bounded on supp f. *Then it follows:*

$$A f(x) = \mathcal{A} f(x) = \sum_i b_i(x) \frac{\partial f}{\partial x_i} + \frac{1}{2} \sum_{i,j} a_{ik}(x) \frac{\partial^2 f}{\partial x_i \partial x_k}.$$

The infinitesimal generator A of the diffusion process given by stochastic differential equation is identical with the corresponding diffusion operator $\mathcal{A}$.

Remark 4.96

(a) Corollary 4.95 also holds in the nonstationary case.
For $f \in C_K^2$ and $u(t,x) := E_x f(X_t)$ it holds:

$$u(0,x) = f(x), \quad \frac{\partial}{\partial t} u = \mathcal{A} u.$$

u is, therefore, the solution of the **Cauchy problem.** For

$$\begin{aligned}&\frac{1}{u}(E_x f(X_{t+u}) - f(X_t)) \\ &= E_x E\left(\frac{f(X_{t+u}) - f(z)}{u} \,\Big|\, X_t = z\right) \underset{u\to 0}{\longrightarrow} E_x \mathcal{A} f(X_t) = \mathcal{A} u(t,x),\end{aligned}$$

if the exchange of expectation and limit is possible, such as for $f \in C_K^2$.
As a special case we have the (one-dimensional) **heat equation**

$$\frac{\partial u}{\partial t} = \frac{1}{2}\frac{\partial^2 u}{\partial x^2}, \quad u(0,x) = f(x). \tag{4.52}$$

The transition density of the Brownian motion

$$p(t,x,y) = \frac{1}{\sqrt{2\pi t}} \exp(-(x-y)^2/2t)$$

is a fundamental solution of the above equation

$$\frac{\partial p}{\partial t} = \frac{1}{2}\frac{\partial^2 p}{\partial x^2}.$$

If $\int e^{-ax^2} |f(x)| \,\mathrm{d}x < \infty$ then

$$u(t,x) := E_x f(B_t) = \int f(y) p(t,x,y) \,\mathrm{d}y$$

for $0 < t < \frac{1}{2a}$ is a solution of the heat equation (4.52).
Similarly, for $d \geq 1$ one can construct the solution with the multivariate Brownian motion.

(b) Corollary 4.94 also holds for stopping times using the optional sampling theorem
Let $f \in C_K^2$, τ be a stopping time with $E_x \tau < \infty$, then it holds:

$$E_x f(X_\tau) = f(x) + E_x \int_0^\tau \mathcal{A} f(X_s) \,\mathrm{d}s \quad (\textbf{\textit{Dynkin formula}}). \tag{4.53}$$

Examples of the Connection with PDE

Let $\mathcal{A}$ be the semi-elliptic differential operator $\mathcal{A} = \sum a_{ij}(x)\frac{\partial^2}{\partial x_i \partial x_j} + \sum b_i \frac{\partial}{\partial x_i}$, where $a = (a_{ij}) \geq 0$ is positive semidefinite.

(1) Dirichlet problem

Let $D \subset \mathbb{R}^d$ be a bounded domain and consider the Dirichlet problem on D

$$(D) \qquad \begin{cases} \mathcal{A}u = 0 & \text{in } D \\ u = f & \text{on } \partial D \end{cases} \tag{4.54}$$

for a continuous function $f \in C(\partial D)$. To construct a solution to this problem, let X be an Itô diffusion i.e. solution of

$$\mathrm{d}X_t = b(X_t)\,\mathrm{d}t + \sigma(X_t)\mathrm{d}B_t, \quad \frac{1}{2}\sigma\sigma^T = a.$$

We define by analogy with the case of Brownian motion for the classical Dirichlet problem:

$$u(x) := E_x f(X_{\tau_D}),$$

where $\tau_D = \inf\{t; X_t \in \partial D\}$. Assuming that $P_x(\tau_D < \infty) = 1, x \in D$, i.e. X leaves the domain D in finite time, it follows $E_x\tau_D < \infty, \forall x \in D$. Thus it follows as in the classical case:

u is a unique solution of the Dirichlet problem (D).

The uniqueness proof from Remark 4.16 b) using the Itô formula can be transferred analogously. Thus u is a unique solution of the Dirichlet problem.

Sufficient for the existence of a solution is the condition

$$\exists i : \inf_{x \in D} a_{ii}(x) > 0$$

This condition implies in particular, the ellipticity condition, $A > 0$. It follows from the uniform ellipticity : $x^\top A(x)x \geq \delta|x|^2$.

(2) Cauchy problem

Let $f = f(x), g = g(t,x) \geq 0$, and let $v \in C^{1,2}$ be polynomially bounded in x and be a solution of the parabolic differential equation

$$\begin{cases} -\frac{\partial v}{\partial t} + kv = \mathcal{A}_t v + g \text{ in } [0,T] \times \mathbb{R}^d \\ v(T,x) = f(x). \end{cases} \tag{4.55}$$

The parabolic differential equation (4.55) has time-dependent coefficients, a potential kernel k and a Lagrangian term g. The solution v has the **Feynman–Kac representation**

$$v(t,x) = E_{t,x}\Big[f(X_T)\exp\Big(-\int_t^T k(u,X_u)\,\mathrm{d}u\Big) + \int_t^T g(s,X_s)\exp\Big\{-\int_t^T k(u,X_u)\,\mathrm{d}u\Big\}\,\mathrm{d}s\Big].$$

In the special case of the Kolmogorov backward equation with $g = k = 0$ holds

$$v(t,x) = E_{t,x} f(X_T) = E(f(X_T) \mid X_t = x).$$

In particular, as a consequence of the Feynman–Kac representation, the uniqueness of the solution is obtained.

The proof of this representation follows by applying Itô's formula to

$$v(s,X_s)\exp\Big(-\int_t^s k(u,X_u)\,\mathrm{d}u\Big).$$

The solution of parabolic differential equations of the form (4.55) is thus in a unique relation to the determination of expected values for functions of diffusion processes.

(3) Poisson equation

Let $D \subset \mathbb{R}^d$ be a bounded domain. The Poisson equation is a generalization of the Dirichlet problem. It is given by :

$$(P) \qquad \begin{cases} \frac{1}{2}\Delta u = -g & \text{in } D \\ u = f & \text{on } \partial D. \end{cases}$$

The Poisson equation (P) has the solution

$$u(x) = E_x\Big(f(B_{\tau_D}) + \int_0^{\tau_D} g(B_t)\,\mathrm{d}t\Big), \quad x \in D. \tag{4.56}$$

In general, for a diffusion operator $\mathcal{A}$ solutions for equations of the type

$$\begin{cases} \mathcal{A}u = -g & \text{in } D \\ u = f & \text{on } \partial D \end{cases}$$

are obtained in the form (4.56), where the Brownian motion B is to be replaced by a diffusion process X and where $E_x\tau_D < \infty$, $x \in D$ is assumed.

The parabolic differential equation (4.55) [illegible] coefficients, a [illegible] kernel K or [illegible] term [illegible]. The solution [illegible] has the **Feynman–Kac representation**

[illegible]

[illegible] the Rotation [illegible] with [illegible]

[illegible]

In particular, [illegible] of the Feynman–Kac representation, the uniqueness of the solution [illegible]

The proof of this [illegible] follows by applying Itô's formula to

[illegible]

The [illegible] of the form (4.55) is [illegible] the [illegible] values for [illegible] functions of diffusion processes.

(2) Poisson [illegible]

Let [illegible]. The Poisson equation is a generalization of the [illegible] given by

[illegible]

The Poisson equation [illegible] the solution

[illegible]

[illegible]

[illegible]

[illegible] the Brownian motion [illegible] can be replaced [illegible]

5 Option Prices in Complete and Incomplete Markets

Already in the introductory Chap. 1 an introduction is given in a non-technical way to the basic principles of the theory of arbitrage-free prices, the hedging principle and the risk-neutral price measure, which leads from the binomial price formula by approximation to the Black–Scholes formula. The necessary link from processes in discrete time (binomial model) to those in continuous time (Black–Scholes model) is given by approximation theorems for stochastic processes in Chap. 2. Theorems of this type allow an interpretation of continuous financial market models using simple discrete models such as the Cox–Ross–Rubinstein model.

This chapter introduces the basics of general arbitrage-free pricing theory. Fundamental are the first and second fundamental theorems of asset pricing and the associated risk-neutral valuation formula, which is based on valuation by means of equivalent martingale measures. For their construction, Girsanov's theorem proves to be very useful. For the standard options, it can be used to determine the corresponding pricing formulas (Black–Scholes formulas) in a simple way. The determination of the corresponding hedging strategies leads in the Black–Scholes model (geometric Brownian motion) to a class of partial differential equations, the Black–Scholes differential equations, going back to the fundamental contributions of Black and Scholes as well as the connection with stochastic calculus by Merton in 1969. This led in the period 1979–1983 to the development of a general theory of arbitrage-free pricing for continuous-time price processes and the important role of equivalent martingale measures for this purpose by Harrison, Kreps and Pliska.

Central topics of this general theory are the completeness and incompleteness of market models, the determination of associated arbitrage-free price intervals via the equivalent martingale measures and the corresponding sub- or super-hedging strategies (optional decomposition theorem).

L. Rüschendorf, *Stochastic Processes and Financial Mathematics*, Mathematics Study Resources 1, https://doi.org/10.1007/978-3-662-64711-0_5

5.1 The Black–Scholes Model and Risk-Neutral Valuation

We consider a continuous-time financial market model with time horizon T and two securities, a risk-free bond (bond) S_t^0 and a risky security (stock) S_t (Fig. 5.1).

The general task of financial mathematics is to describe the performance of the stock in comparison with the bond and, in particular, to determine a 'correct' price for functionals of the stock (options).

Example 5.1 **(Black–Scholes Model)** In the Black–Scholes model, the bond S^0 is described by

$$\mathrm{d}S_t^0 = r S_0^t \,\mathrm{d}t \quad \text{d.h. } S_t^0 = e^{rt} \; \textit{with a deterministic interest rate } r.$$

The stock S is given by a geometric Brownian motion, i.e., a solution of

$$\mathrm{d}S_t = S_t(\mu\,\mathrm{d}t + \sigma\,\mathrm{d}B_t).$$

S has the explicit representation

$$S_t = S_0 \exp\left(\sigma B_t + \left(\mu - \frac{1}{2}\sigma^2\right)t\right), \qquad t \in [0, T].$$

The geometric Brownian motion model has two parameters, the volatility σ and the drift μ. The 'log returns' $\ln \frac{S_t}{S_0}$ of S are given by a Brownian motion with drift $a = \mu - \frac{1}{2}\sigma^2$, and diffusion coefficient σ.

Thus, the log returns have independent increments, but not the process S itself. The reason for this modeling is that empirically, the relative increments

$$\frac{S_t - S_u}{S_u} = \frac{S_t}{S_u} - 1 \approx \ln \frac{S_t}{S_u}$$

turn out for $u \sim t$ to be independent. This leads to the fact that $\ln \frac{S_t}{S_0}$ is modeled as a process with independent increments. Therefore, a natural modeling of security

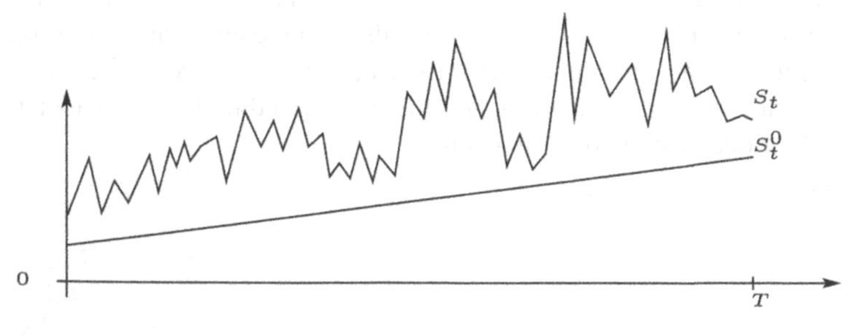

Fig. 5.1 Financial market model

processes is given by the geometric Brownian motion or more generally by a model with exponential Lévy processes $S_t = \exp(X_t)$.

A discrete analogue of geometric Brownian motion is the multiplicative model

$$S^{(n)}_{\frac{k}{n}} = S_0 \prod_{i=1}^{k} Y_i^{(n)}$$
$$= S_0 \exp\Big(\sum_{i=1}^{k} \ln Y_i^{(n)}\Big).$$

If one defines for suitable constants $u_n \geq r \geq d_n$

$$P\left(Y_i^{(n)} = \begin{matrix} u_n \\ d_n \end{matrix}\right) = \begin{cases} p_n, \\ 1 - p_n, \end{cases}$$

then $S^{(n)}$ approximates the model of geometric Brownian motion S if

$$n(p_n \log u_n + (1 - p_n) \log d_n) = n\mu_n \approx a = \left(\mu - \frac{r}{2}\sigma^2\right)$$

and $n\big(p_n \log u_n^2 + (1 - p_n) \log d_n^2 - \mu_n^2\big) \approx \sigma^2$.

These conditions are satisfied, for example, if $u_n = e^{\frac{\sigma}{\sqrt{n}}}$, $d_n = \frac{1}{u_n}$, $p_n = \frac{1}{2}\big(1 + \frac{a}{\sigma\sqrt{n}}\big)$.

This discrete model with one *up* and one *down jump* u_n resp. d_n is the classical Cox–Ross–Rubinstein model used as an approximation for geometric Brownian motion in practice. In the general case, the market model is described by a semimartingale.

We now introduce some *basic terms* of option price theory.

A **trading strategy** Φ, $\Phi = (\Phi_t) = \big((\varphi_t^0, \varphi_t)\big) \in \mathcal{L}(\mathcal{P})$ is a predictable adapted process consisting of two parts:

φ_t^0 gives the share of bonds at time t in the portfolio

φ_t gives the share of stocks at time t in the portfolio

The first general assumption is that φ^0, φ are integrable: $\int_0^T |\varphi_t^0|\, dt < \infty$, $\varphi \in \mathcal{L}^0(S)$.

The **value of the portfolio** at time t then is $V_t(\Phi) = \varphi_t^0 S_t^0 + \varphi_t S_t$.

Self-financing Trading Strategies

In discrete time self-financing trading strategies Φ_n are defined by the property that at time n only a regrouping of the portfolio is undertaken, i.e., the value of the regrouped portfolio is $\Phi_{n+1}^T \cdot S_n$ is identical to the value at time n:

$$\begin{aligned}\Phi_n^T \cdot S_n &= \Phi_{n+1}^T \cdot S_n \\ \Longleftrightarrow V_{n+1}(\Phi) - V_n(\Phi) &= \Phi_{n+1}^T (S_{n+1} - S_n).\end{aligned}$$

The analogous term in continuous time is described by the equation

$$\begin{aligned}\mathrm{d}V_t(\Phi) &= \varphi_t^0 \,\mathrm{d}S_t^0 + \varphi_t \,\mathrm{d}S_t \\ \Longleftrightarrow V_t(\Phi) &= V_0(\Phi) + \int_0^t \varphi_s^0 \,\mathrm{d}S_s^0 + \int_0^t \varphi_u \,\mathrm{d}S_u\end{aligned}$$

Let $\Pi := \{\Phi = (\varphi^0, \varphi) \in \mathcal{L}^0(S);\ \Phi \text{ is a self-financing trading strategy}\}$ be the set of all self-financing trading strategies.

Discounted Price Process
$\widetilde{S}_t := e^{-rt} S_t$ denotes the *discounted price process*, $\widetilde{V}_t(\Phi) := e^{-rt} V_t(\Phi)$ the *discounted value of the portfolio* Φ at time t. The reference variable S_t^0 acts thereby as *'numéraire'*.

Lemma 5.2 *Let $\Phi = (\varphi^0, \varphi) \in \mathcal{L}^0(S)$, then it holds:*

$$\Phi \in \Pi \Longleftrightarrow \widetilde{V}_t(\Phi) = V_0(\Phi) + \int_0^t \varphi_u \,\mathrm{d}\widetilde{S}_u, \quad \forall t > 0.$$

Proof "$\Longrightarrow$": (for S continuous) For $\Phi \in \Pi$ and $\widetilde{V}_t(\Phi) = e^{-rt} V_t(\Phi)$ it holds with partial integration

$$\begin{aligned}\mathrm{d}\widetilde{V}_t(\Phi) &= -r\widetilde{V}_t(\Phi)\,\mathrm{d}t + e^{-rt}\,\mathrm{d}V_t(\Phi) \\ &= -re^{-rt}\big(\varphi_t^0 e^{rt} + \varphi_t S_t\big)\,\mathrm{d}t + e^{-rt}\varphi_t^0\,\mathrm{d}(e^{rt})\varphi_t\,\mathrm{d}S_t \\ &= \varphi_t\big(-re^{-rt}S_t\,\mathrm{d}t + e^{-rt}\,\mathrm{d}S_t\big) = \varphi_t\,\mathrm{d}\widetilde{S}_t.\end{aligned}$$

Thus it follows

$$\widetilde{V}_t(\Phi) = V_0(\Phi) + \int_0^t \varphi_u \,\mathrm{d}\widetilde{S}_u$$

"$\Longleftarrow$": analogously □

Remark 5.3 If $\Phi \in \Pi$ and φ^0, φ are of finite variation, then, with partial integration, it holds

$$\begin{aligned} dV_t(\Phi) &= d(\varphi_t^0 S_t^0) + d(\varphi_t S_t) \\ &= \varphi_t^0 \, dS_t^0 + S_t^0 \, d\varphi_t^0 + \varphi_t \, dS_t + S_t d\varphi_t, \quad \text{as } [\varphi, S] = 0, \\ &= \varphi_t^0 \, dS_t^0 + \varphi_t \, dS_t. \end{aligned}$$

It follows

$$S_t^0 \, d\varphi_t^0 + S_t \, d\varphi_t = 0,$$

i.e. the change in value due to reallocation in the bond is identical to the negative of the change in value due to reallocation in the stock.

A basic principle of price theory is the **no-arbitrage principle,** which states that no risk-free profit is possible in a 'realistic market model'. A fundamental question is to describe price models that satisfy the no-arbitrage principle.

Definition 5.4 $\Phi \in \Pi$ is called an **arbitrage strategy** if

$$V_0(\Phi) = 0, \, V_T(\Phi) \geq 0 \, [P] \text{ and } P(V_T(\Phi) > 0) > 0,$$

i.e., there exists a risk-free profit from an initial value of zero.

Remark 5.5 **(Extended No-Arbitrage-Principle)** For an extended portfolio consisting of securities (stocks, bonds) S^i and derivatives C_i we assign to a strategy $\pi = ((\varphi^i), (a_i))$ with φ^i shares of S^i, a_i shares of C_i, $a_i = a_i(t, \omega)$ the value

$$V_t(\pi) = \varphi_t^0 S_t^0 + \sum \varphi_t^i S_t^i + \sum a_i(t, \cdot) C_i$$

to. The extended no-arbitrage principle states that no-arbitrage strategy exists in this extended market model.

Definition 5.6 Let (S^0, S, P) be a market model. $Q \in M^1(\Omega, \mathfrak{A})$ is an **equivalent (local) martingale measure** to P , if:

(a) $Q \sim P$, the measures Q and P are equivalent.
(b) $\widetilde{S} \in \mathcal{M}_{\text{loc}}(Q)$.

$\mathcal{M}_e = \mathcal{M}_e(P)$ denotes the set of equivalent (local) martingale measures. Elements from $\mathcal{M}_e$ are also called risk neutral measures.

For self-financing trading strategies and equivalent martingale measures $Q \in \mathcal{M}_e$ it holds

$$\widetilde{V}_t(\Phi) = \widetilde{V}_0(\Phi) + \int_0^t \varphi_u \, d\widetilde{S}_u \in \mathcal{M}_{\text{loc}}(Q);$$

the discounted value process is equal to the initial value plus the discounted profit process by trading from the stock. In general $\widetilde{V}(\Phi)$ only a local martingale.

The absence of arbitrage opportunities is a central economic requirement for a stochastic market model. However, already in discrete models arbitrage strategies typically exist for infinite time horizons, as can be seen from doubling strategies. In models in continuous time, such strategies already exist for finite time horizons. We, therefore, need to restrict the set of strategies appropriately.

Definition 5.7

(a) For $Q \in \mathcal{M}_e$, $\Phi \in \Pi$ is called ***Q*-regular,** $\Phi \in \Pi_r(Q)$ if $\widetilde{V}_t(\Phi)$ is a supermartingale concerning w.r.t. Q.

Let $\Pi_r = \bigcap_{Q \in \mathcal{M}_e} \Pi_r(Q)$ be the set of **regular trading strategies**.

(b) Let $Q \in \mathcal{M}_e(P)$ then a trading strategy $\Phi \in \Pi_b := \{\Phi \in \Pi \mid \exists c \in \mathbb{R}, \text{ so that } \widetilde{V}_t(\Phi) \geq c, \ \forall t \in [0, T]\}$ is called ***Q*-admissible**, if $\widetilde{V}_T(\Phi)$ is a martingale with respect to Q,

Let $\Pi_a(Q)$ be the set of *Q-admissible* self-financing trading strategies and let $\Pi_a = \bigcap_{Q \in \mathcal{M}_e(P)} \Pi_a(Q)$ be the set of **admissible strategies**.

Remark 5.8

(a) It holds that $\Pi_b \subseteq \Pi_r$, because for $\Phi \in \Pi_b$, $\widetilde{V}_t(\Phi) \in \mathcal{M}_{\text{loc}}(Q)$, and since $\widetilde{V}_t(\Phi) \geq c$ is bounded below, $\widetilde{V}_t(\Phi)$ is a supermartingale.

(b) For non-negative processes $S^i \geq 0$ it is sufficient to pose instead of the boundedness from below the weaker condition $V(\Phi) \geq -\sum_{i=1}^n c_i S^i$ with $c_i \geq 0$.

The existence of equivalent martingale measures excludes the existence of regular arbitrage strategies.

Theorem 5.9 *Let $\mathcal{M}_e(P) \neq \emptyset$ then no regular arbitrage strategy exists.*

Proof Let $Q \in \mathcal{M}_e(P)$, $\Phi \in \Pi_r$ then the discounted value process $\widetilde{V}_t(\Phi)$ is a supermartingale with respect to Q. It follows that:

$$E_Q \widetilde{V}_T(\Phi) \leq E_Q \widetilde{V}_0(\Phi). \tag{5.1}$$

Suppose Φ would be an arbitrage strategy, then $\widetilde{V}_0(\Phi) = 0$, $\widetilde{V}_T(\Phi) \geq 0$, but this implies according to (5.1)

$$E_Q \widetilde{V}_T(\Phi) \leq 0 \quad \text{also} \quad \widetilde{V}_T(\Phi) = 0[Q].$$

From this also follows $\widetilde{V}_T(\Phi) = 0\,[P]$ and thus a contradiction to the definition of an arbitrage strategy. □

In the discrete case, also the converse of Theorem 5.9 holds, i.e.,

$$\text{no-arbitrage is equivalent to } \mathcal{M}_e(P) \neq \emptyset.$$

In the continuous case, one needs a modification of the no-arbitrage notion.

Definition 5.10

(a) $\Phi \in \Pi$ is **δ-admissible**, $\delta > 0$ if

$$V_0(\Phi) = 0 \, and \, V_t(\Phi) \geq -\delta, \forall t \in [0, T].$$

(b) S fulfills the **NFLVR condition** (***N**o **f**ree **l**unch with **v**anishing **r**isk*)
$\Longleftrightarrow \forall \delta_n \downarrow 0 : \forall \Phi_n \in \Pi$ that are δ_n-admissible, it holds $V_T(\Phi_n) \xrightarrow{P} 0$.

Theorem 5.11 (First Fundamental Theorem of Asset Pricing) *In the market model (S^0, S, P) let (S, P) be a locally bounded semimartingale. Then it holds:*

$$S \text{ fulfills the NFLVR condition} \Longleftrightarrow \mathcal{M}_e(P) \neq \emptyset, \quad \text{e.g. there is an equivalent martingale measure.}$$

Proof "$\Longleftarrow$" As in Theorem 5.9.
"$\Longrightarrow$" Proof idea

(1) In the first step, reduction to bounded semimartingales by means of a 'change of numeraire' argument using Girsanov's theorem.
(2) Approximation by a discrete semimartingale model.
(3) In the third step, an equivalent martingale measure is constructed using Hahn–Banach's separation theorem (cf. Delbaen and Schachermayer (2006)). □

Remark 5.12 In the case of general (non-locally bounded) semimartingales, the class $\mathcal{M}_e(P)$ of the equivalent martingale measures must be replaced by the larger class $\mathcal{M}_\sigma(P)$ of equivalent σ martingale measures. A *d-dimensional* semimartingale S in $\mathcal{M}^d$ is called **σ-martingale** with respect to Q if a φ in $\mathcal{L}(\mathcal{P})$ with values in $\mathbb{R}_+$ exists such that

$$\varphi \cdot S = (\varphi \cdot S^i) \in \mathcal{M}(Q). \tag{5.2}$$

Now the following characterization of the NFLVR condition holds by the existence of an equivalent σ-martingale measure (ESMM) (cf. Delbaen and Schachermayer (1998)):

Theorem 5.13 ((General) First Fundamental Theorem of Asset Pricing:) *Let $S \in \mathcal{S}^d$, then the following are equivalent:*

1. *(ESMM): $\exists\, Q \sim P : S \in \mathcal{M}_\sigma(Q)$*
2. *NFLVR: S satisfies the NFLVR condition.*

In particular, it holds that $\mathcal{M}_{\mathrm{loc}} \subset \mathcal{M}_\sigma$.

We now show, as an application of Theorem 5.11 that the Black–Scholes model of geometric Brownian motion is arbitrage-free.

So let $\widetilde{S}_t = e^{-rt} S_t = S_0 e^{X_t}$ with $X_t = \sigma B_t + (\mu - r - \frac{\sigma^2}{2})t$, volatility σ^2 and drift $\mu - r - \frac{\sigma^2}{2}$. To establish that S is arbitrage-free, we show that there exists an equivalent martingale measure.

Proposition 5.14 *Define $Q \in M^1(\Omega, \mathfrak{A})$ by the Radon–Nikodým density.*

$$\frac{\mathrm{d}Q}{\mathrm{d}P} := L_T = \exp\left(\frac{r-\mu}{\sigma} B_T - \frac{(r-\mu)^2}{2\sigma^2} T\right).$$

Then it holds:

(a) $\widetilde{S} \in \mathcal{M}(Q)$ i.e. $Q \in \mathcal{M}_e(P)$.
(b) (X,Q) is a Brownian motion with drift $-\frac{\sigma^2}{2}$ and with volatility σ^2.

Proof B is a Brownian motion with respect to P and L is the exponential martingale. Therefore, by Girsanov's theorem, it follows:

$$\overline{B}_t := B_t - \frac{r-\mu}{\sigma} t \text{ is a Brownian motion with respect to } Q.$$

$$X_t = \sigma B_t + \left(\mu - r - \frac{\sigma^2}{2}\right) t = \sigma \overline{B}_t - \frac{\sigma^2}{2} t$$

is, therefore, a Brownian motion with volatility σ^2 and with drift $-\frac{\sigma^2}{2}$ with respect to Q. It follows that

$$\widetilde{S}_t = S_0 e^{\sigma \overline{B}_t - \frac{\sigma^2}{2} t} \text{ is an exponential martingale with respect to } Q,$$

and thus $\widetilde{S} \in \mathcal{M}(Q)$. □

Thus, there is no regular arbitrage strategy in the extended sense in the Black–Scholes model.

5.1.1 Risk-Neutral Valuation of Options

An option is a measurable functional of the process, i.e. an element of $L(\mathfrak{A}_T)$. The basic options

$$C = (S_T - K)_+ = \text{European Call} \quad \text{and} \quad C = (K - S_T)_+ = \text{European Put}$$

are called vanilla options. For the European options, the (acquired) right to buy and sell can only be exercised at the end of the term (validity) of the option, in this case at time T.

In contrast "American options", such as an American Call $C = (S_\tau - K)_+$ respectively an American Put $C = (K - S_\tau)_+$ can be exercised during the entire term at any stopping time τ.

Definition 5.15

(a) An option $C \in L(\mathfrak{A}_T)$ is **duplicable**

$$\Longleftrightarrow \exists\, \Phi \in \Pi_r \text{ with } V_T(\Phi) = C.$$

Each such Φ is called a **hedge** of C.

(b) A hedge Φ of C is called **admissible** or **martingale hedge** of C, if $\Phi \in \Pi_a$, i.e.

$$\exists\, Q \in \mathcal{M}_e(P) \quad \text{with } \widetilde{V}(\Phi) \in \mathcal{M}(Q).$$

(c) C is called **strongly hedgeable,** if there is a martingale hedge of C.

Remark 5.16 If Φ is an admissible hedge strategy of C, then

$$\widetilde{V}_T(\Phi) = e^{-rT} C \in \mathcal{L}^1(Q).$$

For what follows, it is of fundamental importance that two admissible hedge strategies for an option C have the same initial cost.

Proposition 5.17 *Let $\Phi, \Psi \in \Pi_{be}$ admissible hedge strategies of the option C, then it holds:*

$$V_0(\Phi) = V_0(\Psi).$$

Proof Let $Q_i \in \mathcal{M}_e(P)$ and $\Phi \in \Pi_a(Q_1)$, $\Psi \in \Pi_a(Q_2)$ with $x := V_0(\Phi)$ and $y := V_0(\Psi)$ the initial costs of these strategies. Then it holds:

$$0 = E_{Q_1}\Big(\underbrace{\widetilde{V}_T(\Phi)}_{=\widetilde{C}} - \underbrace{\widetilde{V}_T(\Psi)}_{=\widetilde{C}}\Big) = E_{Q_1}\widetilde{V}_T(\Phi) - E_{Q_1}\widetilde{V}_T(\Psi).$$

It is $E_{Q_1}\widetilde{V}_T(\Phi) = E_{Q_1}\widetilde{V}_0(\Phi) = x$.

Further $\Psi \in \Pi_r$ and, therefore, $\widetilde{V}(\Psi)$ is a supermartingale w.r.t. Q_1. From this follows

$$E_{Q_1}\widetilde{V}_T(\Psi) \leq E_{Q_1}\widetilde{V}_0(\Psi) = y.$$

Altogether, as a result we obtain

$$0 = E_{Q_1}\widetilde{V}_T(\Phi) - E_{Q_1}\widetilde{V}_T(\Psi) \geq x - y.$$

So $x \leq y$. Conversely, one obtains analogously $y \leq x$ and thus $x = y$. □

Remark 5.18

(a) More generally, it holds for admissible Φ, Ψ:

$$V_T(\Phi) \geq V_T(\Psi) \text{ implies } V_t(\Phi) \geq V_t(\Psi), \quad 0 \leq t \leq T.$$

(b) If $\Phi, \Psi \in \Pi_a(Q)$ are martingale hedges of C with the same martingale measure Q, then even without the assumption of regularity

$$V_0(\Phi) = V_0(\Psi).$$

According to these preparations, we can now introduce the basic no-arbitrage price as a consequence of the no-arbitrage principle.

Definition 5.19 (No-arbitrage Price) Let $Q \in \mathcal{M}_e(P)$ and let $C \in \mathcal{L}^1(\mathcal{A}_T, Q)$ be a claim. If $\Phi \in \Pi_a(Q)$ is an admissible hedge strategy of C, then

$$p(C) := V_0(\Phi) = \varphi_0^0 \underbrace{S_0^0}_{=1} + \varphi_0 S_0$$

is called **no-arbitrage price (Black–Scholes price)** of C (at the time t).

$p(C,t) := V_t(\Phi) = \varphi_t^0 e^{r \cdot t} + \varphi_t S_t^0$ is called no-arbitrage price at time t.

Remark 5.20

(a) According to Proposition 5.17, the no-arbitrage price of a hedgeable claim is uniquely defined. If $Q_1, Q_2 \in \mathcal{M}_e(P)$ and $C \in \mathcal{L}^1(Q_1) \cap \mathcal{L}^1(Q_2)$ have admissible hedge strategies Φ, Ψ, then $V_0(\Phi) = V_0(\Psi)$. Thus $p_{Q_1}(C) = p_{Q_2}(C)$ is defined independently of Q_i.

Thus, the no-arbitrage price is defined for **strongly hedgeable claims** $C \in L(\mathfrak{A}_T)$, i.e. there exists a $Q \in \mathcal{M}_e(P)$ and a $\Phi \in \Pi_a(Q)$ such that $C = V_T(\Phi)$.

(b) **Economic justification of the no-arbitrage price.**

The **generalized no-arbitrage principle**, which also includes non-hedgeable claims with their prices, states that in a realistic market model no risk-free profit is possible.

If the price $p(C)$ of a hedgeable option C were not equal to the no-arbitrage price $V_0(\Phi)$, $p(C) \neq V_0(\Phi)$, then there would be a risk-free profit:

If $p(C) < V_0(\Phi)$, then a risk-free profit strategy consists of the following steps:

(1) short selling from the stock portfolio,
(2) buying the option C,
(3) investing the difference without risk.

This strategy leads to risk-free profit.

If $p(C) > V_0(\Phi)$, then the analogous dual strategy leads to risk-free profit:

(1) short selling from the claim,
(2) buying from the stock portfolio,
(3) investing the difference without risk.

This again results in a risk-free profit as opposed to the no-arbitrage principle.

An equivalent formulation of the no-arbitrage principle is: Two financial instruments with the same payoff at time T are equally expensive at time $t, \forall t \leq T$.

Of fundamental importance is the following risk-neutral valuation formula. The no-arbitrage price can be determined as the expected value of the option with respect to an equivalent martingale measure without having to explicitly determine hedge strategies.

Theorem 5.21 (Risk Neutral Valuation Formula) *Let $Q \in \mathcal{M}_e(P)$ and let $C \in L(\mathfrak{A}_T, Q)$ be a duplicable option with admissible hedge $\Phi \in \Pi_a(Q)$. Then it holds:*

$$p(C) = E_Q \widetilde{C}, \quad \widetilde{C} = e^{-rT} C.$$

More generally: $p(C,t) = e^{-r(T-t)} E_Q(C \mid \mathfrak{A}_t) = e^{rt} E_Q(\widetilde{C} \mid \mathfrak{A}_t)$.

Remark 5.22 The expected value is with respect to the martingale measure Q, but **not** with respect to the underlying statistical model measure P. It turns out, however, that real markets behave 'approximately' like risk-neutral markets. To calculate the price of an option one does not need to know the hedging strategy.

Proof Let $\Phi \in \Pi_a(Q)$ be a hedge strategy of C. Then

$$V_T(\Phi) = V_0(\Phi) + \int_0^T \varphi_s \, dS_s + \int_0^T \varphi_s^0 \, dS_s^0 = C.$$

Equivalent is

$$\widetilde{V}_T(\Phi) = V_0(\Phi) + \int_0^T \varphi \, d\widetilde{S} = \widetilde{C} \quad \text{(cf. Lemma 5.2).}$$

Now

$$\widetilde{V}_t(\Phi) = V_0(\Phi) + \int_0^t \varphi_u \, d\widetilde{S}_u \in \mathcal{M}(Q), \text{ as } \Phi \text{ is admissible with respect to } Q.$$

It follows

$$E_Q\widetilde{C} = E_Q\widetilde{V}_T(\Phi) = E_Q V_0(\Phi) = V_0(\Phi) = p(C).$$

Similarly it follows

$$\begin{aligned} p(C, t) = V_t(\Phi) = e^{rt}\widetilde{V}_t &= e^{rt} E_Q(\widetilde{V}_T(\Phi) \mid \mathfrak{A}_t) \\ &= e^{rt} E_Q(C \mid \mathfrak{A}_t). \end{aligned}$$

□

Basic Questions

(1) For which models is each claim strongly hedgeable? This question leads to the notion of a complete market model.
(2) How is a meaningful price determination for non-hedgeable claims possible?

With the no-arbitrage principle, one obtains only an interval of compatible prices in general (not complete) models. We will describe this so-called no-arbitrage interval in more detail below.

Proposition 5.23 (Trivial Price Bounds) *Let $\mathcal{M}_e(P) \neq \emptyset$, $S \geq 0$, $K \geq 0$, and let $C := (S_T - K)_+$ be a European Call. Then, for any no-arbitrage price $p(C)$ in the sense of the generalized no-arbitrage principle, it holds:*

$$(S_0 - K)_+ \leq p(C) \leq S_0.$$

Proof If $p(C) > S_0$ then consider the following strategy:

(1) At time $t = 0$ buy the stock,
(2) shortselling of option C and
(3) investing the difference risk-free in bonds.

At time $t = T$ the value of the stock is $S_T \geq (S_T - K)_+$. The profit is then > 0 because of the risk-free investment of the difference in bonds.

If $0 \leq p(C) < (S_0 - K)_+$, then an analogous argument yields a risk-free profit again. □

Remark 5.24 **Forwards** are OTC contracts (OTC—over the counter), i.e. contracts between parties that only meet when the contract is concluded. There is no more trading in the further course until maturity. There is no no-arbitrage argument with trading strategies for the valuation of forwards. However, the extended no-arbitrage principle can be used for valuation in an analogous manner.

Example Two parties, one receiving at time T the underlying S^1 the other receives at time T the forward price K. Let K be such that the contract costs nothing at time 0.

The payoff at maturity is $C := S^1_T - K$ at time T from the perspective of party 1.

Financing occurs through the underlying security and a bond maturing in T, i.e., a security S^0, which at time T pays out the value 1 (a freely traded **zero coupon bond**).

The constant portfolio $\varphi = (\varphi^0, \varphi^1) = (-K, 1)$ has at maturity the value C. The initial value of this portfolio is $V_0(\Phi) = -K S^0_0 + S^1_0$. From the extended no-arbitrage principle, it follows for the 'fair price':

$$V_0(\Phi) = 0 \text{ or equivalently: } K = \frac{S^1_0}{S^0_0} \text{ is the unambiguous fair}$$

$$\text{forward price of the base paper } S^1.$$

For the Black–Scholes model, all options can be uniquely priced by the no-arbitrage price.

Theorem 5.25 *In the Black–Scholes model with interest rate r and with equivalent martingale measure Q it holds:*

(a) Each claim $C \in \mathcal{L}^1(\mathfrak{A}_T, Q)$ *is strongly hedgeable.*
(b) $\forall\, C \in \mathcal{L}^2(\mathfrak{A}_T, Q)$ *there exists exactly one martingale hedge of* C.

Proof The discounted price process $\widetilde{S}$ has the representation

$$\widetilde{S}_t = e^{-rt} S_t = S_0 e^{\sigma \overline{B}_t - \frac{\sigma^2}{2} t},$$

where $\overline{B}$ is a Brownian motion with respect to Q and $\mathfrak{A}^B = \mathfrak{A}^{\overline{B}}$. For $C \in \mathcal{L}^1(\mathfrak{A}_T, Q)$ and with $\widetilde{C}_t = E_Q(\widetilde{C}|\mathfrak{A}_t)$, the martingale generated by $\widetilde{C} = e^{-rT} C$, then it holds by the martingale representation theorem: $\exists\, f_s \in \mathcal{L}^0(\overline{B})$ $(\exists! f \in \mathcal{L}^2(\overline{B})$, if $C \in \mathcal{L}^2(Q))$ such that

$$\widetilde{C}_t = E_Q \widetilde{C} + \int_0^t f_s \, \mathrm{d}\overline{B}_s.$$

It holds $d\widetilde{S}_t = \sigma \widetilde{S}_t d\overline{B}_t$ since $\widetilde{S}$ is exponential martingale with respect to Q.

We define:

$$\varphi_t := \frac{f_t}{\sigma \widetilde{S}_t}, \quad \varphi_t^0 := E_Q \widetilde{C} + \int_0^t f_s \, \mathrm{d}\overline{B}_s - \varphi_t \widetilde{S}_t, \quad \text{and} \quad \Phi := (\varphi^0, \varphi).$$

Then $\Phi \in \Pi_a(Q)$ is an admissible hedge of C, because:

(1) $\widetilde{V}_0(\Phi) = \varphi_0^0 + \varphi_0 \widetilde{S}_0 = E_Q \widetilde{C}$ according to the definition of φ^0.

(2) $\widetilde{V}_t(\Phi) = \widetilde{C}_t \in \mathcal{M}(Q)$ because

$$\begin{aligned} \widetilde{V}_t &= e^{-rt} V_t(\Phi) = e^{-rt}(\varphi_t^0 e^{rt} + \varphi_t S_t) = \varphi_t^0 + \varphi_t \widetilde{S}_t \\ &= E_Q \widetilde{C} + \int_0^t f_s \, \mathrm{d}\overline{B}_s - \varphi_t \widetilde{S}_t + \varphi_t \widetilde{S}_t = E_Q \widetilde{C} + \int_0^t f_s \, \mathrm{d}\overline{B}_s \\ &= \widetilde{C}_t \in \mathcal{M}_c(Q), \text{ and } \widetilde{V}_t \in \mathcal{M}_c^2(Q), \text{ if } C \in \mathcal{L}^2(\mathfrak{A}_T, Q). \end{aligned}$$

$\widetilde{V}(\Phi)$ is, therefore, a continuous martingale.

(3) Φ is regular, because $\widetilde{V}_t(\Phi) = \widetilde{C}_t \in \mathcal{M}(Q)$ for $C \in \mathcal{L}^1(\mathfrak{A}_T, Q)$.

(4) Φ is self-financing, because with the help of the above representation of $\widetilde{V}$ it follows from the Doléans–Dade equation for $\widetilde{S}$

$$\begin{aligned} \widetilde{V}_t(\Phi) &= V_0(\Phi) + \int_0^t f_s \, \mathrm{d}\overline{B}_s \\ &= V_0(\Phi) + \int_0^t \frac{f_s}{\sigma \widetilde{S}_s} \underbrace{\sigma \widetilde{S}_s \, \mathrm{d}\overline{B}_s}_{\mathrm{d}\widetilde{S}_s} \\ &= V_0(\Phi) + \int_0^t \varphi_s \, \mathrm{d}\widetilde{S}_s. \end{aligned}$$

So Φ is according to Lemma 5.2 self-financing.

(5) Φ is a martingale hedge for C, because $\widetilde{V}_T(\Phi) = \widetilde{C}_T = \widetilde{C}$ and, therefore, it follows $V_T = C$.
From (1) to (5) it follows: $\Phi \in \Pi_a(Q)$.

The uniqueness statement in (b) follows from the corresponding uniqueness statement for a Brownian motion. □

As a consequence of the above theorem, we obtain: Every contingent claim $C \in \mathcal{L}^1(Q)$ in the Black–Scholes model is strongly hedgeable. This statement applies analogously to claims $C \geq c$ bounded from below.

In particular, it can be applied to the example of the "**European Call**".

Let $C = (S_T - K)_+$ be a European Call with strike K and term T. As a consequence of the risk-neutral valuation formula in Proposition 5.21 we obtain the classical Black–Scholes formula for the valuation of a European Call option.

Theorem 5.26 (Black–Scholes Formula) *The no-arbitrage price of a European Call $C = (S_T - K)_+$ with strike K and running time T is given by*

$$\begin{aligned} p(C) &= E_Q e^{-rT} C \\ &= S_0 \Phi(d_1) - K e^{-rT} \Phi(d_2) =: p(S_0, T, K, \sigma). \end{aligned}$$

Here Φ is the distribution function of the standard normal distribution $N(0, 1)$,

$$d_2 := \frac{\log \frac{S_0}{K} + (r - \frac{\sigma^2}{2})}{T\sigma\sqrt{T}} \quad \text{and } d_1 := d_2 + \sigma\sqrt{T}.$$

Proof Using the risk-neutral valuation formula in Theorem 5.21 the following holds for the no-arbitrage price $p\ (C)$

$$\begin{aligned} p(C) &= E_Q e^{-rT}(S_T - K)_+ = E_Q e^{-rT}(S_0 e^{\sigma \overline{B}_T + (r - \frac{\sigma^2}{2})T} - K)_+ \\ &= E_Q e^{-rT}(S_0 e^Z - K)_+ \quad \text{with } Z := \sigma \overline{B}_T + \Big(r - \frac{\sigma^2}{2}\Big)T. \end{aligned}$$

Z is normally distributed, $Q^Z = N((r - \frac{\sigma^2}{2})T, \sigma^2 T)$.

In general, for a normally distributed random variable $Z \sim N(a, \gamma^2)$ the following formula holds:

$$E(be^Z - c)_+ = be^{a+\frac{\gamma^2}{2}} \Phi\left(\frac{\log \frac{b}{c} + a + \gamma^2}{\gamma}\right) - c\Phi\left(\frac{\log \frac{b}{c} + a}{\gamma}\right).$$

From this, by specialization, the assertion follows. □

Remark 5.27

(a) Analogously it follows for the price process: $p(C, t) = p(S_t, T - t, K, \sigma)$. For the proof, substitute in the above formula S_0 with S_t, T by the residual time $T - t$.

(b) **Approximation formula** For the calculation of the Black–Scholes formula, in practice a (very good) approximation of the distribution function Φ is used: $\Phi(x) \approx 1 - \varphi(x)(a_1 k + a_2 k^2 + \cdots + a_5 k^5)$, where $k := \frac{1}{1+\gamma x}$, $\gamma = 0.23167$, $a_1 = 0.31538$, $a_2 = \ldots$

This approximation yields six-digit precision for the price p.

Using the following call-put parity yields the no-arbitrage price for the European Put option.

Proposition 5.28 (Call-Put Parity) *Let $c_T := p((S_T - K)_+)$ and $p_T := p((K - S_T)_+)$ be the prices of the European Call and Put. Then it holds:*

$$p_T = c_T - S_0 + K e^{-rT}.$$

Proof It holds:

$$(K - S_T)_+ = (S_T - K)_+ - S_T + K.$$

Therefore, it follows from the no-arbitrage valuation formula with the equivalent martingale measure Q

$$\begin{aligned} p_T &= p((K - S_T)_+) \\ &= E_Q e^{-rT}(K - S_T)_+ \\ &= E_Q e^{-rT}(S_T - K)_+ - \underbrace{E_Q e^{-rT} S_T}_{S_0} + K e^{-rT} \\ &= c_T - S_0 + K e^{-rT}. \end{aligned}$$

□

5.1.2 Discussion of the Black–Scholes Formula

(a) **Dividend payments**

A rule of thumb for price changes in dividend payments is: Following a dividend payment of size d causes the share price to fall by 80% of the dividend paid, i.e., by 0.8 d.

For valuation purposes, the stock price can be decomposed into two components.

Into a risk-free component, which is equal to the sum of the discounted dividend payments multiplied by 0.8; and into the remaining risk component, which is valued by the Black–Scholes formula.

(b) Hedging portfolio—hedge strategy

Let $p(x, t, K, \sigma)$ be the price of a call $C = (S_T - K)_+$ with market value x, remaining (residual) time t and strike K. p depends only on the volatility parameter σ but is independent of the drift parameter μ of the model. For $t \to 0$, i.e., the remaining term approaches zero, the following applies

$$p(x, t, K, \sigma) \to (x - K)_+.$$

The Black–Scholes formula

$$p(x, t, K, \sigma) = x\Phi(d_1) - e^{-rt} K\Phi(d_2)$$

can also be interpreted as a **hedging portfolio** : The first component $x\Phi(d_1)$ corresponds to the value of a portfolio with $\Phi(d_1)$ shares in stocks at the time t, the second component corresponds to a short position of $K\Phi(d_2)$ shares in bonds, i.e. $(\Phi(d_1), -K\Phi(d_2))$ is a hedging portfolio for the option C with residual maturity t. It remains to show that these strategies are self-financing, i.e. $\varphi_t^0 = \Phi(d_1)$, $\varphi_t = -K\Phi(d_2)e^{-r(T-t)}$ defines a hedging strategy for the call C.

(c) The 'Greeks'

In the following, we discuss the dependence of the Black–Scholes price $p(x, t, K, \sigma)$ on the parameters x = market value, t = residual maturity, K = strike and σ = volatility. These relationships have names with Greek letters and are called the 'Greeks'.

Useful for this is the following relation for the density φ of the standard normal distribution. It is valid with $d_i = d_i(x, t, K, \sigma)$ according to the Black–Scholes formula:

$$x\varphi(d_1) - Ke^{-rt}\varphi(d_2) = 0. \tag{5.3}$$

(c1) From (5.3) and $\frac{\partial d_1}{\partial x} = \frac{\partial d_2}{\partial x}$ it follows

$$\frac{\partial p}{\partial x} = \Phi(d_1) + \underbrace{\left(x\varphi(d_1) - Ke^{-rtT}\varphi(d_2)\right)}_{=0} \frac{\partial d_1}{\partial x} = \Phi(d_1) > 0. \tag{5.4}$$

So p increasing in the spot price x.

$\Delta := \frac{\partial p}{\partial x}$ is the **Delta of the option**. Δ describes the share of equities in the hedging portfolio with residual maturity t (Delta-hedging).

(c2) $\Gamma := \frac{\partial^2 p}{\partial x^2} = \frac{\partial \Delta}{\partial x} = \varphi(d_1)\frac{\partial d_1}{\partial x} > 0$ is the **Gamma of the option** and describes the sensitivity in the hedging portfolio as a function of the market value x. The Black–Scholes price p is strictly convex in x.

(c3)

$$\frac{\partial p}{\partial t} = Ke^{-rt}\Big(r\Phi(d_2) + \frac{\sigma\varphi(d_2)}{2\sqrt{t}}\Big) > 0. \tag{5.5}$$

The price p is monotonically increasing in the residual time t.

(c4) $\Lambda = \frac{\partial p}{\partial \sigma} = x\varphi(d_1)\sqrt{t} > 0$ is the **Lambda of the option**. The price p is increasing in the volatility σ.

(c5)

$$\frac{\partial p}{\partial r} = Kte^{-rt}\Phi(d_2) > 0. \tag{5.6}$$

p is monotonically increasing in the rate of interest r.

(c6)

$$\frac{\partial p}{\partial K} = -e^{-rt}\Phi(d_2) < 0, \tag{5.7}$$

p is monotonically decreasing in the strike K.

(d) The volatility σ is to be estimated from market data. To do this, one uses historical data, such as the data of the last 30–180 trading days. Now, let $\overline{\sigma}$ be the implied volatility, i.e., the solution of the equation

$$p(x, t, K, \overline{\sigma}) = \overline{p}$$

where $\overline{p}$ denotes the current market price. Then, if the Black–Scholes model correctly describes the market, it should be that $\overline{\sigma}$ is independent of the strike K of the traded options. However, in empirical data one actually observes a dependence in the form of a 'smile' (cf. Fig. 5.2), an indication that the Black–Scholes model does not describe stock prices correctly, in particular that it underestimates tail probabilities:

For $x \approx K$ (at the money) is $\overline{\sigma} < \sigma$, for $x \gg K$ and $x \ll K$ (out of the money) is $\overline{\sigma} > \sigma$.

That is, stocks with a strike $K \gg x$ or $K \ll x$ are more expensive than indicated by the Black–Scholes model. The BS model underestimates the large jumps in prices, i.e., the tails of the distributions.

(e) Examples of options

The basic options (vanilla options) are the European Call and Put. With the help of these basic options, a wealth of other options (derivatives) can be constructed and valued by simple combinations.

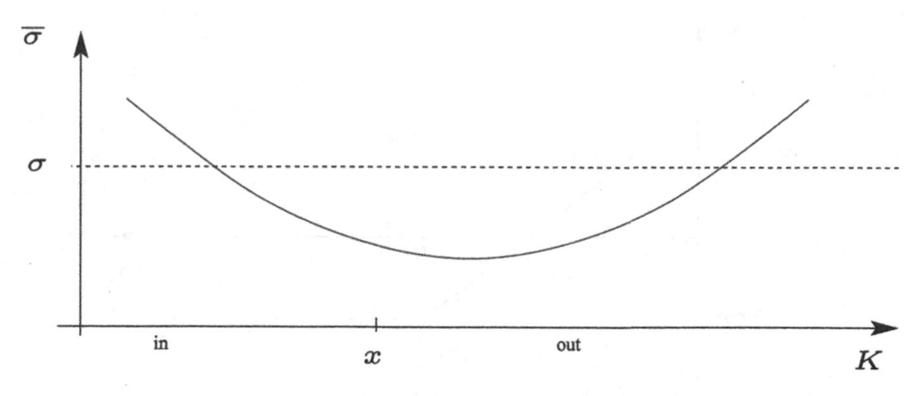

Fig. 5.2 Smile effect

(e1) **European Call:** $(S_T - K)_+ = f(S_T)$

Let $c = c_T$ be the no-arbitrage price, then the buyer's profit is $V_T = f(S_T) - c_T$ as in the adjacent form. Thus, a call provides a profit if the price of the underlying security rises.

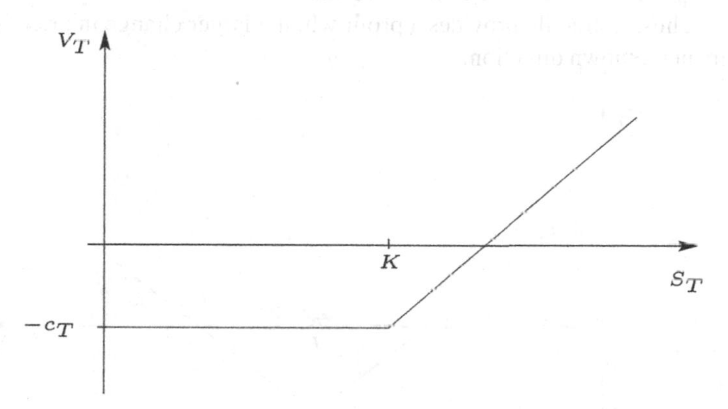

(e2) **European Put:** $f(S_T) = (K - S_T)_+$

$p_T = c_T - S_0 + Ke^{-rT}$ is the price of the European Put. Thus the profit of the buyer is

$$V_T = f(S_T) - p_T$$

as in the adjacent form. A put thus yields a profit if the price of the underlying security falls.

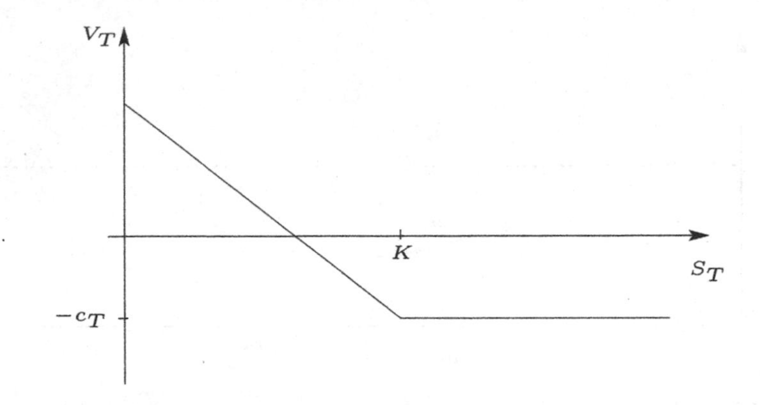

(e3) A **straddle** is a combination of a call and a put with the same strike K. The cost of the straddle is $c = c_T + p_T$ and with

$$f(S_T) = |S_T - K|$$

the profit is V_T is of the adjacent form.

Thus, a straddle provides a profit when a larger change in price is expected in an unknown direction.

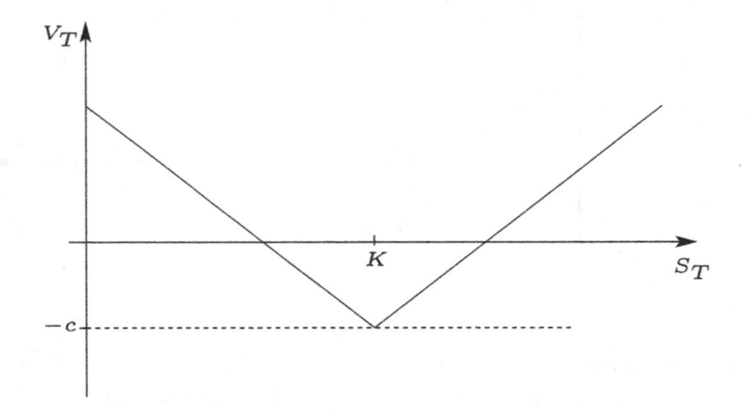

(e4) A **strangle** is a combination of a call and a put with different strikes. This leads to the profit

$$V_T = (S_T - K_2)1_{(S_T > K_2)} \\ + (K_1 - S_T)1_{(S_T < K_1)} - c.$$

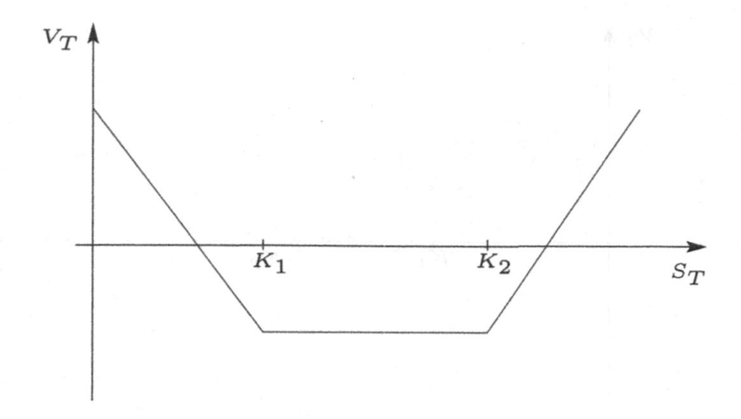

(e5) A **strap** is a combination of two calls with one put. This allows the slope of the profit to be increased in the event of an expected price rise, while also providing a hedge in the event of a price fall. The profit has the form

$$V_T = 2(S_T - K)_+ + (K - S_T)_+ - c.$$

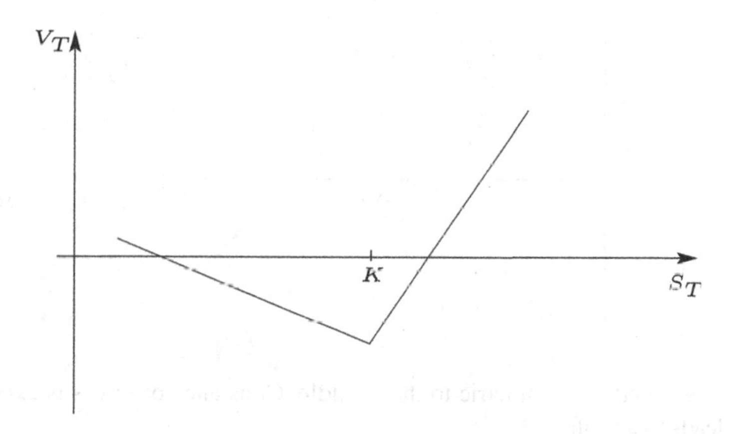

(e6) A **bull spread** is the purchase of a call option with strike K_1 and the sale of a call option with strike $K_2 > K_1$. This leads to a profit profile of the form

$$V_T = (K_2 - K_1)1_{(S_T \geq K_2)} + (S_T - K_1)1_{(K_1 < S_T < K_2)} - c.$$

A rise is expected, losses (and also profits) are bounded.

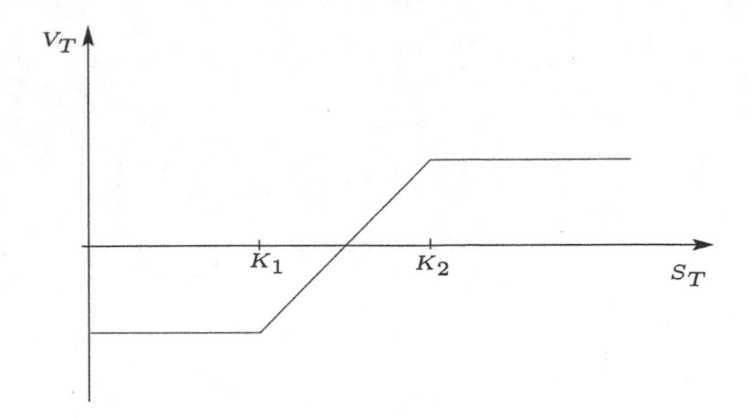

(e7) A **bear spread** is symmetrical to the bull spread is the purchase of a call option with strike K_2 and selling a call option with strike $K_1 < K_2$. A fall in prices is expected, losses are limited.

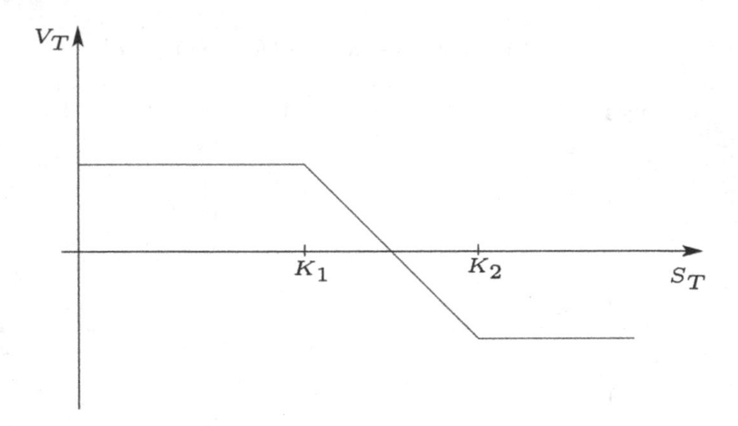

(e8) A **butterfly** is symmetric to the straddle. Constancy of prices is expected and leads to a profit.

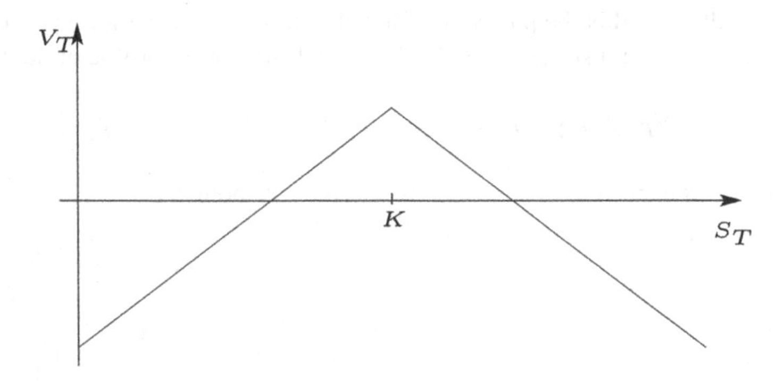

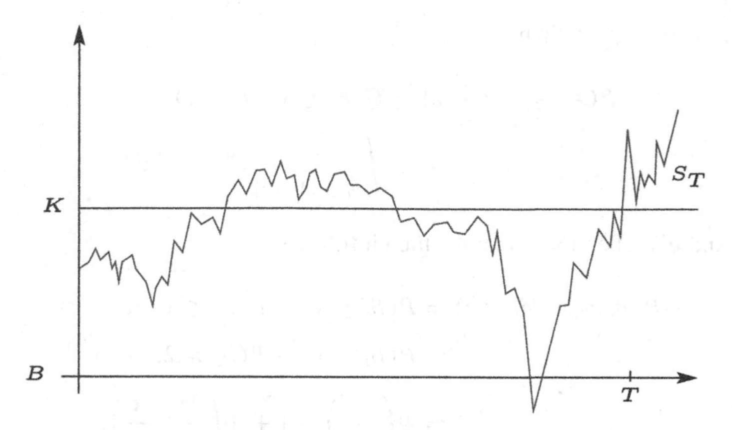

Fig. 5.3 Barrier option

Example 5.29 (**Barrier Option**) A barrier option (Fig. 5.3) expires when the price falls below a barrier B,

$$C_B := (S_T - K)_+ 1_{\{\inf_{0\le t\le T} S_t > B\}}.$$

To calculate the no-arbitrage price of the barrier option C_B, i.e., $E_Q e^{-rT} C_B$ we need the following proposition about the joint distribution of a Brownian motion and its supremum.

Proposition 5.30 *Let X be a Brownian motion with $\sigma = 1$ and drift a and let $Z_t := \sup_{0\le s\le t} X_s$. Then*

$$P(X_t \le x, Z_t < z) = \Phi\left(\frac{x - at}{\sqrt{t}}\right) - e^{2az}\varphi\left(\frac{x - 2z - at}{\sqrt{t}}\right)$$

Proof Let B be a standard Brownian motion with respect to P. By the stochastic exponential

$$L_t := e^{aB_t - \frac{a^2}{2}t}$$

and the measure transformation, $\frac{dQ}{dP} := L_t$, then it follows according to Girsanov's theorem

$$(B_s)_{0\le s\le t} \text{ is w.r.t. } Q \text{ a Brownian motion with drift } a \text{ and volatility } 1.$$

With $M_t := \sup_{s \le t} B_t$ then

$$P(X_t \le x, Z < z) = Q(B_t \le x, M_t < z)$$
$$= \int_{\{B_t \le x, M_t < z\}} e^{aB_t - \frac{a^2}{p}t} \, dP.$$

Using André's reflection principle, then it follows:

$$\begin{aligned} P(B_t \le x, M_t < z) &= P(B_t \le x) - P(B_t \le x, M_t \ge z) \\ &= P(B_t \le x) - P(B_t > 2z - x) \\ &= \Phi\Big(\frac{x}{\sqrt{t}}\Big) - 1 + \Phi\Big(\frac{2z - x}{\sqrt{t}}\Big). \end{aligned}$$

From this it follows:

$$P(B_t \le x | M_t < z) = \begin{cases} 1, & x > z, \\ \dfrac{\Phi(\frac{x}{\sqrt{t}}) - 1 + \Phi\left(\frac{2z-x}{\sqrt{t}}\right)}{P(M_t < z)}, & x \le z. \end{cases}$$

This conditional distribution has the density

$$h(y) = \frac{1}{\sqrt{t} P(M_t < z)} \Big(\varphi\Big(\frac{y}{\sqrt{t}}\Big) - \varphi\Big(\frac{2z - y}{\sqrt{t}}\Big)\Big), \quad y \le z.$$

This gives

$$P(X_t \le x, Z_t < z) = \int f(y) h(y) \, dy \, P(M_t < z)$$

with

$$\begin{aligned} f(y) &= \mathbb{1}_{(-\infty, x]}(y) e^{ay - \frac{a^2}{2}t} \\ &= \int_{-\infty}^{z} \frac{1}{\sqrt{t}} \Big(\varphi\Big(\frac{y}{\sqrt{t}}\Big) - \varphi\Big(\frac{2z - y}{\sqrt{t}}\Big)\Big) \mathbb{1}_{(-\infty, x]}(y) e^{-ay - \frac{a^2}{2}t} \, dy \\ &= \int_{-\infty}^{z} e^{ay - \frac{a^2}{2}t} \frac{1}{\sqrt{2\pi t}} \Big(e^{-\frac{y^2}{2t}} - e \frac{(2z - y)^2}{2t} \Big) \, dy \\ &= \Phi\Big(\frac{x - at}{\sqrt{t}}\Big) - e^{2az} \Phi\Big(\frac{x - 2z - at}{\sqrt{t}}\Big). \end{aligned}$$

□

Theorem 5.31 (Price of a Barrier Option) *The no-arbitrage price of the barrier option*

$$C_B = (S_T - K)^+ \mathbb{1}_{\{\inf_{0\le t\le T} S_t > B\}}$$

in the Black–Scholes model is

$$p(C_B) = E_Q e^{-rT} C_B = p(S_0, T, K, \sigma) - \frac{e^{2\alpha\beta}}{\gamma} p(S_0, T, \gamma K, \sigma).$$

Here p is the Black–Scholes price of the European Call, with $\alpha := \frac{\sigma}{2} - \frac{r}{\sigma}$, $\beta := \log\left(\frac{S_0}{B}\right)\frac{1}{\sigma}$, and $\gamma := \left(\frac{S_0}{B}\right)^2$.

Proof It is $S_t = S_0 e^{\sigma \overline{B}_t + \left(r - \frac{\sigma^2}{2}\right)t}$ with respect to Q, with a Brownian motion $(\overline{B}, Q)$. Thus it follows by Proposition 5.30

$$\begin{aligned}
&E_Q \underbrace{e^{-rT}(S_T - K)_+}_{=:f(S_T)} \mathbb{1}_{\{\inf_{t\le T} S_t > B\}} \\
&\quad = \int_{\{S_T > B\}} f(S_T)\mathrm{d}Q - e^{2\alpha\beta} \int_{\{S_T > B\}} f\left(\frac{S_T}{\gamma}\right)\mathrm{d}Q \\
&\quad = E_Q e^{-rT}(S_T - K)^+ - \frac{e^{2\alpha\beta}}{\gamma} E_Q e^{-rT}(S_T - \gamma K)_+ \\
&\quad = p(S_0, T, K, \sigma) - \frac{e^{2\alpha\beta}}{\gamma} p(S_0, T, \gamma K, \sigma).
\end{aligned}$$

For the second equality it is used that $Q(S_t \ge x, \inf_{s\le t} S_s > B) = Q(S_t \ge x) - e^{2\alpha\beta} Q(\frac{S_t}{\gamma} \ge x)$. □

Example 5.32 **(American Option)** Execution at stopping times $0 \le \tau \le T$.

Let $Z = (Z_t)$ be an adapted process and let Z_t be the payoff at time t. The payoffs available are $C(Z, \tau) := Z_\tau$ with the value $p(C(Z, \tau))$. The optimal payoff is given by

$$\begin{aligned}
p(Z) &:= \sup_{\substack{\tau \le T, \\ \tau \text{ a stopping time}}} p(C(Z, \tau)) \\
&:= \sup_{\tau} E_Q e^{-r\tau} Z_\tau.
\end{aligned}$$

Thus, determining the American Call leads to an optimal stopping problem.

Proposition 5.33 (American Call) *The no-arbitrage price of the American Call option is identical to the price of the European Call. The following, therefore, holds*

$$\sup_{\tau \leq T} E_Q e^{-r\tau}(S_\tau - K)^+ = E_Q e^{-rT}(S_T - K)^+$$
$$= p(S_t, T, K, \sigma).$$

Proof We show that $e^{-rt}(S_t - K)_+$ a submartingale with respect to Q. The function $f(x) = (x - K)_+$ is convex. So according to the Jensen inequality it follows

$$\begin{aligned} E_Q(e^{-rt}(S_t - K)_+ \mid \mathfrak{A}_s) &\geq e^{-rt}(E_Q(S_t \mid \mathfrak{A}_s) - K)_+ \\ &= (E_Q(\widetilde{S}_t \mid \mathfrak{A}_s) - e^{-rt}K)_+ \\ &= (\widetilde{S}_s - e^{-rt}K)_+ \geq e^{-rs}(S_s - K)_+, \end{aligned}$$

since $(x - \alpha K)_+ \geq (x - K)_+$ for $\alpha \leq 1$, $K \geq 0$.

It follows that $e^{-rt}(S_t - K)_+$ is a submartingale. For all stopping times $\tau \leq T$ the following applies according to the optional sampling theorem

$$E_Q e^{-r\tau}(S_\tau - K)_+ \leq E_Q e^{-rT}(S_T - K)_+.$$

The price of the American Call is, therefore, identical to the price of the European Call. □

For the American Put, however, the price is different from that of the European Put. To get an idea of the price of an American (or European) Put, consider an example from Geske and Johnson (1984) with period $T = 7$ months, $S_0 = 40$ €, the strike $K = 45$, the interest rate $r = 4.88\%$ and the volatility $\sigma = 0.3$. In this case, the Black–Scholes formula yields the price of the European Put $p^E(C) = 5.97$ €, whereas the price of the American Put is $p^A(C) = 6.23$ €.

5.1.3 Hedging Strategies and Partial Differential Equations

For the European Call, a hedging portfolio for the call had resulted as a consequence of the Black–Scholes formula:

$$p(x, t, K, \sigma) = x \underbrace{\Phi(d_1(x, t, K, \sigma))}_{=:h(x,t)=\frac{\partial p}{\partial x}=\Delta} - e^{-rt} \underbrace{K\Phi(d_2(x, t, K, \sigma))}_{=:g(x,t)}.$$

Here, t is the residual time (to be replaced by $T - t$ for the price at time t). $h(x, t) = \frac{\partial p}{\partial x} = \Delta$ the delta of the option indicates the proportion of shares in the hedging portfolio. In the following proposition, we give general conditions for a trading strategy $(h(S_t, t), g(S_t, t))$ to be self-financing.

Proposition 5.34 *In the Black–Scholes model let $h, g \in C^{2,1}$ be solutions of the differential equation*

$$\begin{cases} xh_x + e^{rt} g_x &= 0 \\ \frac{1}{2}\sigma^2 x^2 h_x + xh_t + e^{rt} g_t &= 0. \end{cases} \tag{5.8}$$

Let $\varphi := h(S_t, t)$, $\varphi_t^0 := g(S_t, t)$. Then the trading strategy is $\Phi = (\varphi^0, \varphi)$ is self-financing, i.e. $\Phi \in \Pi$.

Proof Define $f(x, t) = g(x, t)e^{rt} + xh(x, t)$, then

$$V_t = V_t(\Phi) = f(S_t, t).$$

Using the Itô formula, it follows

$$\begin{aligned} \mathrm{d}V_t &= f_x(S_t, t)\mathrm{d}S_t + f_t(S_t, t)\,\mathrm{d}t + \frac{1}{2} f_{xx}(S_t, t)\,\mathrm{d}\langle S\rangle_t \\ &= f_x(S_t, t)\,\mathrm{d}S_t + \left(f_t(S_t, t) + \frac{1}{2} f_{xx}(S_t, t)\sigma^2 S_t^2\right)\mathrm{d}t. \end{aligned}$$

To be shown is the self-financing condition:

$$\mathrm{d}V_t = h(S_t, t)\,\mathrm{d}S_t + re^{rt} g(S_t, t)\,\mathrm{d}t.$$

So to show is: $f_x = h$, $f_t + \frac{1}{2} f_{xx}\sigma^2 x^2 = re^{rt} g$.

(1) $f_x = h + \underbrace{xh_x + e^{rt} g_x}_{=0} = h$ using (5.8).

(2) $$\begin{aligned} f_t + \tfrac{1}{2} f_{xx}\sigma^2 x^2 &= xh_t + g_t e^{rt} + rge^{rt} + \tfrac{1}{2}\sigma^2 x^2 h_x && \text{by (1)} \\ &= rge^{rt} && \text{by assumption.} \end{aligned}$$

It follows that the trading strategy Φ is self-financing. □

Remark 5.35 The above conditions in (5.8) an g, h hold for the hedging portfolio in the Black–Scholes model for European options. It follows that: The delta hedge strategy is self-financing and hence an admissible hedge strategy.

More generally, the following theorem for options of type $C = f(S_T, T)$ has an analytical approach to determine self-financing trading strategies.

Theorem 5.36 (Black–Scholes Differential Equation) *In the Black–Scholes model let $f \in C^{2,1}$ be a solution of the Black–Scholes differential equation*

$$\frac{1}{2}\sigma^2 x^2 f_{xx} + rxf_x + f_t - rf = 0. \tag{5.9}$$

Let $g := e^{-rt}(f - xf_x)$, $h := f_x$, then

$$\Phi = (\varphi^0, \varphi) \quad \text{with } \varphi_t^0 := g(S_t, t), \varphi_t := h(S_t, t),$$

self-financing with value process $V_t = f(S_t, t)$.

Proof The value process $V_t(\Phi)$ has the form

$$\begin{aligned} V_t(\Phi) &= g(S_t, t)e^{rt} + h(S_t, t)S_t \\ &= (f(S_t, t) - S_t f_x(S_t, t)) + f_x(S_t, t)S_t \\ &= f(S_t, t). \end{aligned}$$

To prove $\Phi \in \Pi$ we verify the conditions from Proposition 5.34 to:

$$xh_x + e^{rt}g_x = xf_{xx} + e^{rt}(e^{-rt}(f_x - f_x - xf_{xx})) = 0$$

and

$$\begin{aligned} &\frac{1}{2}\sigma^2 x^2 h_x + xh_t + e^{rt}g_t \\ &\quad = \frac{1}{2}\sigma^2 x^2 f_{xx} + xf_{xt} + e^{rt}(-r)e^{-rt}(f - xf_x) + e^{rt}e^{-rt}(f_t - xf_{xt}) \\ &\quad = \frac{1}{2}\sigma^2 x^2 f_{xx} + xf_{xt} - rf + rxf_x + f_t - xf_{xt} \\ &\quad = \frac{1}{2}\sigma^2 x f_{xx} - rf + rxf_x + f_t = 0 \end{aligned}$$

by the Black–Scholes differential equation.

□

Consequence
In order to evaluate and hedge an option $C = f(S_T, T)$, the following steps are necessary:

(1) Solve the Black–Scholes differential equation (5.9) with boundary condition $f(x, T)$.

This is a Cauchy boundary value problem. Then the no-arbitrage price of C is:

$$p(C) = f(S_0, 0) \text{ the no-arbitrage price}$$

(2) Determine the hedge strategy: $\varphi_t^0 = g(S_t, t)$, $\varphi_t = h(S_t, t)$ with $g = e^{-rt}(f - xf_x)$, $h = f_x$ from Theorem 5.36. For the explicit solution of the Black–Scholes equation in the case of a European Call $C = (x - K)_+ = f(x, T)$ the Black–Scholes equation is transformed into the heat equation. This can then be solved explicitly.

Remark 5.37

(1) The above PDE approach for the Black–Scholes formula was used in the original Black–Scholes paper (1973). The elegant martingale argument goes back to the work of Harrison and Pliska (1981). The no-arbitrage argument for determining prices was introduced by Merton (1973). Shortly after the Black–Scholes work (since 1973), options were traded on the Chicago Board of Trade (CBO).
(2) Let now be more general $C = f(X_T, T)$ an option with value process $V_T = E_Q(C \mid \mathfrak{A}_t)$ and let (X, Q) be a diffusion process with infinitesimal producer A.

Let h be a solution of the boundary value problem

$$\left(\frac{\partial}{\partial t} + A\right)h = 0, \quad h(\cdot, T) = f(\cdot, T), \tag{5.10}$$

then it follows with the Itô formula:

$$h(X_t, t) = f(X_0, 0) + \int_0^t h_x(X_s, s)\,\mathrm{d}X_s.$$

From this, analogously to Theorem 5.36:
$\varphi_s^C = h_x(X_s, s)$ is a self-financing hedge strategy for $C = f(X_T, T)$. The corresponding value process is $V_s = h(X_s, s)$.

5.2 Complete and Incomplete Markets

We consider a market model $S = (S^0, S^1, \dots, S^k)$ with a bond S^0 continuous, increasing, $S_t^0 > 0$, $S_0^0 = 1$ and with semimartingales S_t^i. Let $(\mathfrak{A}_t) = (\mathfrak{A}_t^S)$ be the filtration $(S^0, S^1, \dots, S^k)$ and let $\widetilde{S}_t^i = \frac{S_t^i}{S_t^0}$ be the discounted stocks.

Let $\vartheta = (x, \pi)$ be a trading strategy with initial capital x, $\pi = (\pi^0, \dots, \pi^k)$, π^i the shares of security i in the portfolio.

ϑ is called **self-financing**, if for the discounted value process $\widetilde{V}_t$ holds:

$$\widetilde{V}_t(\vartheta) = x + \sum_{i=1}^{k} \int_0^t \pi_u^i \, d\widetilde{S}_u^i.$$

We make the general assumption : $\mathcal{M}_e(P) \neq \emptyset$. This assumption implies the arbitrage-free nature of the market model.

Definition 5.38 The market model S is called **complete,** if for all $Q \in \mathcal{M}_e(P)$ and for all claims $C \in L^1(\mathfrak{A}_T, Q)$ a martingale hedge $\vartheta \in \Pi_a(C)$ exists, i.e.

$$\widetilde{V}_t(\vartheta) \in \mathcal{M}(Q) \text{ and } V_T(\vartheta) = C.$$

Proposition 5.39 *Let $S \in \mathcal{S}^k$ be a semimartingale model, then it holds:*
S is complete

$\iff$ *$\forall\, Q \in \mathcal{M}_e(P)$, $\forall\, M \in \mathcal{M}(Q)$ there exists a representation of the form* $M_t = E_Q M_T + \sum_{i=1}^k \int_0^t \varphi_u^i \, d\widetilde{S}_u^i$, $\varphi \in \mathfrak{L}^0(\widetilde{S})$.

$\iff$ *$\forall\, Q \in \mathcal{M}_e(P)$, $\forall\, F \in L(\mathfrak{A}_T, Q)$ there exists an integral representation* $\widetilde{F} = E_Q F + \sum_{i=1}^k \int_0^T \varphi_u^i d\widetilde{S}_u^i$, $\varphi \in \mathfrak{L}^0(\widetilde{S})$.

Proof The proof is similar to the proof of Theorem 4.27.

For $F \in L(Q, \mathfrak{A}_t)$ let M be the corresponding generated martingale, $M_t := E_Q(\widetilde{F} \mid \mathfrak{A}_t)$. If F is bounded, then the hedge property $V_T(\vartheta) = F$ is equivalent to $\widetilde{V}_T(\vartheta) = \widetilde{F}$. Because of the denseness of $B(\mathfrak{A}_T)$ in $L^1(\mathfrak{A}_T, Q)$ it follows by approximation the representability for $F \in L^1(\mathfrak{A}_T, Q)$ from the bounded case. The approximation argument uses stopping times and continuity. □

Definition 5.40 $C \in L(\mathfrak{A}_T)$ is a **contingent claim,** if $C \geq 0$ and $\exists\, Q \in \mathcal{M}_e(P)$ such that $\widetilde{C} \in L^1(Q)$.

Corollary 5.41 *If S is complete and C is a contingent claim, then C is strongly hedgeable, i.e. $\exists\, Q \in \mathcal{M}_e(P)$, $\exists\, \pi \in \Pi_a(Q)$ so that $\widetilde{C} = \widetilde{V}(\pi)$.*

Proof Let M_t be the martingale generated by $\widetilde{C}$,

$$M_t := E_Q(\widetilde{C} \mid \mathfrak{A}_t), \quad M \in \mathcal{M}(Q).$$

According to Proposition 5.39 M has a stochastic integral representation

$$M_t = V_t(\pi) = V_0(\pi) + \int_0^t \pi_u \, d\widetilde{S}_u.$$

π is, therefore, a martingale hedge and π is regular, since $M_t \geq 0$. □

Remark 5.42 Thus, in complete markets, all contingent claims can be assigned a unique no-arbitrage price.

For $\widetilde{C} = x + \int_0^t \pi_u \, \mathrm{d}\widetilde{S}_u = \widetilde{V}_T(\pi)$ with $\widetilde{C} \in L^1(Q)$ holds:

$$p(C) = x = E_Q \widetilde{C}.$$

Example 5.43 **(Diffusion Model)** Let S be a multivariate diffusion model such that $\mathrm{d}S_t = S_t \diamond \big(b(t, S_t)\,\mathrm{d}t + \sigma(t, S_t)\,\mathrm{d}B_t\big)$, where $\diamond$ represents the component-wise product, $B = (B^1, \dots, B^k)$, is a *k-dimensional* Brownian motion and $S_t^0 = e^{rt}$. Then S has the explicit representation

$$S_t = S_0 \diamond \exp\left(\int_0^t \left(b_s - \frac{1}{2} \Big(\sum_j \sigma_{ij}^2(s)\Big) \right) \mathrm{d}s + \int_0^t \sigma_s \cdot \mathrm{d}B_s \right), \quad b_s = b(s, \cdot).$$

Assumption: Let $a(t, x) := \sigma(t, x) \cdot \sigma(t, x)^T$ be positive definite, let $R_T := \exp\big(\int_0^T f(u, \widetilde{S}_u) \cdot \mathrm{d}B_u - \frac{1}{2} \sum \int f_i^2(u, \widetilde{S}_u)\,\mathrm{d}u \big)$ be the exponential martingale with

$$f_i(u, \widetilde{S}_u) := \sum_{i=1}^k (\sigma^{-1})_{ij} (b_j(u, \widetilde{S}_u) - r). \tag{5.11}$$

Let $E R_T = 1$ and define $\frac{\mathrm{d}Q}{\mathrm{d}P} = R_T$. Then by Girsanov's theorem follows: $\overline{B}_t := B_t - \int_0^t f(u, \widetilde{S}_u)\,\mathrm{d}u$ is a Brownian motion with respect to Q. According to (5.11), the relation

$$b - r \cdot 1 = \sigma f$$

follows, and thus

$$\begin{aligned} \mathrm{d}\widetilde{S}_t &= \widetilde{S} \diamond \{(b - r \cdot 1 - \sigma f)\,\mathrm{d}t + \widetilde{\sigma}(t, \widetilde{S}_t) \cdot \mathrm{d}\overline{B}_t \\ &= \widetilde{S} \diamond \widetilde{\sigma}(t, \widetilde{S}_t) \cdot \mathrm{d}\overline{B}_t \qquad \text{with respect to } Q. \end{aligned}$$

So $\widetilde{S}$ is an exponential martingale with respect to Q, $\widetilde{S} \in \mathcal{M}(Q)$ and $\mathfrak{A}^{\widetilde{S}} = \mathfrak{A}^{\overline{B}}$ since σ is positive definite.

If C is a contingent claim, $C \in L^1(\mathfrak{A}_T, Q)$ and $M_t := E_Q(\widetilde{C} \mid \mathfrak{A}_t^{\widetilde{S}}) = E_Q(\widetilde{C} \mid \mathfrak{A}_t^{\overline{B}})$ the generated martingale, then it follows by according to the martingale representation theorem (for *k-dim* martingales)

$$M_t = E_Q \widetilde{C} + \int_0^t g_s \, \mathrm{d}\overline{B}_s \quad \text{with } g \in \mathfrak{L}^0(\overline{B})$$

and according to (5.11) it holds:

$$d\overline{B}_t^i = \sum_{j=1}^k (\sigma^{-1})_{ij} \frac{d\widetilde{S}_t^j}{\widetilde{S}_t^j}.$$

From this follows

$$M_t = E_Q\widetilde{C} + \sum_{i=1}^k \int_0^t \Phi_u^i \, d\widetilde{S}_u^j \quad with\ \Phi_u^i = \sum_{j=1}^k g_u^i (\sigma^{-1})_{ij} \frac{1}{S_u^j}.$$

This results in the representation

$$\widetilde{C} = M_T = \widetilde{V}_T(\vartheta) = x + \sum_j \int_0^T \Phi_u^j \, d\widetilde{S}_u^j$$

$$\text{with } \vartheta = (x, \Phi^1, \dots, \Phi^k),\ x = E_Q(\widetilde{C}).$$

So $\widetilde{C}$ is duplicable by a regular martingale hedge. Thus, the diffusion model is complete.

For semimartingale models, completeness is closely coupled to the existence of a unique equivalent martingale measure. The following theorem deals with the case of continuous models.

Theorem 5.44 (Second Fundamental Theorem of Asset Pricing, Completeness) *Let S be a continuous martingale model, $\mathcal{M}_e(P) \neq \emptyset$, and $S = (S^0, \dots, S^k)$ with a bond S^0, and k stocks $S^1, \dots, S^k$. Then it holds:*

$$S\ is\ complete \iff |\mathcal{M}_e(P)| = 1.$$

Proof "$\Longrightarrow$": Let $Q^0, Q^1 \in \mathcal{M}_e(P)$ be equivalent martingale measures, then $Q := \frac{1}{2}(Q^0 + Q^1) \ll Q^0$ and Q is an equivalent martingale measure, and $Q \in \mathcal{M}_e(P)$.
Let $R := \frac{dQ}{dQ^0}$, then the assertion is equivalent to: $R \equiv 1$.
Let $R_t := E_{Q^0}(R \mid \mathfrak{A}_t^S) \in \mathcal{M}(Q)$ be the martingale generated by R. Because of completeness and according to Proposition 5.4 it holds:

$$\exists \varphi \in L^0(\widetilde{S}) \text{ so that } R_t = 1 + \int_0^t \varphi_u \, d\widetilde{S}_u$$

By stopping with $\tau_m := \inf\{t \geq 0; |\widetilde{S}_t| \geq m\} \wedge T$ we can w.l.g. assume that $\widetilde{S}^i$ are martingales with respect to Q and Q^0, $(\widetilde{S}^i) \in \mathcal{M}(Q) \cap \mathcal{M}(Q^0)$.

From this follows

$$\forall B \in \mathfrak{A}_s^S \quad \text{and} \quad \forall i \text{ holds } E_Q \mathbb{1}_B \widetilde{S}_t^i = E_Q \mathbb{1}_B \widetilde{S}_s^i.$$

However, it is:

$$E_Q \mathbb{1}_B \widetilde{S}_t^i = E_{Q^0} \mathbb{1}_B \widetilde{S}_t^i R_t \quad \text{and} \quad E_Q \mathbb{1}_B \widetilde{S}_s^i = E_{Q^0} \mathbb{1}_B \widetilde{S}_s^i R_s.$$

Therefore, it follows:

$$E_{Q^0} \mathbb{1}_B \widetilde{S}_t^i R_t = E_{Q^0} \mathbb{1}_B \widetilde{S}_s^i R_s \text{ for } B \in \mathfrak{A}_s^S.$$

Therefore, $(\widetilde{S}_t^i R_t)$ is a martingale with respect to Q^0, $(\widetilde{S}_t^i R_t) \in \mathcal{M}(Q^0)$.

The martingales $\widetilde{S}^i$ and R are, therefore, orthogonal with respect to Q^0 or equivalently, the predictable quadratic covariation is zero, $\langle \widetilde{S}^i, R \rangle_t = 0, \forall t\ [Q^0]$. It follows that the quadratic variation of R is zero

$$\langle R \rangle_T = \sum_{i=1}^k \int_0^T \varphi_u^i \, \mathrm{d}\langle \widetilde{S}^i, R \rangle_u = 0, \text{ i.e. } R_t^2 \in \mathcal{M}(Q^0).$$

It follows that: $R_t = R_T = 1\ [Q^0]$, so also Q a.s. and P almost surely.

It follows

$$Q = RQ^0 = Q^0,$$

i.e. there exists only one equivalent martingale measure.

"$\Longleftarrow$" Suppose there exists only one equivalent martingale measure Q, i.e. $\mathcal{M}_e(P) = \{Q\}$.

To show: $\forall C \in B(\mathfrak{A}_T^S)$ with $|C| \leq K$ is $\widetilde{C}$ is representable by a martingale hedge.

For this let $M_t := E_Q(\widetilde{C} \mid \mathfrak{A}_t^S)$ be the martingale generated by $\widetilde{C}$ in $\mathcal{H}^2$.

By Kunita–Watanabe's theorem, M has a representation of the form

$$M_t = E_Q \widetilde{C} + \int_0^t \varphi_u \, \mathrm{d}\widetilde{S}_u + Z_t,$$

with $\varphi \in \overline{\mathcal{L}}^2(\widetilde{S})$ and an L^2-martingale Z with respect to Q, with $E_Q Z_t = 0$ and $\langle \widetilde{S}^i, Z \rangle = 0, \forall i$.

To show $Z = 0$ let σ_m be the stopping time defined by

$$\sigma_m := \inf \left\{ t \geq 0 : \left| \int_0^t \varphi_u \, \mathrm{d}\widetilde{S}_u \right| \geq m \right\} \wedge T.$$

Because of the continuity of $\widetilde{S}^i$ is $Z_t^{\sigma_m} = Z_{t\wedge\sigma_m}$ is bounded by $K^1 := 2K + m$. Further $\langle \widetilde{S}^i, Z^{\sigma_m}\rangle_t = \langle \widetilde{S}^i, Z\rangle_t^{\sigma_m} = 0, \forall m$ and, therefore, $\widetilde{S}^i Z^{\sigma_m}$ is a martingale, $\widetilde{S}^i \cdot Z^{\sigma_m} \in \mathcal{M}(Q)$.

By $R_t := \underbrace{\frac{1}{K^1}(K^1 + Z_t^{\sigma_m})}_{\geq 0}$ a non-negative process is defined with $E_Q R_t = 1$, $R_t \geq 0, \forall t$ and $\widetilde{S}^i R$ is a Q-martingale, $\widetilde{S}^i R \in \mathcal{M}(Q)$.

With respect to the probability measure Q^0 defined by the Radon–Nikodým density R_T,

$$Q^0 := R_T Q$$

then $\widetilde{S}^i$ is a martingale, $\widetilde{S}^i \in \mathcal{M}(Q^0)$. The argument to this is as in the first part of the proof.

However, this means that Q^0 is also an equivalent martingale measure. By assumption, therefore, it follows $Q^0 = Q$. So $R_t = R = 1\,[Q]$.

According to the definition of R then follows $Z^{\sigma_m} = 0\,[Q]$, a.s. $\forall m$.
σ_m is a sequence of stopping times converging to infinity. From the above Kunita–Watanabe-representation follows then the representability-property of M,

$$M_t = E_Q\widetilde{C} + \int_0^t \varphi_u \,\mathrm{d}\widetilde{S}_u,$$

thus the completeness of S. □

General Second Fundamental Theorem The second fundamental theorem can be generalized to general not necessarily continuous semimartingales. This requires the notion of **predictable representation property (PRP).**

$M \in \mathcal{M}^2_{\text{loc}}$ has the PRP if the generated **stable subspace** $\mathcal{F}(M) := \{\varphi \cdot M; \varphi \in \mathcal{L}^2(M)\} = \mathcal{M}^2$ is, i.e. every element $H \in \mathcal{M}^2$ has a representation of the form

$$H = H_0 + \int \varphi \,\mathrm{d}M, \quad \varphi \in \mathcal{L}^2(M).$$

The property $\varphi \in \mathcal{L}^2(M)$ is important. Even for continuous local martingales, representations can be found, e.g., of the form $1 = \int_0^T \varphi_t \,\mathrm{d}M_t$. But necessarily then $\varphi \notin \mathcal{L}^2(M)$, because otherwise $E(\int_0^T \varphi_t \,\mathrm{d}M_t) = 0$.

For a semimartingale model S now it holds:

Theorem 5.45 ((General) Second Fundamental Theorem) *The following statements are equivalent:*

(i) S is complete
(ii) $|\mathcal{M}_e(P)| = 1$
(iii) $\exists\, Q \in \mathcal{M}_e$, such that S has the PRP with respect to Q.

Proof

(i) $\Longrightarrow$ (ii): Because of completeness, there exists to $A \in \mathcal{A}$ a martingale hedge $\vartheta \in \Pi_a(1_A)$, i.e. $1_A = c + \int_0^T \vartheta_t \, \mathrm{d}S_t$. Therefore, for all $Q \in \mathcal{M}_e$,

$$Q(A) = E_Q 1_A = c, \text{ i.e. } Q \text{ is unambiguously determined.}$$

(ii) $\Longrightarrow$ (iii): for this elaborate part, we refer to Protter (1990, Section IV.3).
(iii) $\Longrightarrow$ (i): Let $H \in B(\mathcal{A}_T)$ be a bounded claim. Then for the associated Q-martingale $M_t = E_Q(H \mid \mathcal{A}_T)$ it holds by the PRP a representation of the form

$$M_t = M_0 + \int_0^t \vartheta_n \, \mathrm{d}S_n.$$

Since M is bounded, it follows that M is a martingale and hence $\vartheta \in \Pi_a$ is a martingale hedging strategy.

□

It is remarkable that not only continuous semimartingales such as diffusion processes, e.g. Brownian motion or Brownian bridge, are complete. There are also (purely) discontinuous processes which are complete.

In complete models, a contingent claim is assigned a unique price by the no-arbitrage argument. In non-complete models, prices are no longer uniquely determined by the no-arbitrage argument.

Definition 5.46 Let $S = (S^0, \dots, S^k)$ be a market model with $S_t^0 = e^{rt}$ and let $C \in L_+(\mathfrak{A}_T)$ be a contingent claim. A price $p = p_C$ for C is called **arbitrage-free** if a $Q \in \mathcal{M}_e(P)$ and an adapted process $S^{k+1} \in \mathcal{M}(Q)$ exist, such that

$$S_0^{k+1} = p, \quad S_t^{k+1} \geq 0, \quad \forall t \leq T \textit{ and } S_T^{k+1} = C, \tag{5.12}$$

and the extended market model $\widehat{S} = (S^0, \dots, S^k, S^{k+1})$ is arbitrage-free in the sense of NFLVR. Let Π_C be the set of arbitrage-free prices for C and let

$$\pi_*(C) := \inf_{p \in \Pi_C} p \textit{ and } \pi^*(C) := \sup_{p \in \Pi_C} p$$

denote the **lower** and the **upper arbitrage-free price**

The following theorem states that the set of arbitrage-free prices forms an interval. Each arbitrage-free price is described by the expected value with respect to an equivalent martingale measure.

Theorem 5.47 (Arbitrage-Free Prices) *Let $\mathcal{M}_e(P) \neq \emptyset$ and $C \in L_+(\mathfrak{A}_T)$ be a contingent claim. Then:*

(a) $\Pi_C = \{E_Q\widetilde{C};\ Q \in \mathcal{M}_e(P), E_Q C < \infty\}$
(b) $\pi_*(C) = \inf_{Q\in\mathcal{M}_e(P)} E_Q\widetilde{C},\ \pi^*(C) = \sup_{Q\in\mathcal{M}_e(P)} E_Q\widetilde{C}$.

Proof

(a) According to the First Fundamental Theorem 5.13, $p \in \Pi_C$ is an arbitrage-free price for C if and only if C can be hedged in the extended arbitrage-free model $\widehat{S}$, and hence, equivalently, there exists for the discounted extended martingale model $\widetilde{\widehat{S}}$ a martingale measure $\widehat{Q} \in \mathcal{M}_e^{\widetilde{\widehat{S}}}(Q)$ with $p = E_{\widehat{Q}}\widetilde{C}$, $\widetilde{S}^i \in \mathcal{M}(\widehat{Q})$, $1 \le i \le k+1$ and $\widetilde{C} = \widetilde{S}_T^{k+1} = p + \int_0^T 1\,\mathrm{d}\widetilde{S}_t^{k+1}$.
Thus, C is hedgeable with a martingale hedge in the extended model $\widehat{S}$. In particular, $\widehat{Q} \in \mathcal{M}_e(P)$ and $p = E_{\widehat{Q}}\widetilde{C}$ by the risk-neutral valuation formula in Theorem 5.21, i.e. $\Pi_C \subset \{E_Q\widetilde{C};\ Q \in \mathcal{M}_e(P), E_Q C < \infty\}$.
For the reverse inclusion, let $p_C = E_Q\widetilde{C} < \infty$ for a $Q \in \mathcal{M}_e(P)$.
Then define a new process S_t^{k+1} as the martingale with respect to Q generated by C

$$S_t^{k+1} = E_Q(C \mid \mathfrak{A}_t), \quad t \le T.$$

It then holds $S_0^{k+1} = p_C$ and $S_T^{k+1} = C$. The extended market model $\widehat{S}$ is arbitrage-free and Q is a martingale measure for the discounted extended model $\widetilde{\widehat{S}}$.
So $p_C \in \Pi_C$ is an arbitrage-free price. Hence the assertion follows

$$\Pi_C = \{E_Q\widetilde{C};\ Q \in \mathcal{M}_e(P), E_Q C < \infty\}.$$

(b) The representation of the lower arbitrage-free price stated in (b), $\pi_*(C) = \inf_{Q\in\mathcal{M}_e(P)} E_Q\widetilde{C}$, follows directly from (a).
The representation of $\pi^*(C)$ needs an additional argument:

We show: If $E_{Q^*}C = \infty$ for a $Q^* \in \mathcal{M}_c(P)$,
then for all $c > 0$ there exists a $p \in \Pi_C$ with $p > c$.

For this, let n be such that $E_{Q^*}C \wedge n =: \widetilde{p} > c$ and define

$$S_t^{k+1} := E_{Q^*}(C \wedge n \mid \mathfrak{A}_t), \quad t \le T.$$

Q^* is a martingale measure for the extended model $\widehat{S} = (S^0, \dots, S^{k+1})$; so this is arbitrage-free.

Let w.l.g. $\Pi_C \neq \Phi$ and let $Q \in \mathcal{M}_e(P)$ and $E_Q C < \infty$. We consider Q as a new underlying measure instead of P.

According to the first fundamental theorem, there exists $\widetilde{Q} \sim Q$ such that $\widehat{S} \in \mathcal{M}(\widetilde{Q})$. The proof of the fundamental theorem is based on a Hahn–Banach separation argument ($L^1 - L^\infty$-duality) and implies that the density $\frac{d\widetilde{Q}}{dQ}$ can be chosen boundedly. It follows:

$$E_{\widetilde{Q}} C < \infty$$

Because of $\widetilde{Q} \in \mathcal{M}_e(S)$ according to part (a) of the proof $p = E_{\widetilde{Q}} C$ is an arbitrage-free price for C. Further it holds:

$$p = E_{\widetilde{Q}} C \geq E_{\widetilde{Q}} C \wedge n = E_{\widetilde{Q}} S_T^{k+1} = S_0^{k+1} = \widetilde{p} > c.$$

From this follows the assertion. □

Remark 5.48

(a) The set of all equivalent martingale measures $\mathcal{M}_e(P)$ is convex. Therefore, it follows, from Theorem 5.47 that the set of arbitrage-free prices of C is also convex, i.e., it is an interval (see Fig. 5.4).
(b) Thus, the arbitrage-free prices determine an interval, the **no-arbitrage interval** $[(\pi_*(C), \pi^*(C))]$. In general, it is not clear whether the lower or upper bound is attained. Prices below $\pi_*(C)$ favour the buyer, prices above $\pi^*(C)$ favour the seller. It turns out that unique prices exist precisely for hedgeable claims. It holds:

$$\text{claim } C \in L_+(\mathfrak{A}_T) \text{ is strongly hedgeable} \iff \pi_*(C) = \pi^*(C).$$

In incomplete models, claims C are generally not hedgeable. The following extension of the hedging idea leads to a fundamental connection with the arbitrage-free price principle and the no-arbitrage interval.

Definition 5.49 Let $S = (S^0, \ldots, S^k)$ be a market model, $\mathcal{M}_e(P) \neq \emptyset$ and let $C \in L_+(\mathfrak{A}_T)$ be a contingent claim.

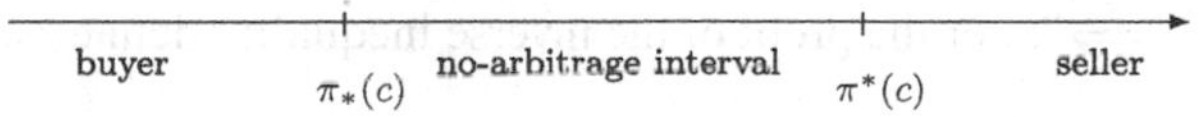

buyer | $\pi_*(c)$ | no-arbitrage interval | $\pi^*(c)$ | seller

Fig. 5.4 Convex price interval

(a) $\vartheta = (x, \pi)$ is a **super-hedging strategy** of C, if

$$\pi \in \Pi_r \text{ and } \widetilde{V}_T(\vartheta) \geq \widetilde{C}\ [P],\ V_0(\vartheta) = x.$$

Let $\mathcal{A}_C^+ := \{\vartheta = (x, \pi);\ \vartheta \text{ is super} - \text{hedging strategy for } C\}$ be the set of super-hedging strategies for C.

(b) $p^\star(C, P) := \inf\{x;\ \exists\,(x, \pi) \in \mathcal{A}_C^+\}$ is called **upper hedging price** from C.

Theorem 5.50 *Let S be a locally bounded semimartingale model with $\mathcal{M}_e(P) \neq \emptyset$.*

Let $C \in L_+(\mathfrak{A}_T)$ and $\pi^\star(C) = \sup_{Q \in \mathcal{M}_e(P)} E_Q\widetilde{C} < \infty$ be the upper arbitrage-free price. Then:

$$\pi^\star(C) = p^\star(C, P)$$

i.e., the upper hedging price of C is identical to the upper arbitrage-free price of C.

Proof The proof is based on the optional decomposition theorem (cf. Kramkov (1996)).

Optional Decomposition Theorem
Let $X \in \mathcal{S}$ be a locally bounded semimartingale and $V = (V_t) \geq 0$ an adapted process to $\mathfrak{A}^X$, then it holds:
V is a supermartingale $\forall\, Q \in \mathcal{M}_e^X(P)$
$\Longleftrightarrow \exists$ decomposition $V_t = V_0 + (\varphi \cdot X)_t - L_t,\ t \geq 0$
with $\varphi \in \mathcal{L}^0(X)$, L_t is an optional increasing process, $L_0 = 0$.

L is an adapted optional process; it can in general not be chosen predictable.

One direction of the proof of Theorem 5.50 is simple:

"$\Longleftarrow$": If $\vartheta = (X, \varphi) \in \mathcal{A}_C^+$ is a super-hedging strategy for C and if $Q \in \mathcal{M}_e(P)$ then it follows that $\widetilde{C} \leq x + \int_0^T \varphi_u\, \mathrm{d}\widetilde{S}_u$.

Because of $\widetilde{C} \geq 0$ $\int_0^t \varphi_u\, \mathrm{d}\widetilde{S}_u$ is a supermartingale. Therefore, it follows:

$$E_Q\widetilde{C} \leq x + E_Q \int_0^T \varphi_u\, \mathrm{d}\widetilde{S}_u \leq x.$$

This gives the inequality

$$\pi^*(C) \leq p^*(C, P).$$

"$\Longrightarrow$": For the proof of the inverse inequality define

$$V_t := \operatorname*{ess\,sup}_{Q \in \mathcal{M}_e(P)} E_Q(\widetilde{C} \mid \mathfrak{A}_t).$$

Here $\mathfrak{A}_t = \mathfrak{A}_t^{\widetilde{S}}$ and ess sup is the essential supremum.

It follows that $V = (V_t)$ is a supermartingale with respect to $Q, \forall\, Q \in \mathcal{M}_e(P) = \mathcal{M}_e^{\widetilde{S}}(P)$. This property is somewhat involved to show and remains here without proof. After the optional decomposition theorem follows the decomposition

$$V_t = x_0 + \int_0^t \varphi_u \, \mathrm{d}\widetilde{S}_u - L_t$$

with $\varphi \in \mathcal{L}^0(\widetilde{S}), 0 \le L_t, L_0 = 0$. L_t is monotonically increasing, adapted and optional.

For $t = 0$ follows from this representation

$$V_0 = x_0 = \sup_{Q \in \mathcal{M}_e(P)} E_Q \widetilde{C} = \pi^\star(C).$$

By definition $V_t \ge 0$ and, therefore, $\vartheta_0 = (x_0, \varphi) \in \Pi_r$.

So $\vartheta_0 = (x_0, \varphi)$ is a super-hedging strategy $\vartheta_0 \in \mathcal{A}_C^+$. This results in the inverse inequality

$$x_0 = \pi^\star(C) \ge p^\star(C, P).$$

$\square$

Remark 5.51

(a) If $|\widetilde{C}| \le K$, then $|V_t| \le K$. Thus $\forall\, Q \in \mathcal{M}_e(P)$: $E_Q L_T \le x_0 + K$. From this follows

$$E_Q \sup_{t \le T} |(\varphi \cdot \widetilde{S})_t| \le x_0 + K + E_Q L_T < \infty,$$

because L_t is monotonically increasing in t. It follows that $\varphi \cdot \widetilde{S} \in \mathcal{M}(Q)$, i.e. $\widetilde{V}_t(\vartheta_0) = x_0 + \int_0^t \varphi_u \, \mathrm{d}\widetilde{S}_u$ is a martingale with respect to Q.

Thus, for bounded options, one can dispense with the regularity assumption for strategies and use as a strategy class $\Pi_a = \bigcap_{Q \in \mathcal{M}_e} \Pi_a(Q)$.

(b) Similarly, the lower hedge price $s_*(C, P)$ can also be determined over the subhedging strategies and by analogy to Theorem 5.50 the equality of the lower hedge price and the lower arbitrage-free price hold:

$$s_*(C, P) = \pi_*(C).$$

In the hedging problem for a claim $H \in L_+(\mathfrak{A}_T)$ also a consumption process C can be introduced as an additional component.

Definition 5.52 For the market model $(S, (\mathfrak{A}_t), P)$ $\widetilde{\pi} = (x, \varphi, C)$ (is called extended) **portfolio process,** if $\varphi \in \mathcal{L}^0(S)$ and the consumption process (C_t) is an adapted increasing process with $C_0 = 0$.

$\widetilde{V}_t := x + \int_0^t \varphi_s \, d\widetilde{S}_s - C_t$ is called (discounted) value process (capital process with consumption) to the portfolio $\widetilde{\pi}$.

$\widetilde{\pi}$ is called **admissible portfolio process,** if $\widetilde{V}_t \geq 0, \forall t$.

Changes in the value process are driven by the portfolio strategy φ and the consumption C.

Remark 5.53 **(Characterization of Capital Processes)** The optional decomposition theorem now has the following interpretation:

Let $V_t \geq 0$ be an adapted process. Then it holds:

$\widetilde{V}$ is a capital process with consumption of a generalized portfolio process.

$\iff \widetilde{V}$ is a Q-supermartingale, $\forall\, Q \in \mathcal{M}_e(P)$.

Correspondingly $\widetilde{V}$ is a capital process (without consumption) of a portfolio process, exactly if $\widetilde{V} \in \mathcal{M}_{\mathrm{loc}}(Q), \forall\, Q \in \mathcal{M}_e(P)$.

Definition 5.54 Let $H \in L_+(\mathfrak{A}_T)$ be a claim.

(a) A portfolio process $\widetilde{\pi} = (x, \varphi, C)$ is a **super-hedging portfolio** for H, if $\widetilde{V}_T \geq H$.

(b) A portfolio process $\widehat{\pi} = (\widehat{x}, \widehat{\varphi}, \widehat{C})$ with capital process $\widehat{V}$ is a **minimal super-hedging portfolio,** if

$$\widehat{V}_t \leq \widetilde{V}_t \,[P], \quad \forall t \leq T \textit{ and for all super-hedging portfolios } (\widetilde{\pi}, \widetilde{V}).$$

Theorem 5.55 *Let $H \in L_+(\mathfrak{A}_T)$ be a contingent claim with bounded price $\sup_{Q\in\mathcal{M}_e(P)} E_Q H < \infty$ and let S be a locally bounded semimartingale. Then there exists a minimal super-hedging portfolio $\widehat{\pi} = (\widehat{x}, \widehat{\varphi}, \widehat{C})$ and the corresponding capital process $\widehat{V}$ has the representation*

$$\widehat{V}_t = \widehat{x} + \left(\widehat{\varphi} \cdot \widetilde{S}\right)_t - \widehat{C}_t = \operatorname*{\textit{ess sup}}_{Q\in\mathcal{M}_e(P)} E_Q(H \mid \mathfrak{A}_t).$$

Proof For all admissible portfolio processes $\widetilde{\pi} = (x, \widetilde{\varphi}, \widetilde{C})$ and $Q \in \mathcal{M}_e(P)$ the associated capital process $\widetilde{V}$ is a supermartingale. Therefore:

$$\widetilde{V}_t \geq \operatorname*{ess\,sup}_{Q\in\mathcal{M}_e(P)} E_Q(\widetilde{V}_T | \mathfrak{A}_t).$$

If $\widetilde{\pi}$ is a super-hedging portfolio, then it follows that

$$\widetilde{V}_t \geq \operatorname*{ess\,sup}_{Q\in\mathcal{M}_e(P)} E_Q(H \mid \mathfrak{A}_t).$$

However, according to the optional decomposition theorem, the lower bound is attained, i.e., there exist $\widehat{x}$, $\widehat{\varphi} \in \mathcal{L}^0(S)$ and $\widehat{C} \geq 0$ adapted (even optionally), such that for the associated value process $\widehat{V}$ with consumption it holds:

$$\widehat{V}_t = \widehat{x} + (\widehat{\varphi} \cdot \widetilde{S})_t - \widehat{C}_t = \operatorname*{ess\,sup}_{Q \in \mathcal{M}_e(P)} E_Q(H \mid \mathfrak{A}_t).$$

□

The boundedness of the semimartingale S can be replaced by the weaker condition of w-admissibility.

Definition 5.56 (w-Admissibility) Let $w \in L_+(\mathcal{A})$, $w \geq 1$, $Q_0 \in \mathcal{M}_\sigma^e$ where $E_{Q_0} w < \infty$ and $a > 0$.

$\varphi \in L(\mathcal{P})$ is called (a, w)-admissible

$$\Longleftrightarrow \forall Q \in \mathcal{M}_\sigma^e, \forall t \geq 0 : (\varphi \cdot S)_t \geq -a E_Q(w \mid \mathcal{A}_t)$$

φ is called w-admissible

$$\Longleftrightarrow \exists a > 0 : \varphi \text{ is } (a, w)\text{-admissible.}$$

Remark 5.57 (**Semimartingale Topology**) On the set of semimartingales $\mathcal{S}^d(P)$ the **Emery metric** of two semimartingales X, Y is defined by

$$\mathrm{d}(X, Y) = \sup_{|\vartheta| \leq 1} \sum_n 2^{-n} E \min(1, |(\vartheta \cdot (X - Y))_n|), \tag{5.13}$$

$(\vartheta \cdot X)_t = \int_0^t \vartheta_s \,\mathrm{d}X_s$ denotes the (vector) stochastic integral over $\vartheta \in \mathcal{L}(\mathcal{P})$ with $|\vartheta| \leq 1$.

Let $\mathcal{L}(S)$ be the space of S-integrable processes ($L(S)$ the set of equivalence classes of such processes). Then it holds:

(a) $\vartheta \in \mathcal{L}(S) \Longleftrightarrow (\vartheta 1_{\{|\vartheta| \leq n\}} \cdot S)$ is a Cauchy sequence with respect to the semimartingale topology on $\mathcal{S}^d(P)$ generated by the Emery metric d.
(b) $\{\vartheta \cdot S; \vartheta \in \mathcal{L}(S)\}$ is closed in $\mathcal{S}^d(P)$ and
(c) $(\mathcal{L}(S), d)$ is a complete topological vector space (cf. Memin (1980)).

There exist $S \in \mathcal{S}_p$ and $\vartheta \in \mathcal{L}(S)$ such that $\vartheta \cdot S \notin \mathcal{M}_{\mathrm{loc}}$. More exactly it holds for $S \in \mathcal{S}_p$, $S = M + A$:

$$\vartheta \in \mathcal{L}(S) \Longleftrightarrow \vartheta \cdot M \in \mathcal{M}_{\mathrm{loc}} \textit{ and } \vartheta \cdot A \textit{ exists as}$$
$$\textit{Lebesgue–Stieltjes integral}$$

The following theorems by Jacka (1992) and Ansel and Stricker (1994) generalize the super-hedging result in Theorem 5.50.

Theorem 5.58 *Let $S \in \mathcal{S}_p$ and let $C \geq -w$ be a claim, $w \geq 1$ a weight function. Then it holds:*
(a)

$$\sup_{\substack{Q \in \mathcal{M}_\sigma^e \\ E_Q w < \infty}} E_Q C = \inf\{\alpha;\ \exists\, \vartheta \in \mathcal{L}(\mathcal{P})\ w\text{-admissible } C \leq \alpha + (\vartheta \cdot S)_\infty\} \tag{5.14}$$

(b) The inf *in (a) is attained if it is finite.*

More precisely, the following characterization holds for the existence of solutions:

Theorem 5.59 (Attainment Theorem) *Let $S \in \mathcal{S}_p$, let w be an admissible weight function and $C \geq -w$.*
Then it holds:

(1) $\exists\, Q \in \mathcal{M}_\sigma^e,\ E_Q w < \infty$ *and* $E_Q C = \sup\{E_R C;\ R \in \mathcal{M}_\sigma^e,\ E_R w < \infty\}$

$\Longleftrightarrow$ *(2) f can be hedged, i.e. $\exists\, \alpha \in \mathbb{R}$, $Q \in \mathcal{M}_\sigma^e$, $E_Q w < \infty$, $\exists\, \vartheta \in \mathcal{L}(S)$ w-admissible, such that $\vartheta \cdot S$ is a uniformly integrable martingale with respect to Q and*

$$C = \alpha + (\vartheta \cdot S)_\infty.$$

As a consequence, the following general answer to the question of the characterization of hedgeable claims can be given.

Theorem 5.60 *Let $C \in L^0(P)$, $C^- \in L^\infty(P)$, then it holds:*

(1) $\exists\, \vartheta \in \mathcal{L}(\mathcal{P})$, $Q \in \mathcal{M}_e(P)$, $\alpha \in \mathbb{R}$ such that $\vartheta \cdot S \in \mathcal{M}(Q)$ is an uniformly integrable martingale and $C = \alpha + \vartheta \cdot S$. Equivalent to this is:

(2) $\exists\, Q \in \mathcal{M}_e(P)$ such that $E_R C \leq E_Q C$, for all $R \in \mathcal{M}_e(P)$.

If C is bounded, then (2) is equivalent to
(3) $E_R C = c$, for all $R \in \mathcal{M}_e(P)$.

Hedgeable claims are thus characterized by the fact that for them the sup over the valuations by equivalent martingale measures is attained.

6 Utility Optimization, Minimum Distance Martingale Measures, and Utility Indifference Pricing

This chapter gives an introduction to the determination (resp. selection) of option prices via minimum distance martingales as well as to pricing and hedging via utility functions. Furthermore, the problem of portfolio optimization is treated. Using exponential Lévy models, these methods are characterized in detail for a number of standard utility functions. In an incomplete market model, for a contingent claim C there is no unique arbitrage-free price, as in complete models, but a whole (possibly infinite) interval of arbitrage-free prices. Therefore, additional criteria must be used to select a specific price. There are a number of different criteria for this purpose; some of these are discussed below:

(1) An intuitive suggestive approach is to select the price measure for pricing that has the smallest distance to the underlying market measure → minimum distance martingale measure.
(2) Given a utility function appropriate to the buyer, a price is determined that appears fair to the buyer—according to the buyer's judgment → utility indifference price.

This problem leads to utility optimization problems and to their solution by means of duality results (convex duality theory). As a consequence, there is a close connection between the two approaches. The statements in this chapter are largely based on the work of He and Pearson (1991a,b), Kallsen (2000), Goll and Rüschendorf (2001), and Rheinländer and Sexton (2011).

6.1 Utility Optimization and Utility Indifference Pricing

The utility of a rational, risk-averse trader is described by a utility function $u : \mathbb{R} \to \mathbb{R} \cup \{-\infty\}$, i.e. u is strictly monotonically increasing, concave and it holds that

$$u'(\infty) = \lim u'(x) = 0, \quad u'(\overline{x}) = \lim_{x \downarrow \overline{x}} u'(x) = \infty \tag{6.1}$$

for $\overline{x} = \inf\{x \in \mathbb{R} \mid u(x) > -\infty\}$. This implies that $\mathrm{dom}(u) = (\overline{x}, \infty)$ or $\mathrm{dom}(u) = [\overline{x}, \infty)$.

L. Rüschendorf, *Stochastic Processes and Financial Mathematics*, Mathematics Study Resources 1, https://doi.org/10.1007/978-3-662-64711-0_6

Classical examples of utility functions are

- $u(x) = 1 - e^{-px}$ exponential utility function, $p > 0$
- $u(x) = \frac{x^p}{p}$, $x > 0$, $p \in (-\infty, 1) \setminus \{0\}$ power utility function
- $u(x) = \log x$, $x > 0$ logarithmic utility function

The trader's objective is, given the initial capital x, to maximize his utility by investing in the financial market. With a class $\mathcal{E}$ of admissible trading strategies and market price model S, this leads to the utility-based **portfolio optimization problem**

$$V(x) = \sup_{\vartheta \in \mathcal{E}} Eu\left(x + \int_0^T \vartheta_t \, \mathrm{d}S_t\right). \tag{6.2}$$

The optimal hedging of a claim B leads utility-based to the **hedging optimization problem**

$$V(x - B) = \sup_{\vartheta \in \mathcal{E}} Eu\left(x + \int_0^T \vartheta_t \, \mathrm{d}S_t - B\right) \tag{6.3}$$

The solution to these problems is rendered possible by a duality theorem from convex calculus.

Definition 6.1 The **convex conjugate function** $u^* : \mathbb{R}_+ \to \mathbb{R}$ of u is defined by

$$u^*(y) = \sup_{x \in \mathbb{R}} (u(x) - xy). \tag{6.4}$$

It holds:

$$u^*(y) = u(I(y)) - yI(y) \quad \text{with } I := (u')^{-1}. \tag{6.5}$$

With (6.1) it follows: I takes values in $(\overline{x}, \infty)$ an and $I(0) = \infty$.

Remark 6.2 For the power utility function $\frac{x^p}{p}$ the logarithmic utility function $\log x$ and the exponential utility function $1 - e^{-px}$ the corresponding conjugates $u^*(y)$ are given by $-\frac{p-1}{p} y^{\frac{p}{p-1}}$, $-\log y - 1$ and $1 - \frac{y}{p} + \frac{y}{p} \log\left(\frac{y}{p}\right)$. From the definition of u^* it follows a form of Fenchel's inequality

$$-xy \leq u^*(y) - u(x). \tag{6.6}$$

The utility indifference method is based on the comparison between the optimal behavior under the alternative scenarios of selling a claim and receiving a premium for it resp. not selling the claim.

Definition 6.3 (Utility Indifference Price) The **utility indifference price** π (of the seller) of a claim B is defined in a market without trading opportunities as a solution $\pi = \pi(B)$ of the equation

$$u(x) = Eu(x + \pi - B). \tag{6.7}$$

In a market with trading opportunities $\mathcal{E}$ it is defined as a solution $\pi = \pi(B)$ of the equation

$$V(x) = V(x + \pi - B). \tag{6.8}$$

The utility indifference price $\widetilde{\pi}$ of a buyer is defined analogously as the solution of the equation

$$V(x) = V(x - \widetilde{\pi} + B).$$

6.2 Minimum Distance Martingale Measures

A minimum distance martingale measure minimizes the distance of the underlying statistical measure to the class of martingale measures (or a suitable subset of them). For this purpose f-divergences are considered as distances.

Definition 6.4 Let $Q, P \in \mathcal{P}$ with $Q \ll P$ and let $f : (0, \infty) \to \mathbb{R}$ be a convex function. The **f-divergence** between Q and P is defined by

$$f(Q \parallel P) := \begin{cases} \int f\left(\frac{\mathrm{d}Q}{\mathrm{d}P}\right) \mathrm{d}P, & \textit{if the integral exists,} \\ \infty, & \textit{otherwise,} \end{cases}$$

where $f(0) = \lim_{x \downarrow 0} f(x)$.

A detailed treatment of f-divergences can be found in Liese and Vajda (1987). For $f(x) = x \log x$ the f-divergence is given by the **relative entropy**

$$H(Q \parallel P) = \begin{cases} E\left[\frac{\mathrm{d}Q}{\mathrm{d}P} \log\left(\frac{\mathrm{d}Q}{\mathrm{d}P}\right)\right], & \text{if } Q \ll P, \\ \infty, & \text{else.} \end{cases}$$

For $f(x) = -\log x$ the **reverse relative entropy** is obtained,

for $f(x) = |x - 1|$ the total variation distance, and
for $f(x) = -\sqrt{x}$ the Hellinger distance.

Let $\mathcal{K}$ be a convex dominated subset of probability measures on $(\Omega, \mathcal{A})$.

Definition 6.5 A probability measure $Q^* \in \mathcal{K}$ is called ***f*-projection** (or **minimum distance measure**) of P on (in) $\mathcal{K}$ if

$$f(Q^* \parallel P) = \inf_{Q \in \mathcal{K}} f(Q \parallel P) =: f(\mathcal{K} \parallel P). \tag{6.9}$$

f-projections are uniquely determined.

Proposition 6.6 (Uniqueness of *f*-Projections) *Let the function f be strictly convex and $f(\mathcal{K} \parallel P) < \infty$. Then there exists at most one f-projection of P on $\mathcal{K}$.*

Proof Let Q_1 and Q_2 be two f-projections of P on $\mathcal{K}$ with $Q_1 \neq Q_2$. Then $\widehat{Q} := \frac{1}{2}(Q_1 + Q_2)$ is a probability measure. Because of the strict convexity it holds

$$\begin{aligned} f(\mathcal{K} \parallel P) &= \inf_{Q \in \mathcal{K}} f(Q \parallel P) \leq f(\widehat{Q} \parallel P) = \int f\left(\frac{1}{2}\frac{\mathrm{d}Q_1}{\mathrm{d}P} + \frac{1}{2}\frac{\mathrm{d}Q_2}{\mathrm{d}P}\right) \mathrm{d}P \\ &< \frac{1}{2} \int f\left(\frac{\mathrm{d}Q_1}{\mathrm{d}P}\right) + f\left(\frac{\mathrm{d}Q_2}{\mathrm{d}P}\right) \mathrm{d}P = \frac{1}{2}(f(Q_1 \parallel P) + f(Q_2 \parallel P)) \\ &= f(\mathcal{K} \parallel P), \end{aligned}$$

a contradiction and it follows that there is at most one f-projection. □

The existence of f-projections holds under a closure condition on $\mathcal{K}$ (cf. Liese and Vajda 1987).

Theorem 6.7 (Existence of an *f*-Projection) *If $\mathcal{K}$ is closed with respect to the topology of the variation distance and it holds $\lim_{x \to \infty} \frac{f(x)}{x} = \infty$, then there exists an f-projection of P on $\mathcal{K}$.*

The main argument of the proof is to show that the set of probability measures $Q \in \mathcal{K}$ with $f(Q \parallel P) \leq 2a$ is weakly compact.

Denote in the following by $\mathcal{M}^f$ the set of all elements $Q \in \mathcal{K}$ with $f(Q \parallel P) < \infty$.

Let $\widehat{Q} \ll P$ and $F \subset L^1(\widehat{Q})$ be a subvector space with $1 \in F$. Define the by F and $\widehat{Q}$ induced **generalized moment family** $\mathcal{K}_F$ by

$$\mathcal{K}_F := \{Q \ll P : F \subset L^1(Q) \text{ and } E_Q f = E_{\widehat{Q}} f \text{ for all } f \in F\}.$$

In the applications to financial mathematics, $\mathcal{K}_F$ typically represents a class of martingale measures. The characterization of the following set of f-projections from Rüschendorf (1984) generalizes a result for the case of entropy from Csiszár (1975).

Theorem 6.8

(i) Let $Q^* \in \mathcal{M}^f$. Then Q^* is the f-projection of P on $\mathcal{K}$ if and only if

$$\int f'\left(\frac{\mathrm{d}Q^*}{\mathrm{d}P}\right)(\mathrm{d}Q^* - \mathrm{d}Q) \leq 0 \text{ for all } Q \in \mathcal{M}^f.$$

(ii) Let $Q^* \in \mathcal{K}_F$ with $Q^* \in \mathcal{M}^f$ and $f'\left(\frac{\mathrm{d}Q^*}{\mathrm{d}P}\right) \in L^1(Q^*)$. Let Q^* be the f-projection of P on $\mathcal{K}_F$ then

$$f'\left(\frac{\mathrm{d}Q^*}{\mathrm{d}P}\right) \in L^1(F, Q^*), \text{ the closure of } F \text{ in } L^1(Q^*).$$

(iii) Let $Q^* \in \mathcal{K}_F$ and $Q^* \in \mathcal{M}^f$. If $f'\left(\frac{\mathrm{d}Q^*}{\mathrm{d}P}\right) \in F$ then Q^* is the f-projection of P on $\mathcal{K}_F$.

Proof

(i) For $Q \in \mathcal{M}^f$ and $\alpha \in [0, 1]$ we define the function

$$h_\alpha = \frac{1}{\alpha - 1}\left(f\left(\alpha\frac{\mathrm{d}Q^*}{\mathrm{d}P} + (1-\alpha)\frac{\mathrm{d}Q}{\mathrm{d}P}\right) - f\left(\frac{\mathrm{d}Q^*}{\mathrm{d}P}\right)\right).$$

For $\alpha \uparrow 1$ we get according to l'Hôpital

$$\lim_{\alpha\uparrow 1} h_\alpha = f'\left(\frac{\mathrm{d}Q^*}{\mathrm{d}P}\right)\left(\frac{\mathrm{d}Q^*}{\mathrm{d}P} - \frac{\mathrm{d}Q}{\mathrm{d}P}\right).$$

Because of the convexity of f it holds

$$f\left(\frac{\mathrm{d}Q}{\mathrm{d}P}\right) \geq f\left(\frac{\mathrm{d}Q^*}{\mathrm{d}P}\right) + f'\left(\frac{\mathrm{d}Q^*}{\mathrm{d}P}\right)\left(\frac{\mathrm{d}Q}{\mathrm{d}P} - \frac{\mathrm{d}Q^*}{\mathrm{d}P}\right)$$

and, therefore, that $\lim_{\alpha\uparrow 1} h_\alpha \geq f\left(\frac{\mathrm{d}Q^*}{\mathrm{d}P}\right) - f\left(\frac{\mathrm{d}Q}{\mathrm{d}P}\right) = h_0$. Defining $\widetilde{h}_\alpha := (\alpha - 1)h_\alpha$, we see by calculating the second derivative that because of the convexity of f the function $\widetilde{h}_\alpha$ is convex and, therefore, h_α is concave, since $\frac{1}{\alpha-1} < 0$. Because of the concavity of h_α and $\lim_{\alpha\uparrow 1} h_\alpha \geq h_0$, h_α is monotonically increasing in α, i.e. $h_\alpha \geq h_0$ for all $\alpha \in [0, 1]$. More precisely

$h_\alpha \geq f\left(\frac{dQ^*}{dP}\right) - f\left(\frac{dQ}{dP}\right) = h_0$. Thus the theorem of monotone convergence can be applied and as a result

$$\begin{aligned}\frac{1}{\alpha-1}\int\left(f\left(\alpha\frac{dQ^*}{dP}+(1-\alpha)\frac{dQ}{dP}\right)-f\left(\frac{dQ^*}{dP}\right)\right)dP\\ \stackrel{\alpha\uparrow 1}{\longrightarrow}\int f'\left(\frac{dQ^*}{dP}\right)\left(\frac{dQ^*}{dP}-\frac{dQ}{dP}\right)dP\end{aligned} \tag{6.10}$$

and the integral over h_α is monotonically increasing in α. $\alpha\frac{dQ^*}{dP}+(1-\alpha)\frac{dQ}{dP}$ defines the density of a probability measure in the class $\mathcal{K}$. If Q^* is the f-projection of P on $\mathcal{K}$, then the left-hand side of (6.10) $is \leq 0$ for all α, i.e. the limit is also ≤ 0.

Conversely, if the right-hand side of (6.10) is ≤ 0, then one obtains, since h_α is monotonically increasing, that

$$f(Q^* \parallel P) - f(Q \parallel P) = \int h_0\, dP \leq \int\left(\lim_{\alpha\uparrow 1} h_\alpha\right) dP \leq 0$$

holds. So $f(Q^* \parallel P) \leq f(Q \parallel P)$ for $Q \in \mathcal{M}^f$ and, therefore, Q^* is the f-projection of P on $\mathcal{K}$.

(ii) Q^* is by assumption the f-projection on $\mathcal{K}_F$, so it follows by part (i) that

$$\int f'\left(\frac{dQ^*}{dP}\right) dQ^* \leq \int f'\left(\frac{dQ^*}{dP}\right) dQ \text{ for all } Q \in \mathcal{M}^f \cap \mathcal{K}_F \tag{6.11}$$

holds. Since $f'\left(\frac{dQ^*}{dP}\right) \in L^1(Q^*)$ and (6.11) holds, it follows with the help of a separation theorem (cf. Rüschendorf 1984, Proposition 1) that $f'\left(\frac{dQ^*}{dP}\right) \in L^1(F, Q^*)$.

(iii) Since $Q^* \in \mathcal{K}_F$ and $f'\left(\frac{dQ^*}{dP}\right) \in F$ it holds

$$\int f'\left(\frac{dQ^*}{dP}\right) dQ^* = \int f'\left(\frac{dQ^*}{dP}\right) \text{ for } Q \in \mathcal{K}_F,$$

so it follows by (i) that Q^* is the f-projection of P on $\mathcal{K}_F$. □

Using Theorem 6.8 (i) it is possible to prove the equivalence of the f-projection of P if $f'(0) = -\infty$ and if there exists a measure $Q \in \mathcal{K}$ such that $Q \sim P$ with $f(Q \parallel P) < \infty$.

Corollary 6.9 (Equivalence of the f-Projection) *Let $f'(0) = -\infty$ and let there exist a measure $Q \in \mathcal{M}^f \cap \mathcal{K}$. If Q^* is the f-projection of P on $\mathcal{K}$, then $Q^* \sim P$.*

Proof Suppose that $Q^* \sim P$ and that Q^* is the f-projection of P on $\mathcal{K}$. So, because of $Q^* \in \mathcal{K}$ it is $Q^* \ll P$. Thus $P\left(\frac{dQ^*}{dP} = 0\right) > 0$. Since $Q \sim P$ it also holds that $Q\left(\frac{dQ^*}{dP} = 0\right) > 0$. According to Theorem 6.8 (i) it holds

$$\int f'\left(\frac{dQ^*}{dP}\right) dQ^* - \int f'\left(\frac{dQ^*}{dP}\right) dQ \leq 0.$$

But since $Q\left(\frac{dQ^*}{dP} = 0\right) > 0$ and according to the assumption $f'(0) = -\infty$ we get a contradiction. □

As an application, this yields a characterization of the minimum distance martingale measure.

The set of martingale measures can be described as a generalized moment family. Let

$$\begin{aligned} G \quad &:= \left\{(\vartheta \cdot S)_T : \vartheta^i = H^i \mathbb{1}_{(s_i,t_i]}, s_i < t_i, H^i \text{ bounded}, \mathcal{F}_{s_i}\text{-measurable}\right\} \\ &\cup \{\mathbb{1}_B : P(B) = 0\} \end{aligned}$$

and

$$\begin{aligned} G_{\text{loc}} := \Big\{&(\vartheta \cdot S)_T : \vartheta^i = H^i \mathbb{1}_{(s_i,t_i]} \mathbb{1}_{[0,\widehat{T}^i]}, s_i < t_i, \\ &H^i \text{ bounded}, \mathcal{F}_{s_i}\text{-measurable}, \widehat{T}^i \in \gamma^i\Big\} \cup \{\mathbb{1}_B : P(B) = 0\}, \end{aligned}$$

where $\gamma^i := \left\{\widehat{T}^i \text{ stopping time}; (S^i)^{\widehat{T}^i} \text{ is bounded}\right\}$.

The following theorem gives a necessary condition for an f-projection on the set $\mathcal{M}$ of martingale measures.

Theorem 6.10 *Let $Q^* \in \mathcal{M}^f \cap \mathcal{M}$ and $f'\left(\frac{dQ^*}{dP}\right) \in L^1(Q^*)$. If Q^* is the f-projection of P on $\mathcal{M}$, then*

$$f'\left(\frac{dQ^*}{dP}\right) = c + \int_0^T \vartheta_t \, dS_t \quad Q^*\text{-almost surely} \tag{6.12}$$

for a process $\vartheta \in \mathcal{L}(S, Q^*)$ such that $\int_0^{\cdot} \vartheta_t \, dS_t$ is a martingale with respect to Q^*.

Proof The class $\mathcal{M}$ of martingale measures can be characterized as a moment-family generated by G,

$$\mathcal{M} = \{Q \in \mathcal{P} : G \subset L^1(Q) \text{ and } E_Q g = 0 \quad \forall g \in G\}$$

Let F be the vector space generated by 1 and G. The conditions of Theorem 6.8 (ii) are satisfied and it, therefore, follows that

$$f'\left(\frac{dQ^*}{dP}\right) = \xi \quad Q^*\text{-almost surely.}$$

for a $\xi \in L^1(F, Q^*)$, the $L^1(Q^*)$-closure of f. Using a closure theorem for stochastic integrals of Yor (1978, Corollary 2.5.2), (for a multidimensional version, see Delbaen and Schachermayer (1999, Theorem 1.6)) we can identify the closure $L^1(G, Q^*)$ with a class of stochastic integrals which are Q^*-martingales; it holds that $L^1(G, Q^*) \subset \{(\vartheta \cdot S)_T : \vartheta \in \mathcal{L}(S, Q^*),$ such that $\vartheta \cdot S$ is a Q^* martingale$\}$. This result holds even without the assumption of a complete filtration, see Jacod (1979, Proposition 1.1). According to Schaefer (1971, Proposition I.3.3) we obtain the necessary condition since f is generated by 1 and G. □

In Theorem 6.10 it is enough to assume that S is a general $\mathbb{R}^{d+1}$-valued semimartingale. In the next theorem, we show that the necessary condition in Theorem 6.10 also holds for the set of local martingale measures, under the additional condition that the price process S is locally bounded.

Theorem 6.11 (Density of *f*-Projection) *Let S be a locally bounded semimartingale. Let $Q^* \in \mathcal{M}^f \cap \mathcal{M}_{\text{loc}}$ and $f'\left(\frac{dQ^*}{dP}\right) \in L^1(Q^*)$. If Q^* is the f-projection of P on $\mathcal{M}_{\text{loc}}$, then*

$$f'\left(\frac{dQ^*}{dP}\right) = c + \int_0^T \vartheta_t \, dS_t \quad Q^*\text{-almost surely} \tag{6.13}$$

for a process $\vartheta \in \mathcal{L}^1_{\text{loc}}(S, Q^*)$ such that $\int_0^\cdot \vartheta_t \, dS_t$ is a Q^*-martingale.

Proof Consider the following characterization of $\mathcal{M}_{\text{loc}}$ as a moment-family

$$\mathcal{M}_{\text{loc}} = \{Q \in \mathcal{P} : G_{\text{loc}} \subset L^1(Q) \text{ and } E_Q g = 0 \quad \forall g \in G_{\text{loc}}\}.$$

The assertion now follows analogously to the proof of Theorem 6.10. □

Remark

(1) If $Q^* \sim P$ and if additionally $-f'\left(\frac{dQ^*}{dP}\right)$ is bounded below, then according to Goll and Rüschendorf (2001) the representability of $f'\left(\frac{dQ^*}{dP}\right)$ as a stochastic integral holds using Theorem 6.8 (i), the martingale representation result Theorem 3.4 from Jacka (1992) and Theorem 3.2 from Ansel and Stricker (1994).
(2) If S is (locally) bounded, then $\mathcal{M}_{\text{loc}}$ is closed with respect to the variation distance. This can be seen by the characterization of $\mathcal{M}_{\text{loc}}$ with the help

of G_{loc} (see Lemma 7.8 in Rheinländer and Sexton (2011)). If in addition $\lim_{x\to\infty} \frac{f(x)}{x} = \infty$ then by Theorem 6.7 there exists an f-projection from P on $\mathcal{M}_{\text{loc}}$. In particular, this condition is satisfied for the relative entropy.

Next we give some sufficient conditions for an f-projection of P on $\mathcal{M}_{\text{loc}}$.

Proposition 6.12 *Let $Q^* \in \mathcal{M}^f \cap \mathcal{M}_{\text{loc}}$ and let*

$$f'\left(\frac{\mathrm{d}Q^*}{\mathrm{d}P}\right) = c + \int_0^T \vartheta_t \,\mathrm{d}S_t \quad P\text{–almost surely} \tag{6.14}$$

for $\int_0^T \vartheta_t \,\mathrm{d}S_t \in G_{\text{loc}}$. Then Q^* is the f-projection of P on $\mathcal{M}_{\text{loc}}$.

Proof Let F be the vector space that generated by 1 and G_{loc}. Because of the representation of $f'\left(\frac{\mathrm{d}Q^*}{\mathrm{d}P}\right)$ as an integral and because of $\int_0^T \vartheta_t \,\mathrm{d}S_t \in G_{\text{loc}}$ one can apply Theorem 6.8 (iii). □

The condition from Proposition 6.12 can be generalized as follows.

Proposition 6.13 *Let $Q^* \in \mathcal{M}^f \cap \mathcal{M}_{\text{loc}}$ such that for a predictable process $\vartheta \in \mathcal{L}(S)$*

$$f'\left(\frac{\mathrm{d}Q^*}{\mathrm{d}P}\right) = c + \int_0^T \vartheta_t \,\mathrm{d}S_t \; P\text{–almost surely},$$

$$-\int_0^{\cdot} \vartheta_t \,\mathrm{d}S_t \text{ is } P\text{–almost surely bounded below},$$

$$E_{Q^*}\left[\int_0^T \vartheta_t \,\mathrm{d}S_t\right] = 0$$

holds. Then Q^* is the f-projection of P on $\mathcal{M}_{\text{loc}}$.

Proof By assumption $\vartheta \in \mathcal{L}(S)$ is a predictable S-integrable process with respect to P and we obtain by Proposition 7.26 (b) from Jacod (1979) that ϑ in dimension $d = 1$ is S-integrable with respect to any $Q \in \mathcal{M}_{\text{loc}}$. Using Proposition 3 from Jacod (1980) one can extend this result to dimensions $d \geq 1$.

According to the definition of $\mathcal{M}_{\text{loc}} := \{Q \ll P : Q \text{ is a (local) martingale measure}\}$ the pricing process S is a (local) Q-martingale. By Corollary 3.5 from Ansel and Stricker (1994) $-\vartheta \cdot S$ is a local Q-martingale and since it is bounded below it is also a Q-supermartingale for every $Q \in \mathcal{M}_{\text{loc}}$.

So $E_Q(\vartheta \cdot S)_T \geq E_Q(\vartheta \cdot S)_0 = 0$ and it follows

$$E_Q f'\left(\frac{\mathrm{d}Q*}{\mathrm{d}P}\right) = c + E_Q(\vartheta \cdot S)_T \geq c + E_Q \underbrace{(\vartheta \cdot S)_0}_{=0}$$

$$= c + \underbrace{E_{Q^*}(\vartheta \cdot S)_T}_{=0} = E_{Q^*} f'\left(\frac{\mathrm{d}Q*}{\mathrm{d}P}\right).$$

Since $Q \in \mathcal{M}_{\mathrm{loc}}$ was chosen arbitrarily, the assertion follows with Theorem 6.8 (i).□

6.3 Duality Results

The utility maximization problem is related by a duality theorem of convex optimization to the problem of determining the minimum of the f-divergence distance. More precisely, the minimum distance martingale measures with respect to the convex conjugate of the utility function turn out to be equivalent to minimax measures. A characterization of minimum distance martingale measures then leads to a characterization of optimal portfolio strategies.

6.3.1 Minimum Distance Martingale Measures and Minimax Measures

Let $\mathcal{K} \subset \mathcal{M}(P)$ be a convex set of P-continuous martingale measures on $(\Omega, \mathcal{A})$ and let u be an utility function as in Sect. 6.1.

For $Q \in \mathcal{K}$ and $x > \overline{x}$ we define

$$U_Q(x) := \sup\{Eu(Y) : Y \in L^1(Q), E_Q Y \leq x, Eu(Y)^- < \infty\}. \tag{6.15}$$

One can interpret $U_Q(x)$ as the maximum expected utility that can be achieved with a capital stock X if the market prices are computed with Q.

In doing so, one would like to avoid that the expected utility is $-\infty$ that is to say, that ruin occurs.

Lemma 6.14 *Let* $Q \in \mathcal{K}$ *and let* $E_Q I\left(\lambda \frac{\mathrm{d}Q}{\mathrm{d}P}\right) < \infty, \forall \lambda > 0$. *Then it holds*

(i) $U_Q(x) = \inf_{\lambda > 0}\{Eu^*\left(\lambda \frac{\mathrm{d}Q}{\mathrm{d}P}\right) + \lambda x\}$.

(ii) *For the equation* $E_Q I\left(\lambda \frac{dQ}{dP}\right) = x$ *there is a unique solution for* λ *denoted by* $\lambda_Q(x) \in (0, \infty)$ *and it holds*

$$U_Q(x) = Eu\left(I\left(\lambda_Q(x)\frac{dQ}{dP}\right)\right).$$

Proof Let $Y \in L^1(Q)$ with $E_Q Y \le x$ and $Eu(Y)^- < \infty$. Then it holds for $\lambda > 0$:

$$\begin{aligned}
Eu(Y) &\le Eu(Y) + \lambda(x - E_Q Y) \\
&= Eu(Y) - E\left[\lambda \frac{dQ}{dP} Y\right] + \lambda x \\
&\le Eu^*\left(\lambda \frac{dQ}{dP}\right) + \lambda x \\
&\overset{(6.6)}{=} Eu\left(I\left(\lambda \frac{dQ}{dP}\right)\right) - E\left[\lambda \frac{dQ}{dP} I\left(\lambda \frac{dQ}{dP}\right)\right] + \lambda x \\
&= Eu\left(I\left(\lambda \frac{dQ}{dP}\right)\right) + \lambda\left(x - E_Q I\left(\lambda \frac{dQ}{dP}\right)\right).
\end{aligned}$$

Equality holds exactly if $Y = I\left(\lambda_Q(x)\frac{dQ}{dP}\right)$ with $\lambda_Q(x)$ a solution of the equation $E_Q I\left(\lambda \frac{dQ}{dP}\right) = x$.

According to (6.1) follows $I(0) = \infty$ and $I(\infty) = \overline{x}$. Together with the strict concavity of u it follows that I is strictly decreasing. Overall, we obtain with the condition $E_Q I\left(\lambda \frac{dQ}{dP}\right) < \infty$ for all $\lambda > 0$ that $E_Q I\left(\lambda \frac{dQ}{dP}\right)$ is a function continuous in λ and monotonically decreasing with values in $(\overline{x}, \infty)$. Since $x > \overline{x}$ the unique existence of $\lambda_Q(x)$ is guaranteed.

It remains to be verified that $E\left[u\left(I\left(\lambda_Q(x)\frac{dQ}{dP}\right)\right)\right]^- < \infty$. With a reformulation of the Fenchel inequality in (6.6) $u(x) - xy \le u(I(y)) - yI(y)$ one obtains

$$E\left[u\left(I\left(\lambda \frac{dQ}{dP}\right)\right) - \lambda \frac{dQ}{dP} I\left(\lambda \frac{dQ}{dP}\right)\right]^- < \infty. \tag{6.16}$$

With the inequality

$$\begin{aligned}
&\left[u\left(I\left(\lambda \frac{dQ}{dP}\right)\right)\right]^- \\
&\quad \le \left[u\left(I\left(\lambda \frac{dQ}{dP}\right)\right) - \lambda \frac{dQ}{dP} I\left(\lambda \frac{dQ}{dP}\right)\right]^- + \left[\lambda \frac{dQ}{dP} I\left(\lambda \frac{dQ}{dP}\right)\right]^-
\end{aligned}$$

we obtain that the condition $E\left[u\left(I\left(\lambda_Q(x)\frac{\mathrm{d}Q}{\mathrm{d}P}\right)\right)\right]^- < \infty$ is fulfilled. □

Remark 6.15

(1) One can interpret $I\left(\lambda_Q(x)\frac{\mathrm{d}Q}{\mathrm{d}P}\right)$ as an optimal claim, which is financeable under the pricing measure Q with the capital x.
(2) If for $Q \in \mathcal{K}$ there exists a $\lambda > 0$ with $Eu^*\left(\lambda\frac{\mathrm{d}Q}{\mathrm{d}P}\right) < \infty$, then we can use Lemma 6.14 (i) to state that $U_Q(x) < \infty$ for all $x > \overline{x}$. If for $Q \in \mathcal{K}$ with $U_Q(x) < \infty$ the assumption $E_Q I\left(\lambda\frac{\mathrm{d}Q}{\mathrm{d}P}\right) < \infty \quad \forall\lambda > 0$ from Lemma 6.14 is satisfied, then $Eu^*\left(\lambda_Q(x)\frac{\mathrm{d}Q}{\mathrm{d}P}\right) < \infty$.
(3) The convex conjugate of the utility function $u(x) = 1 - e^{-x}$ is $u^*(y) = \mathbb{1} - y + y\log y$. The u^*-divergence distance of the exponential utility function is thus

$$Eu^*\left(\frac{\mathrm{d}Q}{\mathrm{d}P}\right) = E\left[1 - \frac{\mathrm{d}Q}{\mathrm{d}P} + \frac{\mathrm{d}Q}{\mathrm{d}P}\log\left(\frac{\mathrm{d}Q}{\mathrm{d}P}\right)\right]$$
$$= E\left[\frac{\mathrm{d}Q}{\mathrm{d}P}\log\left(\frac{\mathrm{d}Q}{\mathrm{d}P}\right)\right] = H(Q \parallel P)$$

the relative entropy distance. Similarly, we see that the divergence distance associated with the logarithmic utility function i.e. the u^*-divergence distance is the reverse relative entropy.
(4) For $\log x$, $\frac{x^p}{p}$, $1-e^{-x}$ the convex conjugate functions are $-\log x - 1$, $-\frac{p-1}{p}x^{\frac{p}{p-1}}$ and $1 - x + x\log x$. Therefore, for $u(x) = 1 - e^{-x}$ the u^*-divergence distance is the relative entropy, for $u(x) = \log x$ the reverse relative entropy and for $u(x) = -x^{-1}$ the Hellinger distance.

Definition 6.16 A measure $Q^* = Q^*(x) \in \mathcal{K}$ is called **minimax measure** for x and $\mathcal{K}$ if it minimizes the mapping $Q \mapsto U_Q(x)$ over all $Q \in \mathcal{K}$, i.e. if holds

$$U_{Q^*}(x) = U(x) := \inf_{Q\in\mathcal{K}} U_Q(x).$$

In general, the minimax measure Q^* depends on x. For standard utility functions such as $u(x) = \frac{x^p}{p}$ $(p \in (-\infty, 1)\backslash\{0\})$, $u(x) = \log x$ and $u(x) = 1 - e^{-px}$ $(p \in (0, \infty))$ the minimax measure is independent of x.

Throughout this chapter we assume that

$$\exists x > \overline{x} \text{ with } U(x) < \infty, \tag{6.17}$$

$$E_Q I\left(\lambda\frac{\mathrm{d}Q}{\mathrm{d}P}\right) < \infty, \quad \forall\lambda > 0, \quad \forall Q \in \mathcal{K}. \tag{6.18}$$

Remark 6.17

(1) The assumption (6.18) is satisfied for $u(x) = \log x$: Here $I(y) = \frac{1}{y}$ and we obtain

$$E_Q I\left(\lambda \frac{dQ}{dP}\right) = E_Q\left[\frac{1}{\lambda}\frac{dP}{dQ}\right] = \frac{1}{\lambda} < \infty.$$

If for any $Q \in \mathcal{K}$, $u^*(Q \| P) < \infty$, then the assumption (6.18) is satisfied likewise for $u(x) = 1 - e^{-px}$ for $p \in (0, \infty)$ and $u(x) = \frac{x^p}{p}(p \in (-\infty, 1)\backslash\{0\})$. For $u(x) = 1 - e^{-px}$ is $I(y) = -\frac{1}{p}\log\left(\frac{y}{p}\right)$ and we get

$$\begin{aligned} E_Q I\left(\lambda \frac{dQ}{dP}\right) &= E_Q\left[-\frac{1}{p}\log\left(\frac{\lambda}{p}\frac{dQ}{dP}\right)\right] \\ &= -\frac{1}{p}\log\left(\frac{\lambda}{p}\right) - \frac{1}{p}\underbrace{E_Q\left[\log\left(\frac{dQ}{dP}\right)\right]}_{<\infty} < \infty. \end{aligned}$$

Analogously, we see the same result for $u(x) = \frac{x^p}{p}$. For the exponential and the power utility function, we can replace the set $\mathcal{K}$ by the convex subset $\{Q \in \mathcal{K}; u^*(Q \mid P) < \infty\}$.

(2) Assumption (6.18) implies, by Remark 6.15 that

$$\{Q \in \mathcal{K}; U_Q(x) < \infty\} = \{Q \in \mathcal{K}; \forall \lambda > 0 : u^*_\lambda(Q \| P) < \infty\}.$$

We denote by $\partial U(x)$ the subdifferential of the function U at x. For $f(x) = u^*(\lambda_0 x)$ we denote the corresponding f-divergence by $u^*_{\lambda_0}(\cdot \| \cdot)$.

The following proposition establishes an important connection between minimax measures and minimum distance martingale measures.

Proposition 6.18 *Let $x > \overline{x}$, $\lambda_0 \in \partial U(x)$, $\lambda_0 > 0$. Then it holds*

(i) $U(x) = u^*_{\lambda_0}(\mathcal{K} \| P) + \lambda_0 x$.

(ii) If $Q^* \in \mathcal{K}$ is a $u^*_{\lambda_0}$-projection of P on $\mathcal{K}$, then Q^* is a minimax measure and $\lambda_0 = \lambda_{Q^*}(x)$.

(iii) If $Q^* \in \mathcal{K}$ is a minimax measure, then Q^* is a $u^*_{\lambda_{Q^*}(x)}$-projection of P on $\mathcal{K}$, $\lambda_{Q^*}(x) \in \partial U(x)$ and it holds

$$U_{Q^*}(x) = \inf_{Q \in \mathcal{K}} U_Q(x) = \sup\{Eu(Y); \sup_{Q \in \mathcal{K}(x)} E_Q Y \le x\}$$

where $\mathcal{K}(x) := \{Q \in \mathcal{K}; U_Q(x) < \infty\}$.

The proposition shows that minimax measures can be described by minimum distance measures. The converse is also true that $u^*_{\lambda_0}$-projections can be interpreted in terms of utility maximization.

Proof

(i) By Lemma 6.14 it holds

$$\begin{aligned} U(x) &= \inf_{Q\in\mathcal{K}} U_Q(x) = \inf_{Q\in\mathcal{K}} \inf_{\lambda>0} \left\{ Eu^*\left(\lambda \frac{\mathrm{d}Q}{\mathrm{d}P}\right) + \lambda x \right\} \\ &= \inf_{\lambda>0} \left\{ u^*_\lambda(\mathcal{K} \parallel P) + \lambda x \right\}. \end{aligned} \tag{6.19}$$

Define $H : (0,\infty) \to \mathbb{R} \cup \{\infty\}$by $H(\lambda) := u^*_\lambda(\mathcal{K} \parallel P)$. By Remark 6.15, the assumptions (6.17) and (6.18) guarantee that there is a $\lambda > 0$ with $H(\lambda) < \infty$. Therefore, from the representation of $U(x)$ in (6.19), it follows that $U(x) < \infty$ for all $x \in \operatorname{dom}(u) = \operatorname{dom}(U)$. Thus we obtain $H(\lambda) < \infty$ for all $\lambda > 0$.

Now to show is: H is a convex function. Let $\varepsilon > 0$ and let $Q_1, Q_2 \in \mathcal{K}$ such that

$$H(\lambda_1) + \varepsilon \geq Eu^*\left(\lambda_1 \frac{\mathrm{d}Q_1}{\mathrm{d}P}\right) \text{ and}$$

$$H(\lambda_2) + \varepsilon \geq Eu^*\left(\lambda_2 \frac{\mathrm{d}Q_2}{\mathrm{d}P}\right).$$

Herewith and with the convexity of u^* it holds:

$$\begin{aligned} &\gamma H(\lambda_1) + (1-\gamma)H(\lambda_2) + 2\varepsilon \\ &\geq \gamma Eu^*\left(\lambda_1 \frac{\mathrm{d}Q_1}{\mathrm{d}P}\right) + (1-\gamma)Eu^*\left(\lambda_2 \frac{\mathrm{d}Q_2}{\mathrm{d}P}\right) \\ &\geq Eu^*\left(\gamma\lambda_1 \frac{\mathrm{d}Q_1}{\mathrm{d}P} + (1-\gamma)\lambda_2 \frac{\mathrm{d}Q_2}{\mathrm{d}P}\right) \\ &= Eu^*\left((\gamma\lambda_1 + (1-\gamma)\lambda_2)\right. \\ &\qquad \left.\cdot\left(\frac{\gamma\lambda_1}{\gamma\lambda_1 + (1-\gamma)\lambda_2} \frac{\mathrm{d}Q_1}{\mathrm{d}P} + \frac{(1-\gamma)\lambda_2}{\gamma\lambda_1 + (1-\gamma)\lambda_2} \frac{\mathrm{d}Q_2}{\mathrm{d}P}\right)\right) \\ &\overset{(*)}{\geq} \inf_{Q\in\mathcal{K}} Eu^*\left((\gamma\lambda_1 + (1-\gamma)\lambda_2) \frac{\mathrm{d}Q}{\mathrm{d}P}\right) = H\left(\gamma\lambda_1 + (1-\gamma)\lambda_2\right). \end{aligned}$$

To $(*)$: Since $\mathcal{K}$ is a convex set and since for $Q_1, Q_2 \in \mathcal{K}$ it holds

$$\frac{\gamma\lambda_1}{\gamma\lambda_1 + (1-\gamma)\lambda_2} \frac{\mathrm{d}Q_1}{\mathrm{d}P} + \frac{(1-\gamma)\lambda_2}{\gamma\lambda_1 + (1-\gamma)\lambda_2} \frac{\mathrm{d}Q_2}{\mathrm{d}P} \in \mathcal{K}.$$

Since H is a convex function, we can use the equivalence (a) $\iff$ (b) from Theorem 23.5 from Rockafellar (1970) and so $\inf_{\lambda>0}\{H(\lambda)+\lambda x\}$ takes its infimum in λ at $\lambda=\lambda_0$ if and only if $-x\in\partial H(\lambda_0)$ or equivalently exactly if $\lambda_0\in\partial U(x)$. We thus obtain by Eq. (6.19) that $U(x)=u^*_{\lambda_0}(\mathcal{K}\,\|\,P)+\lambda_0 x$.

(ii) This statement follows from Lemma 6.14:

$$\begin{aligned} U_{Q^*}(x) &= \inf_{\lambda>0}\left\{Eu^*\left(\lambda\frac{\mathrm{d}Q^*}{\mathrm{d}P}\right)+\lambda x\right\} \\ &= \inf_{\lambda>0}\left\{u^*_\lambda(\mathcal{K}\,\|\,P)+\lambda x\right\} \\ &= U(x). \end{aligned}$$

(iii) If $Q^*\in\mathcal{K}$ is a minimax measure, it follows by Lemma 6.14 (i) that Q^* a $u^*_{\lambda_{Q^*}(x)}$-projection of P on $\mathcal{K}$, since one obtains similarly to Eq. (6.19):

$$\begin{aligned} U(x) &= \inf_{Q\in\mathcal{K}} U_Q(x) = \inf_{Q\in\mathcal{K}}\inf_{\lambda>0}\left\{Eu^*\left(\lambda\frac{\mathrm{d}Q}{\mathrm{d}P}\right)+\lambda x\right\} \\ &= \inf_{\lambda>0}\left\{u^*_\lambda(\mathcal{K}\,\|\,P)+\lambda x\right\} \\ &= U_{Q^*}(x). \end{aligned}$$

For the second part of the statement it holds:

$$\left(u^*_\lambda\right)'(x) = -\lambda I(\lambda x), \tag{6.20}$$

since

$$\begin{aligned} \left(u^*_\lambda\right)'(x) &= \left(u^*(\lambda x)\right)' = \left(u(I(\lambda x)) - \lambda x I(\lambda x)\right)' \\ &= u'(I(\lambda x))I'(\lambda x)\lambda - \lambda I(\lambda x) - \lambda x I'(\lambda x)\lambda \\ &= u'((u')^{-1}(\lambda x))I'(\lambda x)\lambda - \lambda I(\lambda x) - \lambda x I'(\lambda x)\lambda \\ &= \lambda x I'(\lambda x)\lambda - \lambda I(\lambda x) - \lambda x I'(\lambda x)\lambda \\ &= -\lambda I(\lambda x). \end{aligned}$$

By Remark 6.17 it holds

$$\mathcal{K}(x) = \{Q\in\mathcal{K} : U_Q(x)<\infty\} = \{Q\in\mathcal{K} : u^*_{\lambda_{Q^*}(x)}(Q\,\|\,P)<\infty\}.$$

That is, the conditions of Theorem 6.8 are satisfied. Because of the first part of (iii) Q^* is a $u^*_{\lambda_{Q^*}(x)}$-projection and with Theorem 6.8 (i) and Eq. (6.20) it follows.

$$E_{Q^*}I\left(\lambda_{Q^*}(x)\frac{\mathrm{d}Q^*}{\mathrm{d}P}\right) \geq E_Q I\left(\lambda_{Q^*}(x)\frac{\mathrm{d}Q^*}{\mathrm{d}P}\right).$$

With Lemma 6.14 (ii) it follows $U_{Q^*}(x) = \sup\{Eu(Y) : \sup_{Q\in\mathcal{K}(x)} E_Q Y \leq x\}$.

□

As a corollary of Proposition 6.18 we obtain

Corollary 6.19 *Let $x > \overline{x}$, $\lambda_0 \in \partial U(x)$ and $\lambda_0 > 0$. Further, let U be differentiable in x. Then Q^* is a minimax measure if and only if Q^* is the $u^*_{\lambda_0}$-projection of P on $\mathcal{K}$, where $\lambda_0 = \nabla U(x)$.*

Proof Because of the assumption of differentiability, the subdifferential consists of only one point, which is the derivative at the point x. Now the statement follows directly from Proposition 6.18. □

In general one knows $U(x)$ often not explicitly and, therefore, the verification of the condition $\lambda_0 \in \partial U(x)$ is difficult. In the case of the standard utility functions $u(x) = 1 - e^{-px}(p \in (0, \infty))$, $u(x) = \frac{x^p}{p}(p \in (-\infty, 1)\backslash\{0\})$ and $u(x) = \log x$ the minimax measure does not depend on X and accordingly the u^*_λ-projection does not depend on λ and so it does not pose the problem to determine λ.

Proposition 6.20 *Let $\overline{x} = 0$ and let u be bounded from above. Then U is differentiable in every $x > 0$.* □

Proof Sketch According to Theorem 26.3 from Rockafellar (1970), the main argument is to show that $H(\lambda) = u^*_\lambda(\mathcal{K} \parallel P)$ is strictly convex. The strict convexity of $H(\lambda)$ is shown by means of the convexity of $\mathcal{K}$, Alaoglu's theorem, the theorem of dominated convergence and the strict convexity of u^*. □

The differentiability condition from Corollary 6.19 is satisfied if for all $\lambda > 0$ there exists a u^*_λ-projection.

Proposition 6.21 *Let $\mathcal{K}$ be closed with respect to the topology of the variational distance, and let* $\mathrm{dom}(u) = (-\infty, \infty)$. *Then for all $\lambda > 0$ there is a u^*_λ-projection of P on $\mathcal{K}$.*

Proof According to Theorem 6.7 it suffices to show that $\lim_{x\to\infty} \frac{u^*(\lambda x)}{x} = \infty$. Since $u^*(\lambda x) = \sup_y\{u(y) - \lambda xy\} \geq u\left(-\frac{n}{\lambda}\right) + nx$ it follows with $u\left(-\frac{n}{\lambda}\right) > -\infty$, that $\lim_{x\to\infty} \frac{u^*(\lambda x)}{x} \geq n$ for each $n \in \mathbb{N}$. □

The sufficient conditions for projections from Sect. 6.2 provide a way to determine the parameter $\lambda_0 \in \partial U(x)$ and hence also the f-divergence distance associated with a minimax measure.

Proposition 6.22 *Let $Q^* \in \mathcal{M}(\mathcal{M}_{\text{loc}})$, $\lambda > 0$ where $u_\lambda^*(Q^* \parallel P) < \infty$ such that for an S-integrable process ϑ it holds:*

$$I\left(\lambda \frac{dQ^*}{dP}\right) = x + \int_0^T \vartheta_t \, dS_t \quad P\text{-almost surely.}$$

If, in addition, the sufficient conditions for a u_λ^*-projection from Proposition 6.12 hold (Theorem 6.13), then Q^* is a minimax measure for x and $\lambda \in \partial U(x)$.

Remark 6.23 Note that the conditions in Proposition 6.12 and in Proposition 6.13 are formulated for $(u_\lambda^*)'(x) = -\lambda I(\lambda x)$.

Proof In the following $\mathcal{K}$ stands either for $\mathcal{M}$ or for $\mathcal{M}_{\text{loc}}$. If the conditions from Theorem 6.13 hold, then it is already true that $E_{Q^*}\left[\int_0^T \vartheta_t \, dS_t\right] = 0$. If the conditions from Proposition 6.12 hold then Theorem 6.10 or Theorem 6.11 can be applied and it follows that $\int_0^\cdot \vartheta_t \, dS_t$ is a martingale under Q^*. So it holds $E_{Q^*} I\left(\lambda \frac{dQ^*}{dP}\right) = x$ and with Lemma 6.14 (ii) we obtain that $\lambda = \lambda_{Q^*}(x)$. Because of Proposition 6.12 (Theorem 6.13) is Q^* the u_λ^*-projection of P on $\mathcal{K}$, i.e. the assumptions of Theorem 6.8 are satisfied. From Theorem 6.8 and from Eq. (6.20) it follows for all measures $Q \in \mathcal{K}$ with $u_\lambda^*(Q \parallel P) < \infty$, that

$$\begin{aligned} & E_{Q^*} \left(u_\lambda^*\right)' \left(\frac{dQ^*}{dP}\right) \le E_Q \left(u_\lambda^*\right)' \left(\frac{dQ^*}{dP}\right) \\ \iff \quad & E_Q I\left(\lambda \frac{dQ^*}{dP}\right) \le E_{Q^*} I\left(\lambda \frac{dQ^*}{dP}\right) = x. \end{aligned}$$

That is, the claim $I\left(\lambda \frac{dQ^*}{dP}\right)$ is financeable with respect to all measures $Q \in \mathcal{K}$. The maximum expected utility of claims which are w.r.t. $Q \in \mathcal{K}$ financeable is at least as large as the expected utility of the claim $I\left(\lambda \frac{dQ^*}{dP}\right)$.

It holds $U_Q(x) \ge Eu\left(I\left(\lambda \frac{dQ^*}{dP}\right)\right) = U_{Q^*}(x)$ for all $Q \in \mathcal{K}$.

According to Remark 6.15 it holds because of assumption (6.18) that

$$\{Q \in \mathcal{K} : u_\lambda^*(Q \parallel P) < \infty\} = \{Q \in \mathcal{K} : U_Q(x) < \infty\}.$$

Thus $U_{Q^*}(x) < \infty$ and it follows that Q^* is the minimax measure for X and $\mathcal{K}$. By Proposition 6.18 it thus follows $\lambda = \lambda_{Q^*}(x) \in \partial U(x)$. □

6.3.2 Relationship to Portfolio Optimization

The topic of this subsection is to show a connection of the results in Sect. 6.3.1 on minimum distance martingale measures and minimax measures to the portfolio optimization problem in (6.2). In general, we assume the conditions (6.17) and (6.18) for $\mathcal{K} = \mathcal{M}_{\mathrm{loc}}$ as well as for $\overline{x} > -\infty$. For a solution of the portfolio optimization problem, admissibility conditions must be imposed on the class of strategies to exclude doubling strategies.

Definition 6.24 (Admissible Trading Strategies) The set of **admissible trading strategies** is given by

$$\Theta_0 = \left\{ \vartheta \in L(S) \,\middle|\, \int \vartheta \, \mathrm{d}S \textit{ is uniformly bounded from below} \right\},$$

$$\Theta_1 = \left\{ \vartheta \in L(S) \,\middle|\, \int \vartheta \, \mathrm{d}S \textit{ is a } Q^*\textit{-supermartingale} \right\},$$

$$\Theta_2 = \left\{ \vartheta \in L(S) \,\middle|\, \int \vartheta \, \mathrm{d}S \textit{ is a } Q\textit{-supermartingale for all} Q \in \mathcal{M}^f \right\}.$$

Furthermore, let the sets Θ_1' and Θ_2' be defined by replacing in Θ_1 respectively Θ_2 the condition 'supermartingale' by 'martingale'.

Θ_0 represents the class of admissible trading strategies for a trader with finite credit limit. This definition is used by Goll and Rüschendorf (2001) and here it holds $\sum_{i=0}^d \vartheta_t^i S_t^i = x + \int_0^t \vartheta \, \mathrm{d}S$ for each $t \in [0, T]$. The sets Θ_1' and Θ_2' are used for the exponential utility in Rheinländer and Sexton (2011). A discussion of admissible strategies can be found in Schachermayer (2003).

For $\overline{x} = -\infty$ the optimal portfolio is in general not bounded below and thus the optimal portfolio strategy is not included in Θ_0. The following theorem relates minimum distance martingale measures to minimax measures and to optimal portfolio strategies.

Theorem 6.25 *Let $Q^* \in \mathcal{M}^e_{\mathrm{loc}}$ such that $u^*_{\lambda_0}(Q^* \,\|\, P) < \infty$ and $I\left(\lambda_0 \frac{\mathrm{d}Q^*}{\mathrm{d}P}\right) \in L^1(Q^*)$, let S be (locally) bounded, and let $\lambda_0 \in \partial U(x)$. Then*

(i) The following statements are equivalent:

(a) Q^* is a (local) minimum distance martingale measure.

(b) $E_Q I\left(\lambda_0 \frac{\mathrm{d}Q^*}{\mathrm{d}P}\right) \le E_{Q^*} I\left(\lambda_0 \frac{\mathrm{d}Q^*}{\mathrm{d}P}\right)$, $\forall Q \in \mathcal{M}_{\mathrm{loc}}$ with $u^*_{\lambda_0}(Q \,\|\, P) < \infty$.

(c) $I\left(\lambda_0 \frac{\mathrm{d}Q^*}{\mathrm{d}P}\right) = x + \int_0^T \widehat{\vartheta} \, \mathrm{d}S$ and $\widehat{\vartheta} \in \Theta_1'$.

(ii) If (c) holds, then for $i = 0, 1, 2$ *it holds*

$$\sup_{\vartheta \in \Theta_i} Eu\left(x + \int_0^T \vartheta \, dS\right) = Eu\left(x + \int_0^T \widehat{\vartheta} \, dS\right) = U_{Q^*}(x) = U(x),$$

and $\widehat{\vartheta} \in \Theta_i$ *for* $i = 0, 1, 2$. *Furthermore* $\widehat{\vartheta}$ *is the optimal strategy in* Θ_1'.

(iii) If (a) holds then Q^* *is a (local) minimax measure.*

Proof According to Eq. (6.20) it holds $(u_\lambda^*)'(x) = -\lambda I(\lambda x)$.

(i) The equivalence (a) $\Leftrightarrow$ (b) follows by Theorem 6.8 (i): $Q^* \in \mathcal{M}_{\text{loc}}^e$ with $u_{\lambda_0}^*(Q^* \parallel P) < \infty$ is a (local) minimum distance martingale measure if and only if

$$E_{Q^*}\left(u_{\lambda_0}^*\right)'\left(\frac{dQ^*}{dP}\right) \leq E_Q\left(u_{\lambda_0}^*\right)'\left(\frac{dQ^*}{dP}\right), \quad \forall Q \in \mathcal{M}_{\text{loc}}$$

with $u_{\lambda_0}^*(Q \parallel P) < \infty$. By Eq. (6.20) we get

$$E_{Q^*}\left(u_{\lambda_0}^*\right)'\left(\frac{dQ^*}{dP}\right) \leq E_Q\left(u_{\lambda_0}^*\right)'\left(\frac{dQ^*}{dP}\right)$$
$$\Longleftrightarrow E_{Q^*}I\left(\lambda_0\frac{dQ^*}{dP}\right) \geq E_Q I\left(\lambda_0\frac{dQ^*}{dP}\right).$$

(a) $\Rightarrow$ (c): If Q^* is a (local) minimum distance martingale measure, then from Theorem 6.10 or Proposition 6.12 and Eq. (6.20) the representation of $I\left(\lambda_0\frac{dQ^*}{dP}\right)$ follows with a Q^*-martingale $\int_0^\cdot \widehat{\vartheta} \, dS$ for a S-integrable, predictable process $\widehat{\vartheta}$, i.e. $\widehat{\vartheta} \in \Theta_1'$. So it remains to show the direction (c) $\Rightarrow$ (a).

From $I : \mathbb{R} \to (\overline{x}, \infty)$ and the assumption that $\overline{x} > -\infty$ we obtain that $x + \left(\widehat{\vartheta} \cdot S\right)_T \geq \overline{x} > -\infty$. Moreover, since $\widehat{\vartheta} \cdot S$ is a Q^*-martingale and $Q^* \sim P$, it follows that $-\widehat{\vartheta} \cdot S$ is P-almost surely bounded from below. Thus, we obtain by Theorem 6.13 that Q^* is the $u_{\lambda_0}^*$-projection of P on $\mathcal{M}_{\text{loc}}$.

(ii) First, we show that the strategy $\widehat{\vartheta}$ is in the respective set of admissible strategies. According to (c) $\widehat{\vartheta} \in \Theta_1'$ and since $\Theta_1' \subset \Theta_1$ is $\widehat{\vartheta}$ also in Θ_1. As in the proof of (i), for a process $\widehat{\vartheta}$ satisfying condition (c), we can conclude that $\widehat{\vartheta} \cdot S$ is bounded from below and thus $\widehat{\vartheta} \in \Theta_0$. Further, using Corollary 3.5 from Ansel and Stricker (1994) it can be concluded that $\left(\widehat{\vartheta} \cdot S\right)_T$ is a Q-supermartingale for all $Q \in \mathcal{M}^f$ and thus it is in Θ_2.

It remains to show that $\widehat{\vartheta}$ is the optimal strategy. According to Corollary 3.5 from Ansel and Stricker (1994) $\vartheta \cdot S$ is a local Q^* -martingale and for each $\vartheta \in \Theta_0$ it is also bounded from below, and thus is a Q^* -supermartingale. For $\vartheta \in \Theta_1, \Theta_2$ and $\vartheta \in \Theta_1'$ the supermartingale property follows directly.

Because of the supermartingale property it holds

$$E_{Q^*}[x + (\vartheta \cdot S)_T] \le E_{Q^*}[x + (\vartheta \cdot S)_0] = x.$$

By analogy with Lemma 6.14 (ii), for $i = 0, 1, 2$ we can conclude that

$$\sup_{\vartheta \in \Theta_i} Eu\left(x + \int_0^T \vartheta \, \mathrm{d}S\right) = Eu\left(x + \int_0^T \widehat{\vartheta} \, \mathrm{d}S\right)$$

and thus $\widehat{\vartheta}$ is an optimal portfolio strategy. In the same way, it follows that $\widehat{\vartheta}$ is the optimal strategy in Θ_1'. Furthermore, according to (c).

$$Eu\left(x + \int_0^T \widehat{\vartheta} \, \mathrm{d}S\right) = Eu\left(I\left(\lambda_0 \frac{\mathrm{d}Q^*}{\mathrm{d}P}\right)\right) = U_{Q^*}(x).$$

Therefore, according to Proposition 6.18, it follows that $U_{Q^*}(x) = U(x)$.

(iii) Since Q^* is a (local) minimum distance martingale measure and the preconditions of Proposition 6.18 are satisfied, we obtain by applying Proposition 6.18 (ii) that Q^* is a (local) minimax measure. □

Remark 6.26

(1) If $\overline{x} = -\infty$, then in (i) and (ii) only $x + \left(\widehat{\vartheta} \cdot S\right)_T \ge -\infty$ holds, from which it does not follow in part (i) that $-\widehat{\vartheta} \cdot S$ is P-almost surely bounded from below. Furthermore, in part (ii) we cannot conclude that $\widehat{\vartheta} \in \Theta_0$. Consequently, it is not sufficient in case $\overline{x} = -\infty$ to use Θ_0 as the set of admissible strategies. In this case, one uses the larger class Θ_2 (cf. Schachermayer 2003).
(2) For the exponential utility, the optimal strategy $\widehat{\vartheta}$ is also in Θ_2'. For general utility functions, however, it is not in Θ_2'.

Corollary 6.27 *Let Q^* be a minimax measure for X and $\mathcal{M}_{\mathrm{loc}}$.*

(i) If $u^*\left(\widehat{Q} \parallel P\right) < \infty$ for a measure $\widehat{Q} \in \mathcal{M}^e_{\mathrm{loc}}$ then $Q^* \sim P$.
(ii) If $Q^* \sim P$ and S is (locally) bounded, then

$$I\left(\lambda_{Q^*}(x) \frac{\mathrm{d}Q^*}{\mathrm{d}P}\right) = x + \int_0^T \widehat{\vartheta} \, \mathrm{d}S,$$

where $\widehat{\vartheta}$ is an optimal portfolio strategy and it holds

$$U(x) = U_{Q^*}(x) = \sup\{Eu(Y) : E_Q Y \le x \ for \ all \ Q \in \mathcal{M}_{\mathrm{loc}}\}.$$

Proof

(i) Proposition 6.18 (iii) shows that Q^* is a $u^*_{\lambda_{Q^*}(x)}$-projection. Since the u^*-distance is finite, $u^*\left(\widehat{Q} \parallel P\right) = Eu^*\left(\frac{d\widehat{Q}}{dP}\right) < \infty$ and the assumption (6.18) is satisfied, we obtain, by Remark 6.15, $U_{\widehat{Q}}(x) < \infty$ and thus according to Remark 6.17, that $u^*_{\lambda_{Q^*}(x)}\left(\widehat{Q} \parallel P\right) < \infty$. Because of Eq. (6.20) and $I(0) = \infty$ it holds $\left(u^*_{\lambda_{Q^*}(x)}\right)'(0) = -\lambda_{Q^*}(x) I\left(\lambda_{Q^*}(x) \cdot 0\right) = -\infty$. Now the conditions of Corollary 6.9 are satisfied and it follows $Q^* \sim P$.

(ii) By Proposition 6.18 (iii) it follows that Q^* is a $u^*_{\lambda_{Q^*}(x)}$-projection, $\lambda_{Q^*}(x) \in \partial U(x)$ with $U_{Q^*}(x) < \infty$ and thus $u^*_{\lambda_{Q^*}(x)}(Q^* \parallel P) < \infty$. With Theorem 6.25 (i), therefore, it follows that

$$I\left(\lambda_{Q^*}(x)\frac{\mathrm{d}Q^*}{\mathrm{d}P}\right) = x + \left(\widehat{\vartheta} \cdot S\right)_T, \tag{6.21}$$

where $\widehat{\vartheta} \in \Theta_1'$. With Theorem 6.25 (ii) it then follows that

$$\sup_{\vartheta \in \Theta_i} Eu\left(x + \int_0^T \vartheta \,\mathrm{d}S\right) = Eu\left(x + \int_0^T \widehat{\vartheta} \,\mathrm{d}S\right) = U_{Q^*}(x) = U(x)$$

for $i = 0, 1, 2$ and $\widehat{\vartheta} \in \Theta_i$ for $i = 0, 1, 2$. Likewise, $\widehat{\vartheta}$ is the optimal strategy in Θ_1'. So it remains to show that $U_{Q^*}(x) = \sup\{Eu(Y) : E_Q Y \le x \text{ for all } Q \in \mathcal{M}_{\text{loc}}\}$.

According to Corollary 3.5 from Ansel and Stricker (1994) $\widehat{\vartheta} \cdot S$ is a local Q-martingale and because of the representation (6.21) and since $I : \mathbb{R} \to (\overline{x}, \infty)$ it is a Q-supermartingale for each $Q \in \mathcal{M}_{\text{loc}}$. Thus, for all $Q \in \mathcal{M}_{\text{loc}}$ we obtain

$$E_Q\left[x + \left(\widehat{\vartheta} \cdot S\right)_T\right] \le E_Q\left[x + \left(\widehat{\vartheta} \cdot S\right)_0\right] = x.$$

Therefore, the optimal claim from (6.21) is financeable with respect to any price measure $Q \in \mathcal{M}_{\text{loc}}$.

By Lemma 6.14 it holds that $U_{Q^*}(x) = Eu\left(I\left(\lambda_{Q^*}(x)\frac{\mathrm{d}Q^*}{\mathrm{d}P}\right)\right)$. Suppose there was a $\widetilde{Y} \neq I\left(\lambda_{Q^*}(x)\frac{\mathrm{d}Q^*}{\mathrm{d}P}\right)$ with $E_Q\widetilde{Y} \le x$ for all $Q \in \mathcal{M}_{\text{loc}}$ and $Eu(\widetilde{Y}) = \sup\{Eu(Y) : E_Q Y \le x \text{ for all } Q \in \mathcal{M}_{\text{loc}}\}$. Since $Q^* \in \mathcal{M}_{\text{loc}}$ it also holds $E_{Q^*}\widetilde{Y} \le x$ and thus as a consequence $U_{Q^*}(x) = Eu(\widetilde{Y})$ which is a contradiction to the fact that $I\left(\lambda_{Q^*}(x)\frac{\mathrm{d}Q^*}{\mathrm{d}P}\right)$ is an optimal claim. So it holds

$$U_{Q^*}(x) = \sup\left\{Eu(Y) : E_Q Y \le x \text{ for all } Q \in \mathcal{M}_{\text{loc}}\right\}.$$

□

Remark 6.28 If one computes derivative prices using a minimax or a minimum-distance martingale measure Q^*, then Theorem 6.25 shows, that the optimal claim, i.e., the solution to problem (6.15) for $Q = Q^*$ can be hedged by a portfolio strategy $\widehat{\vartheta}$. Thus, no trading in derivatives increases the maximum expected utility relative to an optimal portfolio if prices are calculated by Q^*. It then holds

$$Eu\left(x + (\widehat{\vartheta} \cdot S)_T\right) \geq Eu(Y)$$

for all claims Y such that $E_{Q^*}Y \leq x$, i.e. for all claims that can be financed under the price measure Q^*.

Davis proposes another method for pricing a claim. If the price of a contingent claim is defined by Davis' fair price, then one cannot increase the maximum expected utility of the final asset by taking an infinitesimal long or short position of the derivative in comparison to an optimal portfolio. Under regularity assumptions the **fair price of Davis** of a contingent claim B is obtained as a function of the initial financial x by

$$p(B) = \frac{E\left[u'\left(x + (\widehat{\vartheta} \cdot S)_T\right) B\right]}{\text{const}}, \tag{6.22}$$

where $\widehat{\vartheta}$ is the optimal portfolio strategy. If one puts in formula (6.22) the representation of the density of the minimum distance martingale measure from Theorem 6.25 (i), then one obtains Davis' fair price as the expected value of the claim B with respect to the minimum distance martingale measure. Thus, neither infinitesimal nor general trading increases the maximum expected utility.

Corollary 6.29 *Let the assumptions from Theorem 6.25 hold. If Q^* is the minimum distance martingale measure, then* ***Davis' fair derivative price*** *is given by*

$$p(B) = E_{Q^*}B. \tag{6.23}$$

If we consider the exponential utility, then the duality results also hold for Θ_2'. For this we consider for $p > 0$ the portfolio optimization problem (6.2) for $i = 0, 1, 2$ and likewise for Θ_1' and Θ_2', i.e.

$$\sup_{\vartheta \in \Theta_i} E\left[1 - \exp\left(-p\left(x + \int_0^T \vartheta_t \,\mathrm{d}S_t\right)\right)\right] \tag{6.24}$$

and the associated minimum distance problems, respectively. For these, we obtain:

Theorem 6.30 *It holds*

$$\inf_{\vartheta \in \Theta_i'} E \exp\left(-p \int_0^T \vartheta_t \,\mathrm{d}S_t\right) = \exp\left(-H(Q^* \parallel P)\right), \tag{6.25}$$

where Q^* is the minimum entropy martingale measure and the infimum is attained by $-\frac{1}{p}\widehat{\vartheta} \in \Theta_i'$ for $i = 1, 2$.

Proof Sketch First, the result for Θ_1' is shown. Here the representation

$$\frac{\mathrm{d}Q^*}{\mathrm{d}P} = \exp\left(x + \int_0^T \widehat{\vartheta}_t \,\mathrm{d}S_t\right), \quad \widehat{\vartheta} \in \Theta_1',$$

of the density of the minimal entropy martingale measure (see Rheinländer and Sexton 2011, Theorem 7.14) is fundamental. As an additional tool, the Jensen inequality is used.

Then the result is extended to Θ_2'. The same procedure as for the first case yields Eq. (6.25). Then it must be shown that $-\frac{1}{p}\widehat{\vartheta} \in \Theta_2'$. To do this, one shows using Barron's inequality, see Proposition 7.21 from Rheinländer and Sexton (2011), the uniform integrability of $\left\{\int_0^\tau \widehat{\vartheta}_t \,\mathrm{d}S_t\right\}_\tau$, where τ is from the set of stopping times with values in $[T, T]$. It then follows with a result of Chou et al. (1980) that $\int \widehat{\vartheta}\,\mathrm{d}S$ is a Q-martingale for all $Q \in \mathcal{M}^f$.

Thus, for the exponential utility, the entropic duality is valid for Θ_2', the set of admissible strategies such that $(\widehat{\vartheta} \cdot S)_T$ is a Q-martingale for all $Q \in \mathcal{M}^f$. The duality results for Θ_2', however, cannot be extended to general utility functions $u : \mathbb{R} \to \mathbb{R}$ because the class Θ_2' is too small to contain the optimal strategy (cf. Schachermayer 2003, Proposition 4). □

6.4 Utility-Based Hedging

For the utility-based hedging problem (6.3)

$$\sup_{\vartheta \in \mathcal{E}} Eu\left(x + \int_0^T \vartheta_t \,\mathrm{d}S_t - B\right)$$

with $\mathcal{E} = \Theta_i$, $i = 0, 1, 2$ of a contingent claim $B \geq 0$ there is a similar duality theory as for the corresponding portfolio optimization problem in Sect. 6.3.2. We are looking for an admissible hedging strategy which, given the initial capital x hedges the claim B with respect to the cost function u as well as possible. u describes the risk aversion of an investor.

If $\overline{x} := \inf\{x \in \mathbb{R};\, u(x) > -\infty\} \geq 0$, then the criterion (6.3) only admits super-hedging strategies, since in this case $x + \int_0^T \vartheta \,\mathrm{d}S - B \geq 0$. To allow for more general strategies, suppose that

$$-\infty < \overline{x} < 0 \quad \text{and} \quad x - \overline{x} > \sup_{Q \in \mathcal{M}^e_{\mathrm{loc}}} E_Q B.$$

Here $\sup_{Q \in \mathcal{M}^e_{\text{loc}}} E_Q B$ is the minimum cost of a super-hedging strategy. We choose $\overline{x}$ small enough to not exclusively allow super-hedging strategies, and large enough to not exclude super-hedging strategies.

Analogously to Sect. 6.3.2 we define for $Q \in \mathcal{K}$ and $x > \overline{x}$

$$U_Q^B(x) := \sup\{Eu(Y - B);\, E_Q Y \le x,\, Eu(Y - B)^- < \infty\}.$$

$U_Q^B(x)$ corresponds to the maximum utility to be expected from utility-based hedging of the claim B, which can be achieved with a capital x that can be achieved when market prices are computed by Q. The following lemma generalizes Lemma 6.14, and the proof is similar.

Lemma 6.31 *Let $Q \in \mathcal{M}_{\text{loc}}$ and $E_Q\left[I\left(\lambda \frac{dQ}{dP}\right) + B\right] < \infty \quad \forall \lambda > 0$. Then it holds*

(i) $U_Q^B(x) = \inf_{\lambda > 0}\left(Eu^*\left(\lambda \frac{dQ}{dP}\right) + \lambda\left(x - E_Q B\right)\right)$.

(ii) For the equation $E_Q\left[I\left(\lambda \frac{dQ}{dP}\right) + B\right] = x$ there exists a unique solution for λ denoted by $\lambda_Q(x) \in (0, \infty)$ and it holds

$$U_Q^B(x) = Eu\left(I\left(\lambda_Q(x)\frac{dQ}{dP}\right)\right).$$

Proof Let $I\left(\lambda \frac{dQ}{dP}\right) + B \in L^1(Q)$ with $E_Q Y \le x$ and $Eu(Y - B)^- < \infty$. Then $\lambda > 0$

$$\begin{aligned}
Eu(Y - B) &\le Eu(Y - B) + \lambda(x - E_Q Y)\\
&= Eu(Y - B) + \lambda\left(x - E_Q\left[Y - B + B\right]\right)\\
&= Eu(Y - B) - E\left[\lambda \frac{dQ}{dP}(Y - B)\right] + \lambda\left(x - E_Q B\right)\\
&\le Eu^*\left(\lambda \frac{dQ}{dP}\right) + \lambda\left(x - E_Q B\right)\\
&= E\left(I\left(\lambda \frac{dQ}{dP}\right)\right) - E\left[\lambda \frac{dQ}{dP} I\left(\lambda \frac{dQ}{dP}\right)\right] + \lambda\left(x - E_Q B\right)\\
&= Eu\left(I\left(\lambda \frac{dQ}{dP}\right)\right) + \lambda\left(x - E_Q\left[I\left(\lambda \frac{dQ}{dP}\right) + B\right]\right).
\end{aligned} \tag{6.26}$$

Equality holds if and only if $Y - B = I\left(\lambda_Q(x)\frac{dQ}{dP}\right)$ with $\lambda_Q(x)$ is a solution of the equation $E_Q\left[I\left(\lambda\frac{dQ}{dP}\right) + B\right] = x$. The unambiguous existence of $\lambda_Q(x)$ and the fact that

$$E\left[u\left(I\left(\lambda_Q(x)\frac{dQ}{dP}\right)\right)\right]^- < \infty$$

can be shown analogously to the proof of Lemma 6.14. Thus:

$$U_Q^B(x) = Eu\left(I\left(\lambda_Q(x)\frac{dQ}{dP}\right)\right) \tag{6.27}$$

with $E_Q\left[I\left(\lambda_Q(x)\frac{dQ}{dP}\right) + B\right] = x$. By (6.26) we get

$$U_Q^B(x) = \inf_{\lambda>0}\left\{Eu^*\left(\lambda\frac{dQ}{dP}\right) + \lambda\left(x - E_Q B\right)\right\}.$$

□

Define

$$U^B(x) := \inf_{Q\in\mathcal{M}_{\mathrm{loc}}} \sup_{E_Q Y\le x} Eu(Y - B). \tag{6.28}$$

The measure Q^* which minimizes the mapping $Q \mapsto U_Q^B$ is the analogue of the minimax measure in the utility-based hedging case.

In the following let $U^B(x) < \infty$ and assumption (6.18) be fulfilled for $\mathcal{M}_{\mathrm{loc}}$. Furthermore, let S be a locally bounded semimartingale.

To the hedging problem (6.3), we introduce a dual minimization problem over martingale measures, viz. for $\lambda_0 \in \partial U^B(x)$:

$$\inf_{Q\in\mathcal{M}_{\mathrm{loc}}} E\left[u^*\left(\lambda_0\frac{dQ}{dP}\right) - \lambda_0\frac{dQ}{dP}B\right]. \tag{D}$$

For utility-based hedging, we obtain the following duality result.

Theorem 6.32 *Let $\lambda_0 \in \partial U^B(x)$ and let $Q^* \in \mathcal{M}_{\mathrm{loc}}^e$ be such that $u_{\lambda_0}^*(Q^*||P) < \infty$ and*

$$I\left(\lambda_0\frac{dQ^*}{dP}\right) + B \in L^1(Q^*).$$

Then it holds

(i) The following statements are equivalent:
  (a) Q^* *solves problem (D).*
  (b) $E_Q\left[I\left(\lambda_0 \frac{dQ^*}{dP}\right) + B\right] \le E_{Q^*}\left[I\left(\lambda_0 \frac{dQ^*}{dP}\right) + B\right]$, $\forall Q \in \mathcal{M}_{\text{loc}}$, *with* $E\left[u^*\left(\lambda_0 \frac{dQ}{dP}\right) - \lambda_0 \frac{dQ}{dP} B\right] < \infty$.
  (c) $I\left(\lambda_0 \frac{dQ^*}{dP}\right) + B = x + \int_0^T \widehat{\vartheta}\, dS$ *and* $\widehat{\vartheta} \in \Theta_1'$.

(ii) If (c) holds, then $\widehat{\vartheta}$ *is an optimal hedging strategy.*

The proof essentially uses a result analogously to Theorem 6.8 (i).

Theorem 6.33 *Let* $Q^* \in \mathcal{M}_{\text{loc}}$ *with* $u^*_{\lambda_0}(Q^* \parallel P) < \infty$ *and* $I\left(\lambda_0 \frac{dQ^*}{dP}\right) \in L^1(Q^*)$. *Then* Q^* *solves the dual problem (D) iff*

$$E_Q\left[I\left(\lambda_0 \frac{dQ^*}{dP}\right) + B\right] \le E_{Q^*}\left[I\left(\lambda_0 \frac{dQ^*}{dP}\right) + B\right], \quad \forall Q \in \mathcal{M}_{\text{loc}}$$

with $E\left[u^*\left(\lambda_0 \frac{dQ}{dP}\right) - \lambda_0 \frac{dQ}{dP} B\right] < \infty$.

Proof The proof is analogous to the proof of Theorem 6.8. For $Q \in \mathcal{M}_{\text{loc}}$ with

$$E\left[u^*\left(\lambda_0 \frac{dQ}{dP}\right) - \lambda_0 \frac{dQ}{dP} B\right] < \infty$$

and $\alpha \in [0, 1]$ we define

$$\begin{aligned}
h_\alpha := \frac{1}{\alpha - 1}\Bigg[& u^*\left(\lambda_0\left(\alpha \frac{dQ^*}{dP} + (1-\alpha)\frac{dQ}{dP}\right)\right) \\
&- \lambda_0 B\left(\alpha \frac{dQ^*}{dP} + (1-\alpha)\frac{dQ}{dP}\right) u^*\left(\lambda_0 \frac{dQ^*}{dP}\right) + \lambda_0 \frac{dQ^*}{dP} B\Bigg] \\
= \frac{1}{\alpha - 1}&\left[u^*\left(\lambda_0\left(\alpha \frac{dQ^*}{dP} + (1-\alpha)\frac{dQ}{dP}\right)\right) - u^*\left(\lambda_0 \frac{dQ^*}{dP}\right)\right] \\
&- \lambda_0 B\left(\frac{\alpha}{\alpha - 1}\frac{dQ^*}{dP} + \frac{1-\alpha}{\alpha - 1}\frac{dQ}{dP}\right) + \lambda_0 B \frac{1}{\alpha - 1}\frac{dQ^*}{dP} \\
= -\lambda_0 B&\left(\frac{dQ^*}{dP} - \frac{dQ}{dP}\right) + \frac{1}{\alpha - 1}\left[u^*\left(\lambda_0\left(\alpha \frac{dQ^*}{dP} + (1-\alpha)\frac{dQ}{dP}\right)\right)\right. \\
&\left. - u^*\left(\lambda_0 \frac{dQ^*}{dP}\right)\right].
\end{aligned}$$

With the rule of L'Hôpital we see that h_α for $\alpha \uparrow 1$ converges to the following expression:

$$\begin{aligned}
\lim_{\alpha\uparrow 1} h_\alpha &= \lim_{\alpha\uparrow 1} \left(u^*_{\lambda_0}\right)\left(\alpha\frac{dQ^*}{dP} + (1-\alpha)\frac{dQ}{dP}\right)\left(\frac{dQ^*}{dP} - \frac{dQ}{dP}\right) \\
&\qquad - \lambda_0 B\left(\frac{dQ^*}{dP} - \frac{dQ}{dP}\right) \\
&= \lim_{\alpha\uparrow 1} -\lambda_0 I\left(\lambda_0\left(\alpha\frac{dQ^*}{dP} + (1-\alpha)\frac{dQ}{dP}\right)\right)\left(\frac{dQ^*}{dP} - \frac{dQ}{dP}\right) \\
&\qquad - \lambda_0 B\left(\frac{dQ^*}{dP} - \frac{dQ}{dP}\right) \\
&= -\lambda_0 I\left(\lambda_0\frac{dQ^*}{dP}\right)\left(\frac{dQ^*}{dP} - \frac{dQ}{dP}\right) - \lambda_0 B\left(\frac{dQ^*}{dP} - \frac{dQ}{dP}\right) \\
&= \left(\lambda_0 I\left(\lambda_0\frac{dQ^*}{dP}\right) + \lambda_0 B\right)\left(\frac{dQ}{dP} - \frac{dQ^*}{dP}\right).
\end{aligned}$$

To see that h_α is increasing in α, we show that $h_0 \le \lim_{\alpha\uparrow 1} h_\alpha$ and that h_α is concave. Let

$$\begin{aligned}
h_0 &= -\lambda_0 B\left(\frac{dQ^*}{dP} - \frac{dQ}{dP}\right) - \left[u^*\left(\lambda_0\frac{dQ}{dP}\right) - u^*\left(\lambda_0\frac{dQ^*}{dP}\right)\right] \\
&= \left(u^*\left(\lambda_0\frac{dQ^*}{dP}\right) - \lambda_0 B\frac{dQ^*}{dP}\right) - \left(u^*\left(\lambda_0\frac{dQ}{dP}\right) - \lambda_0 B\frac{dQ}{dP}\right).
\end{aligned}$$

Since $u^*_{\lambda_0}$ is convex, it follows that

$$\begin{aligned}
u^*\left(\lambda_0\frac{dQ}{dP}\right) &\ge u^*\left(\lambda_0\frac{dQ^*}{dP}\right) + \left(u^*_{\lambda_0}\right)'\left(\frac{dQ^*}{dP}\right)\left(\frac{dQ}{dP} - \frac{dQ^*}{dP}\right) \\
&= u^*\left(\lambda_0\frac{dQ^*}{dP}\right) - \lambda_0 I\left(\lambda_0\frac{dQ^*}{dP}\right)\left(\frac{dQ}{dP} - \frac{dQ^*}{dP}\right),
\end{aligned}$$

and it holds

$$u^*\left(\lambda_0\frac{dQ^*}{dP}\right) - u^*\left(\lambda_0\frac{dQ}{dP}\right) \le \lambda_0 I\left(\lambda_0\frac{dQ^*}{dP}\right)\left(\frac{dQ}{dP} - \frac{dQ^*}{dP}\right).$$

From this it follows

$$\begin{aligned} h_0 &= u^*\left(\lambda_0 \frac{\mathrm{d}Q^*}{\mathrm{d}P}\right) - u^*\left(\lambda_0 \frac{\mathrm{d}Q}{\mathrm{d}P}\right) + \lambda_0 B\left(\frac{\mathrm{d}Q}{\mathrm{d}P} - \frac{\mathrm{d}Q^*}{\mathrm{d}P}\right) \\ &\leq \lambda_0 I\left(\lambda_0 \frac{\mathrm{d}Q^*}{\mathrm{d}P}\right)\left(\frac{\mathrm{d}Q}{\mathrm{d}P} - \frac{\mathrm{d}Q^*}{\mathrm{d}P}\right) + \lambda_0 B\left(\frac{\mathrm{d}Q}{\mathrm{d}P} - \frac{\mathrm{d}Q^*}{\mathrm{d}P}\right) \\ &= \lim_{\alpha \uparrow 1} h_\alpha . \end{aligned}$$

To prove the concavity of h_α let $\widetilde{h}_\alpha := (\alpha - 1)h_\alpha$. Then the second derivative of $\widetilde{h}_\alpha \geq 0$ and $\widetilde{h}_\alpha$ is, therefore, convex. It follows that h_α is concave.

With the concavity of h_α and $h_0 \leq \lim_{\alpha\uparrow 1} h_\alpha$ we see that h_α increases monotonically against

$$\left(\lambda_0 I\left(\lambda_0 \frac{\mathrm{d}Q^*}{\mathrm{d}P}\right) + \lambda_0 B\right)\left(\frac{\mathrm{d}Q}{\mathrm{d}P} - \frac{\mathrm{d}Q^*}{\mathrm{d}P}\right).$$

So we can apply Beppo Levi's theorem on monotone convergence and get that $\int h_\alpha \,\mathrm{d}P$ converges against $\int (\lambda_0 I\left(\lambda_0 \frac{\mathrm{d}Q^*}{\mathrm{d}P}\right) + \lambda_0 B)(\mathrm{d}Q - \mathrm{d}Q^*)$.

Since $\mathcal{M}_{\mathrm{loc}}$ is convex, it follows that $\alpha \frac{\mathrm{d}Q^*}{\mathrm{d}P} + (1-\alpha)\frac{\mathrm{d}Q}{\mathrm{d}P}$ in $\mathcal{M}_{\mathrm{loc}}$. If Q^* solves the problem (D), then $h_\alpha \leq 0$ for all $\alpha \in [0, 1]$ and it follows that $\int h_\alpha \,\mathrm{d}P \leq 0$. Further it follows that the limit

$$\int \left(\lambda_0 I\left(\lambda_0 \frac{\mathrm{d}Q^*}{\mathrm{d}P}\right) + \lambda_0 B\right)(\mathrm{d}Q - \mathrm{d}Q^*)$$

is also ≤ 0 and we have shown the first direction.

On the other hand, if $\int \left(\lambda_0 I\left(\lambda_0 \frac{\mathrm{d}Q^*}{\mathrm{d}P}\right) + \lambda_0 B\right)(\mathrm{d}Q - \mathrm{d}Q^*) \leq 0$, then we obtain by the monotonicity of h_α that

$$\begin{aligned} \int h_0 \,\mathrm{d}P &= \int \left[u^*\left(\lambda_0 \frac{\mathrm{d}Q^*}{\mathrm{d}P}\right) - \lambda_0 B \frac{\mathrm{d}Q^*}{\mathrm{d}P} - \left(u^*\left(\lambda_0 \frac{\mathrm{d}Q}{\mathrm{d}P}\right) - \lambda_0 B \frac{\mathrm{d}Q}{\mathrm{d}P}\right)\right] \mathrm{d}P \\ &\leq \int \left(\lim_{\alpha\uparrow 1} h_\alpha\right) \mathrm{d}P \\ &\leq 0. \end{aligned}$$

Thus Q^* solves the dual problem (D). □

Proof of Theorem 6.32

(i) Because of Theorem 6.25, it suffices to show (b) $\Longleftrightarrow$ (c).

(b) $\Longrightarrow$ (c) With the assumption of (b), it follows, according to the projection Theorem 4.18 (ii) that

$$I\left(\lambda_0 \frac{\mathrm{d}Q^*}{\mathrm{d}P}\right) + B \in \overline{L^1(F, Q^*)},$$

the closure of f in $L^1(Q^*)$. Analogously to Theorem 6.11 then follows a representation as a stochastic integral $I\left(\lambda_0 \frac{\mathrm{d}Q^*}{\mathrm{d}P}\right) + B = c + \left(\widehat{\vartheta} \cdot S\right)_T$ for an S-integrable, predictable process $\widehat{\vartheta}$ with $\int_0^{\cdot} \widehat{\vartheta}\, \mathrm{d}S$ a Q^*-martingale, i.e. $\widehat{\vartheta} \in \Theta_1'$. With Lemma 6.31 we obtain for Q^*, the solution of the dual problem (D)

$$\begin{aligned}
U_{Q^*}^B(x) &= \inf_{\lambda>0} \left\{ Eu^*\left(\lambda \frac{\mathrm{d}Q^*}{\mathrm{d}P}\right) + \lambda\left(x - E_{Q^*}B\right) \right\} \\
&= \inf_{\lambda>0} \left\{ E\left[u^*\left(\lambda \frac{\mathrm{d}Q^*}{\mathrm{d}P}\right) - \lambda \frac{\mathrm{d}Q^*}{\mathrm{d}P} B\right] + \lambda x \right\} \\
&= \inf_{\lambda>0} \inf_{Q\in\mathcal{M}_{\mathrm{loc}}} \left\{ E\left[u^*\left(\lambda \frac{\mathrm{d}Q}{\mathrm{d}P}\right) - \lambda \frac{\mathrm{d}Q}{\mathrm{d}P} B\right] + \lambda x \right\} \\
&= \inf_{Q\in\mathcal{M}_{\mathrm{loc}}} \inf_{\lambda>0} \left\{ E\left[u^*\left(\lambda \frac{\mathrm{d}Q}{\mathrm{d}P}\right) - \lambda \frac{\mathrm{d}Q}{\mathrm{d}P} B\right] + \lambda x \right\} \\
&= \inf_{Q\in\mathcal{M}_{\mathrm{loc}}} U_Q^B(x) = U^B(x).
\end{aligned}$$

Altogether it holds

$$\begin{aligned}
U^B(x) &= \sup\left\{Eu(Y - B) : E_{Q^*}Y \le x,\ Eu(Y - B)^- < \infty\right\} \\
&= Eu\left(I\left(\lambda_0 \frac{\mathrm{d}Q^*}{\mathrm{d}P}\right)\right),
\end{aligned}$$

and $E_{Q^*}\left[I\left(\lambda_0 \frac{\mathrm{d}Q^*}{\mathrm{d}P}\right) + B\right] = x$. Since $\widehat{\vartheta} \cdot S$ is a Q^*-martingale it follows $c = x$ and it holds $I\left(\lambda_0 \frac{\mathrm{d}Q^*}{\mathrm{d}P}\right) + B = x + \left(\widehat{\vartheta} \cdot S\right)_T$.

(c) $\Longrightarrow$ (b) With $I : \mathbb{R} \to (\overline{x}, \infty)$ and since B is nonnegative it holds that

$$x + \left(\widehat{\vartheta} \cdot S\right)_T \ge x + \left(\widehat{\vartheta} \cdot S\right)_T - B \ge \overline{x}.$$

As in the proof of Theorem 6.25 $\widehat{\vartheta} \cdot S$ is a Q^*-martingale and with $Q^* \sim P$, it follows that $\widehat{\vartheta} \cdot S$ is P almost surely bounded from below. According to Ansel and Stricker (1994, Corollary 3.5) it follows that $\widehat{\vartheta} \cdot S$ is a local Q-martingale and hence also a Q-supermartingale for every $Q \in \mathcal{M}_{\mathrm{loc}}$. Thus, using assumption (c) and the supermartingale property, we obtain

$$\begin{aligned} E_Q\left[I\left(\lambda_0\frac{\mathrm{d}Q^*}{\mathrm{d}P}\right)+B\right] &= x+E_Q\left(\widehat{\vartheta}\cdot S\right)_T \\ &\leq x+E_Q\left(\widehat{\vartheta}\cdot S\right)_0 = x \\ &= E_{Q^*}\left[I\left(\lambda_0\frac{\mathrm{d}Q^*}{\mathrm{d}P}\right)+B\right]. \end{aligned}$$

(ii) First, we show that the strategy $\widehat{\vartheta}$ is in the respective set of admissible strategies. According to (c) $\widehat{\vartheta} \in \Theta_1' \subset \Theta_1$. For a process $\widehat{\vartheta}$ satisfying (c), we can conclude, as in (i), that $\widehat{\vartheta}\cdot S$ is P-almost surely bounded below, and thus $\widehat{\vartheta} \in \Theta_0$. As above, it follows that $\left(\widehat{\vartheta}\cdot S\right)_T$ is a Q-supermartingale for all $Q \in \mathcal{M}^f$ is and thus $\widehat{\vartheta} \in \Theta_2$.

If $\vartheta \in \Theta_i$ for $i = 0, 1, 2$ respectively $\vartheta \in \Theta_1'$ is an admissible strategy, then

$$\begin{aligned} &Eu\left(x+(\vartheta\cdot S)_T-B\right) \\ &\quad\leq E\left[u\left(x+\left(\widehat{\vartheta}\cdot S\right)_T-B\right)\right. \\ &\qquad\qquad \left.+u'\left(x+\left(\widehat{\vartheta}\cdot S\right)_T-B\right)\left((\vartheta\cdot S)_T-\left(\widehat{\vartheta}\cdot S\right)_T\right)\right] \\ &\quad= E\left[u\left(x+\left(\widehat{\vartheta}\cdot S\right)_T-B\right)+u'\left(I\left(\lambda_0\frac{\mathrm{d}Q^*}{\mathrm{d}P}\right)\right)\left((\vartheta\cdot S)_T-\left(\widehat{\vartheta}\cdot S\right)_T\right)\right] \\ &\quad= E\left[u\left(x+\left(\widehat{\vartheta}\cdot S\right)_T-B\right)+\lambda_0\frac{\mathrm{d}Q^*}{\mathrm{d}P}\left((\vartheta\cdot S)_T-\left(\widehat{\vartheta}\cdot S\right)_T\right)\right] \\ &\quad\leq Eu\left(x+\left(\widehat{\vartheta}\cdot S\right)_T-B\right). \end{aligned}$$

Since u is concave $-u$ convex; hence the first inequality holds. The first inequality holds with the representation $I\left(\lambda_0\frac{\mathrm{d}Q^*}{\mathrm{d}P}\right)+B = x+\int_0^T\widehat{\vartheta}\,\mathrm{d}S$ from (c). The second equality follows with $I := \left(u'\right)^{-1}$.
As before $\vartheta\cdot S-\widehat{\vartheta}\cdot S$ is a local Q^*-martingale. For the last inequality, we differentiate in the argument according to whether we consider the supremum over $\vartheta \in \Theta_0$ or over $\vartheta \in \Theta_i$ for $i = 1, 2$ resp. $\vartheta \in \Theta_1'$ respectively:
If $\vartheta \in \Theta_0$, then $\vartheta\cdot S$ is bounded below. As in part (i), we can conclude that $-\widehat{\vartheta}\cdot S$ is almost surely bounded from below. Thus also $\vartheta\cdot S-\widehat{\vartheta}\cdot S$ is almost surely bounded below and, therefore, since it is also a local Q^*-martingale, it is a Q^*-supermartingale. So it holds $E_{Q^*}\left[(\vartheta\cdot S)_T-\left(\widehat{\vartheta}\cdot S\right)_T\right] \leq E_{Q^*}\left[(\vartheta\cdot S)_0-\left(\widehat{\vartheta}\cdot S\right)_0\right] = 0$.
The last inequality follows, since $\lambda_0 > 0$.
If $\vartheta \in \Theta_i$ for $i = 1, 2$ respectively $\vartheta \in \Theta_1'$ then $\vartheta\cdot S$ a Q^*-supermartingale and it holds

$$E_{Q^*}(\vartheta\cdot S)_T \leq E_{Q^*}(\vartheta\cdot S)_0 = 0.$$

Analogously to part (i) one sees that $-\widehat{\vartheta} \cdot S$ is almost surely bounded below, so it is also a Q^*-supermartingale. Thus it holds

$$E_{Q^*}\left[-\left(\widehat{\vartheta} \cdot S\right)_T\right] \leq E_{Q^*}\left[-\left(\widehat{\vartheta} \cdot S\right)_0\right] = 0$$

and the last inequality follows, since $\lambda_0 > 0$. □

Finally, we conclude this section by giving the duality result for the exponential utility with claim B (cf. Rheinländer and Sexton 2011).

Corollary 6.34 *For any $x \in \mathbb{R}$ holds*

$$\begin{aligned}
&\sup_{\vartheta \in \Theta_i} E\left[1 - \exp\left(-p\left(x + \int_0^T \vartheta_t \,\mathrm{d}S_t - B\right)\right)\right] \\
&\quad = 1 - \exp\left(-p \inf_{Q \in \mathcal{M}^f}\left(\frac{1}{p} H(Q \parallel P) + x - E_Q B\right)\right).
\end{aligned}$$

and the supremum is attained for $\widehat{\vartheta} \in \Theta_2'$.

The proof of this result is based on an application of Theorem 6.30, on measure change and on the representation of the density of the minimal entropy martingale measure.

6.5 Examples in Exponential Lévy Models

Exponential Lévy processes are important models for pricing processes. In this chapter, using the results from Sects. 6.2 and 6.3 optimal portfolios and minimum distance martingales are determined for some standard utility functions.

Exponential Utility Function

The distance to be minimized for the exponential utility function $u(x) = 1 - e^{-px}$ is the relative entropy given by the function $f(x) = x \log x$. Using the necessary and sufficient conditions from the Propositions 6.10–6.13 we obtain for the minimum distance martingale measure a representation of the density of the form

$$\frac{\mathrm{d}Q^*}{\mathrm{d}P} = \frac{1}{\lambda_0} \exp\left(-p\left(x + (\vartheta \cdot S)_T\right)\right). \tag{6.29}$$

Theorem 6.13 provides a sufficient condition: If the stochastic integral from (6.29) is P-a.s. bounded from below and is a Q^*-martingale, then Q^* is the minimum entropy martingale measure, and ϑ is an optimal portfolio strategy. In general, the question is whether $(\vartheta \cdot S)_T$ is bounded from below, i.e., the admissibility of ϑ, is an interesting point.

Let $X = (X^1, \dots, X^d)$ be a $\mathbb{R}^d$-valued Lévy process, $\mathcal{E}$ be the stochastic exponential and let the positive price process $S = (S^1, \dots, S^d)$ be given by

$$S^i = S_0^i \mathcal{E}(X^i). \tag{6.30}$$

These processes also admit a representation of the form $S^i = S_0^i \exp\left(\widetilde{X}^i\right)$ for a $\mathbb{R}^d$-valued Lévy process $\widetilde{X}$ (see Goll and Kallsen 2000).

Let (B, C, f) be the characteristic triplet (the differential characteristic). of X with respect to a truncation function $h : \mathbb{R}^d \to \mathbb{R}^d$. Furthermore, assume there exists a $\gamma \in \mathbb{R}^d$ with the following properties:

1. $$\int |xe^{-\gamma^\top x} - h(x)| F(\mathrm{d}x) < \infty,$$

2. $$b - c\gamma + \int \left(xe^{-\gamma^\top x} - h(x)\right) F(\mathrm{d}x) = 0. \tag{6.31}$$

Let $\vartheta_t^i := \frac{\gamma^i}{S_{t-}^i}$ for $i = 1, \dots, d$, $\quad \vartheta_t^0 := x + \int_0^t \vartheta_s \,\mathrm{d}S_s - \sum_{i=1}^d \vartheta_t^i S_t^i$ for $t \in (0, T]$. Define $Z_t = \mathcal{E}\left(-\gamma^\top X_s^c + \left(e^{-\gamma^\top x} - 1\right) * \left(\mu^X - \nu\right)_s\right)_t$.

The minimum distance martingale measure with respect to the relative entropy is described in the following corollary.

Corollary 6.35 *The measure Q^* defined by $\frac{\mathrm{d}Q^*}{\mathrm{d}P} = Z_T$ is an equivalent local martingale measure. If $\gamma \cdot X$ is bounded below, then Q^* minimizes the relative entropy between P and $\mathcal{M}_{\mathrm{loc}}$, i.e. $H(Q^* \mid P) = H(\mathcal{M}_{\mathrm{loc}} \mid P)$.*

Proof In Kallsen (2000) it is shown that Z as defined above is a martingale and $S^i Z$ is a local martingale with respect to P for $i \in \{1, \dots, d\}$. The density $Z_T = \frac{\mathrm{d}Q^*}{\mathrm{d}P}$ of Q^* with respect to P is a stochastic exponential and has a representation as in (6.29) with $\vartheta_t^i := \frac{\gamma^i}{S_{t-}^i}$. Furthermore $E_{Q^*}(\vartheta \cdot S)_T = 0$. This implies that the relative entropy between Q^* and P is finite, i.e. $E_{Q^*} \log\left(\frac{\mathrm{d}Q^*}{\mathrm{d}P}\right) < \infty$.

If the process is $\vartheta \cdot S = \gamma \cdot X$ is bounded from below, then according to Theorem 6.13 it follows that Q^* minimizes the relative entropy between P and $\mathcal{M}_{\mathrm{loc}}$. □

The condition that $\gamma \cdot X = \vartheta \cdot S$ is bounded from below is not satisfied in general. In the context of portfolio optimization for the unconstrained case, this issue is discussed in Kallsen (2000) and Schachermayer (2001).

Logarithmic Utility Function
To the logarithmic utility function $u(x) = \log x$ the associated f-divergence distance with $f(x) = -\log x$ is the reverse relative entropy.

The characteristic (B, C, ν) of the $\mathbb{R}^d$-valued semimartingale $(S^1, \dots, S^d)$ with respect to a fixed truncation function $h : \mathbb{R}^d \to \mathbb{R}^d$ is given by

$$B = \int_0^{\cdot} b_t \, \mathrm{d}A_t, \quad C = \int_0^{\cdot} c_t \, \mathrm{d}A_t, \quad \nu = A \otimes F, \tag{6.32}$$

where $A \in \mathcal{A}^+_{\mathrm{loc}}$ is a predictable process, B a predictable $\mathbb{R}^d$-valued process, C a predictable $\mathbb{R}^{d\times d}$-valued process whose values are nonnegative symmetric matrices, and f a transition kernel of $(\Omega \times \mathbb{R}_+, \mathscr{P})$ according to $(\mathbb{R}^d, \mathbb{B}^d)$. T hereby $\mathcal{A}^+_{\mathrm{loc}}$ describes the class of locally integrable, adapted, increasing processes, and $\mathscr{P}$ the predictable σ-algebra.

Furthermore, let there exist a $\mathbb{R}^d$-valued, S-integrable process H with the following properties:

1. $1 + H_t^\top x > 0$ for $(A \otimes F)$-almost all $(t, x) \in [0, T] \times \mathbb{R}^d$,
2. $\int \left| \frac{x}{1 + H_t^\top x} - h(x) \right| F_t(\mathrm{d}x) < \infty$ $(P \otimes A)$-almost surely on $\Omega \times [0, T]$,
3. $$b_t - c_t H_t + \int \left(\frac{x}{1 + H_t^\top x} - h(x) \right) F_t(\mathrm{d}x) = 0 \tag{6.33}$$

   $(P \otimes A)$-almost surely on $\Omega \times [0, T]$.

Let for $t \in (0, T]$

$$\vartheta_t^i := x H_t^i \mathcal{E}\left(\int_0^{\cdot} H_s \, \mathrm{d}S_s \right)_{|t-} \quad \text{for } i = 1, \dots d, \ \vartheta_t^0 := x + \int_0^t \vartheta_s \, \mathrm{d}S_s - \sum_{i=1}^d \vartheta_t^i S_t^i.$$

In Theorem 3.1 from Goll and Kallsen (2000) it is shown that ϑ, defined as above, is an optimal portfolio strategy for the logarithmic utility maximization problem. Based on this, Theorem 6.13 gives a characterization of the local martingale measure that minimizes the reverse relative entropy.

Corollary 6.36 *If* $Z_t := \mathcal{E}\left(-H \cdot S_s^c + \left(\frac{1}{1+H^\top x} - 1 \right) * (\mu^S - \nu)_s \right)_t$ *is a martingale, then the associated measure* Q^* *with density* Z_T *is an equivalent local martingale measure and it minimizes the reverse relative entropy.*

Proof According to Goll and Kallsen (2000, proof of Theorem 3.1) (Z_t) is a positive local martingale such that $S^i Z$ is a local martingale for $i \in \{1, \dots, d\}$. Further $\frac{x}{Z_T} = x + (\vartheta \cdot S)_T$ and $\vartheta \cdot S$ is bounded from below. If as assumed Z is a martingale then Z_T is the density of an equivalent local martingale measure Q^*. For Q^* it holds

$$E\left[-\log\left(\frac{\mathrm{d}Q^*}{\mathrm{d}P}\right)\right] = E\log\left(\frac{1}{Z_T}\right) = E\log\left(x + (\vartheta \cdot S)_T\right) - \log x.$$

As in the proof of Lemma 6.14, it follows that

$$E\left[-\log\left(\frac{\mathrm{d}Q^*}{\mathrm{d}P}\right)\right] + \log x = E\log\left(x + (\vartheta \cdot S)_T\right) \leq E\left[-\log\left(\frac{\mathrm{d}Q}{\mathrm{d}P}\right)\right] + x$$

for all $Q \in \mathcal{M}_{\mathrm{loc}}$.

If the reverse relative entropy of Q^* is infinite, it follows from this inequality that all other measures $Q \in \mathcal{M}_{\mathrm{loc}}$ also have infinite reverse relative entropy.

Because of

$$E_{Q^*}[x + (\vartheta \cdot S)_T] = E[Z_T\,(x + (\vartheta \cdot S)_T)] = x$$

it follows by Theorem 6.13 that Q^* minimizes the reverse relative entropy. □

Derivative Pricing by the Esscher Transform

In this subsection, we show that the solution to the utility maximization problem with respect to a special power utility function can be determined using the Esscher transform.

The Esscher transform is a classical means of finding an equivalent martingale measure. Let the price process $S = (S_t)_{t \leq T}$ be given by a Lévy process $X = (X_t)_{t \leq T}$ with $X_0 = 0$ such that $S_t = e^{X_t}$. Let M be the moment generating function of X with $M(u,t) = M(u)^t = Ee^{uX_t}$. Let M exist for $|u| < C$ for a constant $C > 0$. The Esscher transformation defines a set of measures $\left\{Q^\vartheta : |\vartheta| < C\right\}$ by

$$\frac{\mathrm{d}Q^\vartheta}{\mathrm{d}P} = \frac{e^{\vartheta X_T}}{M(\vartheta)^T}.$$

If $\widehat{\vartheta}$ is a solution of

$$0 = \log\left(\frac{M(\vartheta + 1)}{M(\vartheta)}\right), \tag{6.34}$$

then $Q^{\widehat{\vartheta}}$ is an equivalent martingale measure, the Esscher pricing measure (the Esscher transform).

Let u be the power utility function, $u(x) = \frac{x^p}{p}$, $p \in (-\infty, 1)\backslash\{0\}$. Then the condition (c) from Theorem 6.25 (i) for a minimal distance martingale measure and for a strategy $\widehat{\varphi}$ is equivalent to

$$\frac{dQ^*}{dP} = \frac{(x + (\widehat{\varphi} \cdot S)_T)^{p-1}}{\lambda_0} \quad \text{and } \widehat{\varphi} \cdot S \text{ is a } Q^* \text{ martingale.}$$

The Esscher martingale measure $Q^{\widehat{\vartheta}}$, defined by

$$\frac{dQ^{\widehat{\vartheta}}}{dP} = \frac{e^{\widehat{\vartheta} X_T}}{M(\widehat{\vartheta})^T} \quad \text{with } \widehat{p} = \widehat{\vartheta} + 1,$$

with the constant strategy $\widehat{\varphi} = x$ and $\lambda_0 = x^{\widehat{\vartheta}} M(\widehat{\vartheta})$ satisfies condition (c) of Theorem 6.25 (i) respectively, the assumption from Proposition 6.12. Therefore, as a consequence, we have the following corollary.

Corollary 6.37 *In the model of an exponential Lévy process, the Esscher transform $Q^{\widehat{\vartheta}}$ is a minimum distance martingale measure for the power divergence $f(x) = -\frac{\widehat{p}-1}{\widehat{p}} x^{\frac{\widehat{p}}{\widehat{p}-1}}$ if $\widehat{\vartheta}$ solves (6.34) and $\widehat{p} = \widehat{\vartheta} + 1 < 1$. Furthermore, $Q^{\widehat{\vartheta}}$ is a minimax measure for the power utility function $\frac{x^{\widehat{p}}}{\widehat{p}}$.*

Corollary 6.37 shows that the Esscher measure $\mathbb{Q}^{\widehat{\vartheta}}$ is the minimum distance martingale measure to the (special) power utility function $u(x) = \frac{x^{\widehat{p}}}{\widehat{p}}$ where the parameter $\widehat{p}$ is determined such that $K \cdot e^{X_T}$ is equal to the value of the optimal portfolio at time T. The optimal portfolio strategy constantly invests the entire wealth in the risky security.

In the next subsection, we consider the solution to this problem for general power utility functions.

Power Utility Function

Next, we determine the local martingale measure, which minimizes the f-divergence distance for the general power utility function $f(x) = -\frac{p-1}{p} x^{\frac{p}{p-1}}$ ($p \in (-\infty, 1) \setminus \{0\}$) if the discounted price process is $S = (S^1, \dots, S^d)$ is of the form

$$S^i = S_0^i \mathcal{E}(X^i) \tag{6.35}$$

for a $\mathbb{R}^d$-valued Lévy process $X = (X^1, \dots, X^d)$. According to Theorem 6.25 this problem is equivalent to the portfolio optimization problem for the utility function $u(x) = \frac{x^p}{p}$.

Let (B, C, f) be the differential triplet of X with respect to a truncation function $h : \mathbb{R}^d \to \mathbb{R}^d$. Let there exist a $\gamma \in \mathbb{R}^d$ with the following properties:

1. $F(\{x \in \mathbb{R}^d : 1 + \gamma^\top x \leq 0\}) = 0$,

2. $\int \left| \frac{x}{(1+\gamma^\top x)^{1-p}} - h(x) \right| F(\mathrm{d}x) < \infty$,
3.

$$b + (p-1)c\gamma + \int \left(\frac{x}{(1+\gamma^\top x)^{1-p}} - h(x) \right) F(\mathrm{d}x) = 0. \tag{6.36}$$

Let $\vartheta_t^i := \frac{\gamma^i}{S_{t-}^i} \widetilde{V}_{t-}$ for $i = 1, \dots, d$, $\vartheta_t^0 := x + \int_0^t \vartheta_s \, \mathrm{d}S_s - \sum_{i=1}^d \vartheta_t^i S_t^i$ for $t \in (0, T]$, where $\widetilde{V}$ is the value process with respect to ϑ i.e. for $f = u^*$.

Define $Z_t := \mathcal{E}\left((p-1)\gamma^\top X_s^c + \left((1+\gamma^\top x)^{p-1} - 1 \right) * (\mu^X - \nu)_s \right)_t$, then it holds

Corollary 6.38 *The measure Q^* defined by $\frac{\mathrm{d}Q^*}{\mathrm{d}P} = Z_T$ is an equivalent local martingale measure and it minimizes the f-divergence distance for $f(x) = -\frac{p-1}{p} x^{\frac{p}{p-1}}$.*

Proof In the proof of Theorem 3.2 in Kallsen (2000), it is shown that Z is a positive martingale such that $S^i Z$ is a local martingale with respect to P for $i \in \{1, \dots, d\}$. The density process $Z_T = \frac{\mathrm{d}Q^*}{\mathrm{d}P}$ of Q^* with respect to P has the representation

$$Z_T = \frac{(x + (\vartheta \cdot S)_T)^{p-1}}{E\,(x + (\vartheta \cdot S)_T)^{p-1}}$$

$\vartheta_t^i := \frac{\gamma^i}{S_{t-}^i} \widetilde{V}_{t-}$. Furthermore $E_{Q^*}(\vartheta \cdot S)_T = 0$ and, therefore, it follows $f(Q^* \mid P) < \infty$. Since the process $\vartheta \cdot S$ is bounded from below, the assertion follows by Theorem 6.13. □

The portfolio strategy ϑ invests a constant fraction of the relative value process $\frac{\widetilde{V}_t}{S_{t-}^i}$ in the stock. According to Kallsen (2000) ϑ is an optimal portfolio strategy. For locally bounded exponential Lévy processes, this statement also follows from Theorem 6.13.

Remark 6.39 Using the notion of *w-admissible* strategies from Definition 5.56 and the corresponding representation Theorem 5.58 the characterizations of minimum distance martingale measures, minimax measures, and the associated optimal portfolio strategies can also be applied to the case of pricing models that are not necessarily locally bounded (cf. Biagini and Frittelli 2004, 2005).

6.6 Properties of the Utility Indifference Price

The utility indifference price $\pi(B)$ of a claim B for a seller with capital x with trading opportunity is defined in (6.8) as the solution $\pi(B)$ the equation

$$V(x) = V(x + \pi(B) - B);$$

and without trading opportunity as the solution $\pi_0(B)$ of Eq. (6.7)

$$u(x) = Eu(x + \pi_0(B) - B).$$

The utility indifference price is generally a nonlinear function on the set of claims because of the concavity of the utility function.

Let in the following $\mathcal{M}^f$ be non empty, $\mathcal{K} = \mathcal{M}_{\text{loc}}$ and let S be a locally bounded semimartingale. The derivation of the following properties of the utility indifference price makes essential use of the duality theorems of Sect. 6.3, in particular of

$$V(x) = \sup_{\vartheta\in\Theta_2} Eu\left(x + \int_0^T \vartheta_t \, \mathrm{d}S_t\right) = \inf_{\substack{\lambda>0\\ Q\in\mathcal{M}^f}} \left\{u_\lambda^*(Q \parallel P) + \lambda x\right\}, \tag{6.37}$$

$$\begin{aligned} V(x + \pi(B) - B) &= \sup_{\vartheta\in\Theta_2} Eu\left(x + \pi(B) + \int_0^T \vartheta_t \, \mathrm{d}S_t - B\right) \\ &= \inf_{\substack{\lambda>0\\ Q\in\mathcal{M}^f}} \left\{u_\lambda^*(Q \parallel P) + \lambda(x + \pi(B) - E_Q B)\right\}. \end{aligned} \tag{6.38}$$

The following results are based on the work of Owen and Žitković (2009), Biagini et al. (2011), and Rheinländer and Sexton (2011).

Proposition 6.40 (Properties of the Utility Indifference Price) *The utility indifference price is well-defined and satisfies the following properties:*

(1) (Translation invariance) $\pi(B + c) = \pi(B) + c$ *for* $c \in \mathbb{R}$*;*
(2) (Monotonicity) $\pi(B_1) \le \pi(B_2)$ *if* $B_1 \le B_2$*;*
(3) (Convexity) Given two contingent claims B_1 *and* B_2 *then for each* $\gamma \in [0, 1]$

$$\pi(\gamma B_1 + (1 - \gamma)B_2) \le \gamma\pi(B_1) + (1 - \gamma)\pi(B_2);$$

(4) (Pricing of hedgeable claims) For a hedgeable claim $B = b + \int_0^T \vartheta_t \, \mathrm{d}S_t$ *with* $\vartheta \in \Theta_2'$ *it holds:*

$$\pi\left(b + \int_0^T \vartheta_t \, \mathrm{d}S_t\right) = b;$$

(5) *(Pricing with entropic penalty term)* $\pi(B)$ *allows the dual representation*

$$\pi(B) = \sup_{Q \in \mathcal{M}^f} \{E_Q B - \alpha(Q)\}$$

with a penalty term $\alpha : \mathcal{M}^f \to [0, \infty)$, $\alpha(Q) = \inf_{\lambda > 0} \frac{1}{\lambda}(u_\lambda^*(Q \parallel P) + \lambda x - V(x))$

(6) *(Price bounds) Let* Q^* *be the minimum distance martingale measure, then it holds*

$$E_{Q^*} B \le \pi(B) \le \sup_{Q \in \mathcal{M}^f} E_Q B,$$

where $E_{Q^*} B$ *is Davis's fair price;*

(7) *(Strong continuity) If* $(B_n)_{n \ge 0}$ *is a sequence of contingent claims such that*

$$\sup_{Q \in \mathcal{M}^f} E_Q[B_n - B] \to 0 \text{ and } \inf_{Q \in \mathcal{M}^f} E_Q[B_n - B] \to 0,$$

then it holds

$$\pi(B_n) \to \pi(B);$$

(8) *(Fatou property) For a sequence of contingent claims* $(B_n)_{n \ge 0}$ *it holds*

$$\pi\left(\liminf_{n \to \infty} B_n\right) \le \liminf_{n \to \infty} \pi(B_n);$$

(9) *(Continuity from below) If* $(B_n)_{n \ge 0}$ *is a sequence of contingent claims, then it holds*

$$B_n \uparrow B \ P\text{-a.s.} \Rightarrow \pi(B_n) \uparrow \pi(B).$$

Remark

(a) For the utility indifference price $\widetilde{\pi}$ of the buyer according to Definition 6.3 one obtains in part (3) the concavity of the buyer's utility indifference price instead of the convexity. In part (5), we obtain for the buyer's utility indifference price the representation

$$\widetilde{\pi}(B) = \inf_{Q \in \mathcal{M}^f} \{E_Q B + \alpha(Q)\},$$

in part (6) one obtains as price bounds

$$\inf_{Q \in \mathcal{M}^f} E_Q B \le \widetilde{\pi}(B) \le E_{Q^*} B$$

and in part (9) one obtains continuity from above.

(b) As a consequence of strong continuity, in Biagini et al. (2011), norm continuity is shown. The main argument is to apply an extension of the Namioka–Klee theorem (see Biagini and Frittelli 2009, Theorem 1). It also follows from this extension of the Namioka–Klee theorem that π is subdifferentiable.

Proof For the existence and uniqueness of a solution of (6.1) consider $V(x+p-B)$. Analogously to the proof of parts (2) and (3), one can show that this function in P is monotonically increasing and concave. Using the theorem on monotone convergence and the Fenchel inequality, we see that the function $V(x+p-B)$ in P assumes values in $(-\infty, u(+\infty)]$ and well-definedness follows.

(1) Follows directly from the definition and the well-definedness of π.
(2) Using the duality result (6.37) it follows that $V(X)$ is monotonically increasing. Let $x_1 \le x_2$ and Q_2^* be optimal in the dual problem for the initial value x_2. Then it holds

$$\begin{aligned}
V(x_1) &= \inf_{\substack{\lambda>0\\ Q\in\mathcal{M}^f}} \left\{u_\lambda^*(Q \parallel P) + \lambda x_1\right\}\\
&\le \inf_{\lambda>0} \left\{u_\lambda^*(Q_2^* \parallel P) + \lambda x_1\right\}\\
&\le \inf_{\lambda>0} \left\{u_\lambda^*(Q_2^* \parallel P) + \lambda x_2\right\}\\
&= \inf_{\substack{\lambda>0\\ Q\in\mathcal{M}^f}} \left\{u_\lambda^*(Q \parallel P) + \lambda x_2\right\} = V(x_2).
\end{aligned}$$

Thus the assertion follows from the monotonicity of $V(X)$ and the definition of the utility indifference price.
(3) From the duality result (6.37) it further follows that $V(X)$ is concave.
Let $\gamma \in [0,1]$ then it holds

$$\begin{aligned}
&V(\gamma x_1 + (1-\gamma)x_2)\\
&= \inf_{\substack{\lambda>0\\ Q\in\mathcal{M}^f}} \{u_\lambda^*(Q \parallel P) + \lambda(\gamma x_1 + (1-\gamma)x_2)\}\\
&= \inf_{\substack{\lambda>0\\ Q\in\mathcal{M}^f}} \{\gamma(u_\lambda^*(Q \parallel P) + \lambda x_1) + (1-\gamma)(u_\lambda^*(Q \parallel P) + \lambda x_2)\}\\
&\ge \inf_{\substack{\lambda>0\\ Q\in\mathcal{M}^f}} \{\gamma(u_\lambda^*(Q \parallel P) + \lambda x_1)\} + \inf_{\substack{\lambda>0\\ Q\in\mathcal{M}^f}} \{(1-\gamma)(u_\lambda^*(Q \parallel P) + \lambda x_2)\}\\
&= \gamma V(x_1) + (1-\gamma)V(x_2).
\end{aligned}$$

Similarly $V(x, B) := V(x - B) = \sup_{\upsilon \in \Theta_2} Eu(x + \int_0^T \vartheta_t \, dS_t)$ is concave in B. Because for $\gamma \in (0, 1)$ it holds

$$\begin{aligned}
&V(x, \gamma B_1 + (1-\gamma)B_2) \\
&= \sup_{\vartheta \in \Theta_2} Eu(x + (\vartheta \cdot S)_T - (\gamma B_1 + (1-\gamma)B_2)) \\
&= \sup_{\vartheta_1, \vartheta_2 \in \Theta_2} Eu(\gamma(x + (\vartheta_1 \cdot S)_T - B_1) + (1-\gamma)(x + (\vartheta_2 \cdot S)_T - B_2)) \\
&\geq \sup_{\vartheta_1, \vartheta_2 \in \Theta_2} [\gamma E(x + (\vartheta_1 \cdot S)_T - B_1) + (1-\gamma)E(x + (\vartheta_2 \cdot S) - B_2] \\
&= \gamma V(x, B_1) + (1-\gamma)V(x, B_2).
\end{aligned}$$

From this follows

$$\begin{aligned}
&V(\gamma x + (1-\gamma)x + \gamma\pi(B_1) + (1-\gamma)\pi(B_2) - \gamma B_1 - (1-\gamma)B_2) \\
&= V(\gamma(x + \pi(B_1) - B_1) + (1-\gamma)(x + \pi(B_2) - B_2)) \\
&\geq \gamma V(x + \pi(B_1) - B_1) + (1-\gamma)V(x + \pi(B_2) - B_2) \\
&= \gamma V(x) + (1-\gamma)V(x) = V(x).
\end{aligned}$$

Altogether we get

$$\begin{aligned}
V(x) &= V(x + \pi(\gamma B_1 + (1-\gamma)B_2) - (\gamma B_1 + (1-\gamma)B_2)) \\
&\leq V(x + \gamma\pi(B_1) + (1-\gamma)\pi(B_2) - (\gamma B_1 + (1-\gamma)B_2)).
\end{aligned}$$

Because of the monotonicity of $V(X)$, therefore, it holds

$$\pi(\gamma B_1 + (1-\gamma)B_2) \leq \gamma\pi(B_1) + (1-\gamma)\pi(B_2),$$

i.e. π is convex.

(4) Since $\vartheta \in \Theta_2'$ it follows that $\int \vartheta \, dS$ is a Q martingale for all $Q \in \mathcal{M}^f$. Together with the duality result (6.38) we obtain

$$\begin{aligned}
&V\Big(x + \pi\Big(b + \int_0^T \vartheta_t \, dS_t\Big) - b - \int_0^T \vartheta_t \, dS_t\Big) \\
&= \inf_{\substack{\lambda > 0 \\ Q \in \mathcal{M}^f}} \Big\{u_\lambda^*(Q \,\|\, P) + \lambda\Big(x + \pi\Big(b + \int_0^T \vartheta_t \, dS_t\Big) - b\Big)\Big\}.
\end{aligned}$$

In order that

$$V\left(x + \pi\left(b + \int_0^T \vartheta_t \, dS_t\right) - b - \int_0^T \vartheta_t \, dS_t\right) = V(x)$$

holds, it follows because of (6.37) that already $\pi\left(b + \int_0^T \vartheta_t \, dS_t\right) = b$ must hold.

(5) By Definition 6.3 of $\pi(B)$ the dual representation (6.38) implies that

$$\begin{aligned} V(x) &= V(x + \pi(B) - B) \\ &= \inf_{\substack{\lambda > 0 \\ Q \in \mathcal{M}^f}} \{u_\lambda^*(Q \,\|\, P) + \lambda(x + \pi(B) - E_Q B)\} \\ &\leq \inf_{Q \in \mathcal{M}^f} \{u_\lambda^*(Q \,\|\, P) + \lambda(x + \pi(B) - E_Q B)\}, \quad \forall \lambda > 0. \end{aligned}$$

Therefore, $\pi(B) \geq \frac{V(x)}{\lambda} - \inf_{Q \in \mathcal{M}^f} \left\{ \frac{u_\lambda^*(Q \,\|\, P)}{\lambda} + x - E_Q B \right\}, \quad \forall \lambda > 0$

$$\geq \frac{V(x)}{\lambda} - \frac{u_\lambda^*(Q \,\|\, P)}{\lambda} - x + E_Q B, \quad \forall \lambda > 0, \ \forall Q \in \mathcal{M}^f.$$

So it follows that for all $Q \in \mathcal{M}^f$ it holds

$$\pi(B) \geq E_Q B - \inf_{\lambda > 0} \left\{ \frac{u_\lambda^*(Q \,\|\, P)}{\lambda} + x - \frac{V(x)}{\lambda} \right\},$$

with equality for the optimal measure Q^*. Thus it follows

$$\pi(B) = \sup_{Q \in \mathcal{M}^f} \Bigg\{ E_Q B - \underbrace{\inf_{\lambda > 0} \left\{ \frac{u_\lambda^*(Q \,\|\, P)}{\lambda} + x - \frac{V(x)}{\lambda} \right\}}_{=:\alpha(Q)} \Bigg\}.$$

(6) The first inequality follows from the representation in (5) and from $\alpha(Q^*) = 0$:

$$\pi(B) = \sup_{Q \in \mathcal{M}^f} \{E_Q B - \alpha(Q)\} \geq E_{Q^*} B - \alpha(Q^*) \geq E_{Q^*} B.$$

The second inequality follows since the penalty term $\alpha(Q) \geq 0$ and we obtain as an upper bound the weak superreplication price $\sup_{Q \in \mathcal{M}^f} E_Q B$.

(7) This property follows from the representation of the utility indifference price in part (5), since.

$$\begin{aligned}
&\inf_{Q\in\mathcal{M}^f} E_Q[B_n - B] \\
&\quad = - \sup_{Q\in\mathcal{M}^f} E_Q[B - B_n] \\
&\quad = - \sup_{Q\in\mathcal{M}^f} \{(E_Q B - \alpha(Q)) - (E_Q B_n - \alpha(Q))\} \\
&\quad \leq - \sup_{Q\in\mathcal{M}^f} \{E_Q B - \alpha(Q)\} + \sup_{Q\in\mathcal{M}^f} \{E_Q B_n - \alpha(Q)\} \\
&\quad = \pi(B_n) - \pi(B) \leq \sup_{Q\in\mathcal{M}^f} \{(E_Q B_n - \alpha(Q)) - (E_Q B - \alpha(Q))\} \\
&\quad = \sup_{Q\in\mathcal{M}^f} E_Q[B_n - B].
\end{aligned}$$

Thus it follows $\pi(B_n) \to \pi(B)$.

(8) This property follows from the representation of the utility indifference price in part (5) and Fatou's lemma:

$$\begin{aligned}
\pi\left(\liminf_{n\to\infty} B_n\right) &= \sup_{Q\in\mathcal{M}^f} \left\{E_Q\left[\liminf_{n\to\infty} B_n\right] - \alpha(Q)\right\} \\
&\leq \sup_{Q\in\mathcal{M}^f} \liminf_{n\to\infty} \left\{E_Q[B_n] - \alpha(Q)\right\} \\
&\leq \liminf_{n\to\infty} \sup_{Q\in\mathcal{M}^f} \left\{E_Q[B_n] - \alpha(Q)\right\} \\
&= \liminf_{n\to\infty} \pi(B_n).
\end{aligned}$$

(9) This property follows directly from the Fatou property. It holds

$$\pi(B) = \pi\left(\liminf_{n\to\infty} B_n\right) \leq \liminf_{n\to\infty} \pi(B_n).$$

□

Remark

(a) Because of the representation from part (5), the utility indifference price can be viewed as pricing with entropic penalty term.

(b) For the case of the exponential utility function, the utility indifference price does not depend on the initial capital and the utility indifference price $\pi(B; p)$ to the risk aversion parameter P according to (5) can also be represented as

$$\pi(B; p) = \sup_{Q \in \mathcal{M}^f} \left\{ E_Q B - \frac{1}{p} \big(H(Q \parallel P) - H(Q^* \parallel P) \big) \right\}.$$

(cf. Rheinländer and Sexton 2011, Eq. 7.15). In this case, the representation of the utility indifference price is also obtained from the dual representation of the utility maximization problem:

$$\inf_{Q \in \mathcal{M}^f} \frac{1}{p} H(Q \parallel P) = \inf_{Q \in \mathcal{M}^f} \left\{ \frac{1}{p} H(Q \parallel P) + \pi(B; p) - E_Q B \right\}.$$

(c) By Definition 6.3 it holds $\pi(B) = -\widetilde{\pi}(-B)$ with $\widetilde{\pi}$ the buyer's utility indifference price. Because of the convexity of the utility indifference price and $\pi(0) = 0$ it follows that

$$\pi(B) \geq \widetilde{\pi}(B).$$

(d) The utility indifference price is increasing with respect to risk aversion. If the investor is risk averse, he expects a risk premium for a risk is taken. That is, the more risk averse the investor, the higher the risk premium and hence the higher the utility indifference price.

The next topic is the study of volume asymptotics for the mean of the seller's utility indifference price for B units $b > 0$ for $b \to \infty$ and $b \to 0$ of the contingent claim B, i.e. for $\frac{\pi(bB)}{b}$.

Proposition 6.41 (Volume Asymptotics) *Let B be a contingent claim and $b > 0$. Then $\frac{\pi(bB)}{b}$ is a continuous increasing function in b. Furthermore, it holds*

(i) $E_{Q^*} B \leq \frac{\pi(bB)}{b} \leq \sup_{Q \in \mathcal{M}^f} E_Q B;$

(ii) $\lim_{b \uparrow \infty} \frac{\pi(bB)}{b} = \sup_{Q \in \mathcal{M}^f} E_Q B;$

(iii) $\lim_{b \downarrow 0} \frac{\pi(bB)}{b} = E_{Q^*} B.$ □

Proof The continuity in b follows from the strong continuity from Proposition 6.40. Let $0 < b_1 \leq b_2$. Set $\gamma = \frac{b_1}{b_2}$, $B_1 = b_2 B$ and $B_2 = 0$. Because of the convexity of

$\pi(B)$ it holds

$$\frac{\pi(b_1 B)}{b_1} = \frac{1}{b_1}\pi\Big(\underbrace{\frac{b_1}{b_2}}_{=\gamma}\underbrace{b_2 B}_{=B_1} + \underbrace{\Big(1-\frac{b_1}{b_2}\Big)}_{=(1-\gamma)}\underbrace{B_2}_{=0}\Big)$$

$$\leq \frac{1}{b_1}\frac{b_1}{b_2}\pi(b_2 B) + \frac{1}{b_1}\Big(1-\frac{b_1}{b_2}\Big)\underbrace{\pi(0)}_{=0} = \frac{\pi(b_2 B)}{b_2}.$$

(i) Follows from Proposition 6.40, part (6).

(ii) We prove this statement by contradiction. Suppose there exists a $\widetilde{Q} \in \mathcal{M}^f$ such that $E_{\widetilde{Q}}B > \lim_{b\uparrow\infty} \frac{\pi(bB)}{b}$. Then for each $b > 0$,

$$\begin{aligned} V(x) &= V(x + \pi(bB) - bB) \\ &= \inf_{\substack{\lambda > 0 \\ Q \in \mathcal{M}^f}} \{u_\lambda^*(Q \,\|\, P) + \lambda(x + \pi(bB) - bE_Q B)\} \\ &\leq \inf_{\lambda>0} u_\lambda^*(\widetilde{Q} \,\|\, P) + \lambda x + \lambda b\Big(\frac{\pi(bB)}{b} - E_{\widetilde{Q}}B\Big). \end{aligned}$$

The right hand side goes for $b \uparrow \infty$ to $-\infty$ and, therefore, leads to a contradiction.

(iii) Let $\pi'(C, B)$ denote the directional derivative of π at C in direction B, i.e.

$$\pi'(C, B) = \lim_{b\downarrow 0} \frac{\pi(C + bB) - \pi(C)}{b}.$$

For convex functions f on a Banach space X with values in $\mathbb{R} \cup \{\infty\}$ it holds for all continuity points $\overline{x}$ with $f(\overline{x}) < \infty$

$$f'(\overline{x}, d) = \max\{\langle x^*, d\rangle; x^* \in \partial f(\overline{x})\}$$

(cf. Borwein and Zhu 2005). Thus, for the convex price function π and a continuity point C the representation

$$\pi'(C, B) = \sup_{Q\in\partial\pi(C)} E_Q B \text{ follows.}$$

With $C = 0$ and $\pi(0) = 0$ we get

$$\lim_{b\downarrow 0} \frac{\pi(bB)}{b} = \pi'(0, B) = \sup_{Q\in\partial\pi(0)} E_Q B.$$

As a consequence of the dual representation of Proposition 6.40 part (5) (see Biagini et al. 2011, Proposition 4.2, eq. (4.4)), the subdifferential of π at zero i.e. $\partial\pi(0)$ is equal to the minimizing measure of the dual problem (6.37), i.e., equal to Q^* and the statement follows.

□

Remark 6.42

(a) For the limit in part (ii), we obtain the asymptotic behavior of the weak superreplication price for B.
(b) Using Proposition 6.41, it follows for the exponential utility function with Corollary 7.28 (iii) from Rheinländer and Sexton (2011):

$$\lim_{p\uparrow\infty} \pi(B; p) = \sup_{Q\in\mathcal{M}^e} E_Q B, \qquad \lim_{p\downarrow 0} \pi(B; p) = E_{Q^*} B.$$

If the risk aversion parameter P tends to infinity, then the utility indifference price converges to the superreplication price. If the risk aversion parameter P goes towards 0, then the linear pricing rule under the minimum entropy martingale measure Q^* is obtained. The price with respect to the minimum entropy measure is identical to Davis' fair price (cf. Corollary 6.29).

As a corollary of Proposition 6.40 one obtains

Corollary 6.43 *The mapping* $\varrho : B \mapsto \pi(-B)$ *is a convex risk measure on the set of bounded claims.*

Proof Because of the translation invariance, monotonicity and convexity from Propositions 6.40 (1), (2), and (3) and $\pi(0) = 0$ the required properties of a convex risk measure are satisfied. □

7 Variance-Minimal Hedging

This chapter is devoted to the determination of optimal hedging strategies by the criterion of variance-minimal strategies. In incomplete market models, not every claim H is hedgeable. A natural question is, therefore: How good is H hedgebable?

With respect to quadratic deviation, this question was studied in Föllmer and Sondermann (1986), in Föllmer and Schweizer (1991), and in Schweizer (1991). The answer to this question is based on the Föllmer–Schweizer decomposition, a generalization of the Kunita–Watanabe decomposition, and on the associated minimal martingale measure.

Two types of hedging problems can be distinguished. In the first type, the initial capital x_0 of the hedging strategy is fixed. In the second the hedging error of the hedging strategy is minimized over all initial values x.

The criteria for this are given by the following hedging problems:

Definition 7.1 (Mean-Variance Hedging Problem) Let $H \in L_+^2(\mathcal{A}_T, P)$ be a contingent claim.

(a) A strategy φ is called admissible if $\varphi \in \mathcal{L}^2(S)$.
(b) A pair $\Phi = (x_0, \varphi^*)$, φ^* admissible, is called **variance-minimal hedging strategy** for H, if

$$E_P(H - x_o - V_T(\varphi^*))^2 = \inf_{x,\varphi}. \tag{7.1}$$

(c) An admissible strategy $\varphi^* \in \mathcal{L}^2(S)$ is called **variance-minimal for *C* to the initial capital *x*** when

$$E_P(H - x - V_T(\varphi^*))^2 = \inf_{\varphi}. \tag{7.2}$$

L. Rüschendorf, *Stochastic Processes and Financial Mathematics*, Mathematics Study Resources 1, https://doi.org/10.1007/978-3-662-64711-0_7

If (x, φ) is a solution of the hedging problem (7.1), then $x + V_T(\varphi)$ is the hedgeable part of the claim H. x is the associated 'fair' price of H with respect to the variance criterion.

7.1 Hedging in the Martingale Case

In the case of hedging with respect to a martingale measure Q, the Kunita–Watanabe decomposition for martingales provides an optimal hedging strategy. Let $P = Q$ be a martingale measure for S, $S \in \mathcal{M}_{\text{loc}}(Q)$ and $H \in \mathcal{L}^2(\mathcal{A}_T, Q)$. Then H induces the martingale $V_t = E_Q(H \mid \mathcal{A}_t)$, $0 \le t \le T$.

Theorem 7.2 (Mean-Variance Hedging, Martingale Case) *Let $S \in \mathcal{M}_{\text{loc}}(Q)$ and $H \in L^2(\mathcal{A}_T, Q)$ and let*

$$H = x_0 + \int_0^T \varphi_u^* \, \mathrm{d}S_u + L_T \tag{7.3}$$

be the Kunita–Watanabe decomposition of V with $\varphi^* \in \mathcal{L}^2(S)$, $L_0 = 0$, $L \in \mathcal{M}^2$ such that L is strictly orthogonal to V, i.e. $\langle V, L\rangle = 0$.

Then $\Phi^* = (x_0, \varphi^*)$ is a variance-minimal hedging strategy for H.

Proof By the projection theorem for $\mathcal{L}^2(Q)$, $\Phi^* = (x_0, \varphi^*)$ is an optimal hedging strategy

$$\Longleftrightarrow E(H - x_0 - V_T(\varphi^*))(x + V_T(\varphi)) = 0, \quad \forall x, \forall \varphi \in \mathcal{L}^2(S), E = E_Q.$$

Equivalent to this is that $x_0 = E_Q H$ and using the Kunita–Watanabe decomposition that

$$\begin{aligned} E(H - x_0 - V_T(\varphi^*))V_T(\varphi) &= E L_T\Big(\int_0^T \varphi_u^* \, \mathrm{d}S_u\Big) \\ &= E \int_0^T \varphi_n^* \mathrm{d}\langle L, S\rangle_u = 0, \end{aligned}$$

since L is strictly orthogonal to S, i.e. $\langle L, S\rangle = 0$. $\Phi^* = (x_0, \varphi^*)$ is thus a variance-minimal hedging strategy for H. □

An alternative approach to optimal hedging is made possible by considering generalized strategies, not necessarily self-financing, that hedge the claim H.

Definition 7.3 (Generalized Hedging Strategy) A pair $\Phi = (\varphi^0, \varphi)$, $\varphi \in \mathcal{L}^2(S)$, φ^0 an adapted càdlàg process, is called **(generalized) hedging strategy** for $H \in$

$L^2(Q)$ if the value process satisfies

$$V_t(\Phi) = \varphi_t^0 + \varphi_t S_t \in L^2(Q) \quad and \quad V_T(Q) = H. \tag{7.4}$$

Implicitly, the discussed value process $\widetilde{V}_t(\Phi) = \varphi_t^0 + \varphi_t \widetilde{S}_t$ is considered, so that here the interest rate $r_t \equiv 1$ is chosen. It holds

$$V_t(\Phi) = V_o(\Phi) + \int_0^t \varphi_u \, \mathrm{d}S_u \tag{7.5}$$

if and only if Φ is self-financing, i.e. the value process $V_t(\Phi)$ is given by the initial value $V_0(\Phi)$ plus the gain process $G_t(\varphi) := \int_0^t \varphi_u \, \mathrm{d}S_u$.

If one defines the **cost process** as the difference between the value process and the gain process

$$C_t := C_t(\Phi) := V_t(\Phi) - G_t(\Phi), \tag{7.6}$$

then Φ is self-financing, exactly when $C_t = \text{const.} = V_0(\Phi)$. The cost process thus describes how much capital, in addition to trading with Φ has to be raised in order to realize the value process.

The **objective** is then: minimize (quadratic) functionals of the cost process C among all generalized strategies that hedge the claim H, i.e. $V_T(\Phi) = H$.

Under the assumption $H \in L^2(P)$, $S_t \in L^2(P)$, $\forall t$, this leads to the following definition.

Definition 7.4 (Risk Minimizing Strategy)

(a) $\Phi = (\varphi^0, \varphi)$ is called $\mathcal{L}^2$-admissible (hedging) strategy for H, if

$$V_T(\Phi) = H, \quad \varphi \in \mathcal{L}^2(S) \quad \text{and} \quad V_t \in L^2(P), \forall t.$$

(b) **(residual) risk process**

$$R_t(\Phi) := E((C_T - C_t)^2 \mid \mathcal{A}_t) \tag{7.7}$$

is called **(residual) risk process of Φ**.

(c) An $\mathcal{L}^2$-admissible strategy $\widehat{\Phi}$ for H is **risk-minimizing,** if

$$R_t(\widehat{\Phi}) \le R_t(\Phi), \quad \forall t \le T, \; \forall \Phi \; \mathcal{L}^2\text{-}admissible.$$

Among all $\mathcal{L}^2$-admissible strategies for H a risk-minimizing strategy $\widehat{\Phi}$ minimizes the risk process at all times $\le T$. For hedgeable claims H and, in particular, in the case of a complete market model, risk-minimizing is equivalent to φ being self-financing and $R_t(\Phi) = 0, \forall t \leq T$. Risk-minimizing strategies have the attenuated property of being 'self-financing in mean'.

Definition 7.5 (Self-Financing in Mean) An $\mathcal{L}^2$-admissible strategy Φ for H is **self-financing in mean** if $C(\Phi) \in \mathcal{M}(P)$, i.e. for $s \leq t$ holds

$$E(C_t(\Phi) \mid \mathcal{A}_s) = C_s(\Phi).$$

Proposition 7.6 *If Φ is risk-minimizing for H, then Φ is self-financing in mean.*

Proof To $s \in [0, T]$ and $\Phi = (\varphi^0, \varphi)$ let $\widetilde{\varphi} \in \mathcal{L}^2(S)$ and $\widetilde{\eta}$ be such that

$$V_t(\widetilde{\Phi}) = V_t(\Phi) \quad \text{for } 0 \leq t < s$$
$$\text{and } V_t(\widetilde{\Phi}) = E\Big(V_T(\Phi) - \int_t^T \varphi_u \, \mathrm{d}S_u \,\Big|\, \mathcal{A}_t\Big).$$

Then $\widetilde{\Phi}$ is an admissible strategy for H, because $V_T(\widetilde{\Phi}) = V_t(\Phi) = H$ and $C_T(\widetilde{\Phi}) = C_T(\Phi)$.

By construction of $\widetilde{\Phi}$ it holds:

$$C_T(\Phi) - C_s(\Phi) = C_T(\widetilde{\Phi}) - C_s(\widetilde{\Phi}) + E(C_T(\widetilde{\Phi}) \mid \mathcal{A}_s) - C_s(\Phi).$$

Thus it follows

$$R_s(\Phi) = R_s(\widetilde{\Phi}) + \big(E(C_T(\Phi) \mid \mathcal{A}_s) - C_s(\Phi)\big)^2.$$

Since Φ is risk-minimizing it holds

$$R_s(\Phi) - R_s(\widetilde{\Phi}) \leq 0.$$

Thus it follows: $E(C_T(\Phi) \mid \mathcal{A}_s) = C_s(\Phi)$ a.s., i.e. Φ is self-financing in mean. $\square$

The existence and uniqueness of a risk-minimizing strategy is shown in Föllmer and Sondermann (1986). A description of this strategy follows from the Kunita–Watanabe decomposition (KW decomposition) of H

$$H = EH + \int_0^T \varphi_u^H \, \mathrm{d}S_u + L_T^H, \tag{7.8}$$

where $L^H \in \mathcal{M}^2$ is an L^2-martingale orthogonal to S, i.e. $\langle H, S \rangle = 0$. With the value process

$$V_t^H = H_0 + \int_0^t \varphi_u^H \, \mathrm{d}S_u + L_t^H \tag{7.9}$$

it follows $V_t^H = E(H \mid \mathcal{A}_t)$ and

$$\varphi^H = \frac{\mathrm{d}\langle V^H, S\rangle}{\mathrm{d}\langle S\rangle}. \tag{7.10}$$

To prove that φ^H is risk-minimizing, we need one more definition.

Definition 7.7 The process

$$\widehat{R}_t = E((L_T^H - L_t^H)^2 \mid \mathcal{A}_t) = \widehat{R}_t(H), \quad 0 \le t \le T$$

is called **intrinsic risk process** of H.

Theorem 7.8 *There exists a unique risk-minimizing strategy for H, viz.* $\widehat{\Phi} := (V^H - \varphi^H \cdot S, \varphi^H)$ *and it holds*

$$R_t(\widehat{\Phi}) = \widehat{R}_t.$$

Proof By definition $\widehat{\Phi}$ is admissible for H, i.e. $V_T(\widehat{\Phi}) = H$ a.s. Let Φ be an admissible continuation of $\widehat{\Phi}$ at time t, i.e.

$$\Phi_s = \widehat{\Phi}_s \text{ for } s < t \quad \text{and} \quad V_T(\Phi) = V_T(\widehat{\Phi}) = H.$$

Then it holds

$$\begin{aligned} C_T(\Phi) - C_t(\Phi) &= V_T(\Phi) - V_t(\Phi) - \int_t^T \varphi_u \,\mathrm{d}S_u \\ &= \widehat{V}_0 + \int_0^T \varphi_u^H \,\mathrm{d}S_u + L_T^H - V_t(\Phi) - \int_t^T \varphi_u \,\mathrm{d}S_u \\ &= \int_t^T (\varphi_u^H - \varphi_u) \,\mathrm{d}S_u + (L_T^H - L_t^H) + (V_t(\widehat{\Phi}) - V_t(\Phi)) \end{aligned}$$

Since S and L^H are orthogonal it follows

$$\begin{aligned} R_t(\Phi) &= E((C_T(\Phi) - C_t(\Phi))^2 \mid \mathcal{A}_t) \\ &= E\Big(\int_t^T (\varphi_u^H - \varphi_u)^2 \mathrm{d}\langle S_u\rangle \mid \mathcal{A}_t\Big) + R_t(\widehat{\Phi}) + (V_t(\widehat{\Phi}) - V_t(\Phi))^2 \\ &\ge R_t(\widehat{\Phi}); \end{aligned} \tag{7.11}$$

therefore, $\widehat{\Phi}$ a risk-minimal strategy for H.

To prove uniqueness, let $\widetilde{\Phi} = (\widetilde{\eta}, \widetilde{\varphi})$ be another risk-minimizing strategy for H. Then it follows from (7.11) that $\varphi_t^H = \widetilde{\varphi}_t$ a.s. for $0 \leq t \leq T$.

Since by Proposition 7.6 $C(\widetilde{\Phi}) = V(\widetilde{\Phi}) - G(\Phi)$ is a martingale it follows that $V(\widetilde{\Phi})$ is a martingale. Because of $V_T(\widehat{\Phi}) = V_T(\widetilde{\Phi}) = H$ it, therefore, follows $V_t(\widehat{\Phi}) = V_t(\widehat{\Phi}), 0 \leq t \leq T$ and, therefore, $\widehat{\Phi} = \widetilde{\Phi}$. □

In particular, the following corollary is obtained for hedgable claims.

Corollary 7.9 (Hedgable Claims) *The following statements are equivalent:*
The risk-minimizing strategy is self-financing
$\Longleftrightarrow$ *The intrinsic risk of H is 0, $\widehat{R}_t(H) = 0$, for all $t \leq T$*
$\Longleftrightarrow$ *H is hedgeable, i.e. H has a representation of the form*

$$H = EH + \int_0^T \varphi_u^H \, \mathrm{d}S_u \ a.s.$$

However, it is shown below that in the incomplete case the risk-minimizing strategy $\varphi^H = \varphi_Q$ depends on the chosen martingale measure Q. Hence, the question of choosing an 'appropriate' martingale measure arises, as well as the question of hedging under the (statistical) measure P.

7.2 Hedging in the Semimartingale Case

Let now $S = (S_t)_{0 \leq t \leq T}$ a semimartingale with Doob–Meyer decomposition

$$S = S_0 + M + A \tag{7.12}$$

with $M \in \mathcal{M}^2$ and $A \in V$ predictable with bounded variation $|A|$. Then

$$E(S_0^2 + \langle X \rangle_T + |A|_T^2) < \infty \text{ and } \langle X \rangle = \langle M \rangle.$$

The cost process of a strategy $\Phi = (\varphi^0, \varphi)$ is given by

$$C_t(\Phi) = V_t(\Phi) - \int_0^t \varphi_s \, \mathrm{d}M_s - \int_0^t \varphi_s \, \mathrm{d}A_s. \tag{7.13}$$

In contrast to the martingale case, there is no direct generalization of the KW decomposition of H into a stochastic integral and an orthogonal component.

Schweizer (1991) introduced the concept of a local risk-minimizing strategy, i.e. a strategy that minimizes risk under small perturbations. It is shown that local risk-minimizing strategies are self-financing (cf. Sect. 7.2.1). To determine such optimal (= risk-minimizing) strategies, there is a characterization via an extended decomposition, the Föllmer–Schweizer decomposition, which leads to the solution of an optimality equation for the optimal strategy (cf. Sect. 7.2.1). A second approach by Föllmer and Schweizer (1991) reduces the determination of the optimal strategy in the semimartingale case with respect to the statistical measure P to the determination of an optimal strategy in the martingale case, with respect to a suitable equivalent martingale measure, the 'minimum martingale measure' (cf. Sect. 7.2.2).

7.2.1 Föllmer–Schweizer Decomposition and Optimality Equation

To introduce locally risk-minimizing strategies, the notion of a 'small perturbation of a strategy φ is needed.

Definition 7.10 A strategy $\Delta = (\varepsilon, \delta)$ is called (small) perturbation (strategy) if

(1) δ is bounded,
(2) $\int_0^T \delta_s \, \mathrm{d}\,|A|_s$ is bounded, and
(3) $\delta_T = \varepsilon_T = 0$.

In particular, for an admissible strategy Φ and a (small) perturbation Δ it holds that $\Phi + \Delta$ is admissible and the restriction of Δ to a subinterval is again a small perturbation.

For a finite decomposition $\tau = (t_i)_{1\le i\le N}$ of $[0, T]$ with $0 = t_0 < t_1 < \cdots < t_N = T$ let $|\tau| = \max |t_i - t_{i-1}|$. A sequence (τ_n) of decompositions is called increasing if $\tau_n \subset \tau_{n+1}, \forall n$; denomination : $(\tau_n) \uparrow$.

For a perturbation $\Delta = (\varepsilon, \delta)$ let the restriction $\Delta\big|_{(s,t]} = (\varepsilon\big|_{(s,t]}, \delta\big|_{(s,t]})$ be defined by $\delta_u\big|_{(s,t]}(u, w) = \delta_u(w)1_{(s,t]}(u)$ and $\varepsilon\big|_{(s,t]}(u, w) = \varepsilon - u(w)1_{(s,t]}(u)$.

Definition 7.11

(a) For a strategy Φ, a perturbation Δ and a partition τ let

$$r^\tau(\Phi, \Delta)(w, t) := \sum_{t_i\in\tau} \frac{R_{t_i}(\Phi + \Delta\,|_{(t_i,t_{i+1}]} - R_{t_i}(\Phi)}{E(\langle M\rangle_{t_{i+1}} - \langle M\rangle_{t_i} \mid \mathcal{A}_{t_i})}(w)1_{(t_i,t_{i+1}]}(t)$$

be the **risk ratio**.

(b) An admissible strategy Φ for H is called **locally risk-minimizing** if for all perturbations Δ

$$\varliminf_{\substack{n\to\infty\\ (\tau_n)\uparrow,\\ |\tau_n|\to 0}} r^{\tau_n}(\Phi, \Delta) \ge 0. \tag{7.14}$$

The risk ratio describes the relative change in risk when Φ is affected by a perturbation Δ along a partition τ. There is an analogue of Proposition 7.6 for locally risk-minimizing strategies.

Proposition 7.12 *Let for P almost all ω the carrier of the measure $d\langle M\rangle(\omega)$ be equal to $[0, T]$. Then, if Φ is locally risk-minimizing, Φ is self-financing in mean.*

Proof (Idea) The proof is similar to that of Proposition 7.6. Let $\widetilde{\Phi} = (\widetilde{\varphi}_0, \widetilde{\varphi})$ be a self-financing strategy with $\widetilde{\varphi} = \varphi$ and $\widetilde{\varphi}_{0,t} = E(C_T(\Phi) \mid \mathcal{A}_t) + \int_0^t \varphi_u \, \mathrm{d}S_u - \varphi_t S_t$. Then $\Delta = \widetilde{\Phi} - \Phi$ ia a perturbation. Using a similar reasoning as in the proof of Proposition 7.6 it then follows that for a suitable partition τ the risk ratio $r^\tau(\Phi, \Delta)$ is not P a.s. greater than or equal to 0 in contradiction to the assumption that Φ is risk-minimizing. □

In particular, to find a locally risk-minimal (LRM)-strategy only the determination of φ is to be done, since φ^0 then follows unambiguously from the martingale property of $C(\Phi)$.

We now pose the following regularity assumption:

Assumption (A)
(1) A is continuous
(2) $A \ll \langle M\rangle$ with density $\alpha \in L \log^+ L \quad P \times \langle M\rangle$ a.s.
(3) S is continuous in $T P$ a.s.

The following characterization theorem of Schweizer (1991) we state without proof:

Theorem 7.13 (Characterization Theorem of (LRM)) *Under the assumption A) holds for an H-admissible strategy Φ:*

Φ is (LRM) $\Longleftrightarrow$ Φ is self-financing in mean and the martingale $C(\Phi)$ is orthogonal to M.

We refer to (LRM) strategies as optimal using Theorem 7.13.

Definition 7.14 An admissible in mean self-financing strategy Φ for H is called '**optimal**' if the martingale $C(\Phi)$ orthogonal to M.

The following basic theorem gives a characterization of optimal strategies and is an extension of the Kunita–Watanabe decomposition.

Theorem 7.15 (Föllmer–Schweizer Decomposition, (FS Decomposition)) *Theexistence of an optimal strategy $\widehat{\Phi}$ for H is equivalent to the existence of the FS-decomposition*

$$H = H_0 + \int_0^T \varphi_u^H \, \mathrm{d}S_u + L^H \tag{7.15}$$

with $H_0 = E(H \mid \mathcal{A}_0)$, $\varphi^H \in \mathcal{L}^2(S)$ *and* $L_t^H = E(L_T^H \mid \mathcal{A}_t) \in \mathcal{M}^2$ *such that* L^H *is orthogonal to* M *is, i.e.* $\langle L^H, M\rangle = 0$.

Proof

"$\Longleftarrow$": If H has a representation of the form (7.15) and let the strategy $\widehat{\Phi}$ be defined by $\widehat{\Phi} = (V - \varphi^H \cdot S, \varphi^H)$ with $V_t = H_0 + \int_0^t \varphi_u^H \, dS_u + L_t^H$. Then $\widehat{\varphi}$ is admissible for H and the cost process $C_t = H_0 + L_t^H$ is a martingale with respect to P orthogonal to M, i.e. $\widehat{\Phi}$ is optimal for H (cf. Definition 7.14).

"$\Longrightarrow$": If $\widehat{\Phi} = (\varphi_0, \varphi^H)$ is optimal for H, then $C_t = E(C_T \mid \mathcal{A}_t)$ a martingale orthogonal to M. Thus the FS decomposition holds

$$
\begin{aligned}
H = V_T(\widehat{\Phi}) &= C_T + \int_0^T \varphi_u^H \, dS_u \\
&= H_0 + \int_0^T \varphi_u^H \, dS_u + L_T^H
\end{aligned}
$$

with $L_T^H = C_T - C_0$ and L^H is orthogonal to M under P.

□

By Theorem 7.15, determining optimal and hence locally risk-minimizing hedging strategies for H is equivalent to determining the FS decomposition of H. A direct approach to this involves the following three steps:

(1) Application of the KW-decomposition of H with respect to the martingale $M \in \mathcal{M}^2$ provides

$$
H = N_0 + \int_0^T \mu_s \, dM_s + N^H \text{ a.s. with respect to } P \tag{7.16}
$$

with the martingale $N_t^H = E(N_T^H \mid \mathcal{A}_t)$. It is $EN^H = 0$ and N^H is orthogonal to M with respect to P.

(2) Application of the KW decomposition to $\int \varphi^H \, dA$ provides

$$
\int_0^T \varphi_s^H \, dA_s = N_0^\varphi + \int_0^T \mu_s^{\varphi^H} \, dM_s + N^{\varphi^H} \tag{7.17}
$$

5 with $N_t^{\varphi^H} = E(N_T^{\varphi^H} \mid \mathcal{A}_t)$, a martingale with $EN_t^{\varphi^H} = 0$ and N^{φ^H} orthogonal to M.

(3) From the FS decomposition (7.15) and the KW decomposition (7.16) it follows:

$$
H = (H_0 + N_0^{\varphi^H}) + \int_0^T (\varphi_s^H + \mu_s^{\varphi^H}) \, dM_s + (L_T^H + N_T^{\varphi^H}). \tag{7.18}
$$

Since the KW decomposition of H is unique, it follows for an optimal strategy $\widehat{\Phi}$: $\widehat{\Phi}$ is a solution of the **optimality equation:**

$$\varphi_s^H + \mu_s^{\varphi^H} = \mu_s, \tag{7.19}$$

The integrand φ^H of the FS-decomposition thus differs in general from the integrand μ of the KW-decomposition by the compensation term μ^{φ_H}. For some classes of examples this program is carried out in Schweizer (1990, 1991).

7.2.2 Minimal Martingale Measures and Optimal Strategies

The determination of optimal strategies can alternatively be reduced to the martingale case by selecting a suitable martingale measure, the minimal martingale measure.

Definition 7.16 (Minimal Martingale Measure) An equivalent martingale measure $\widehat{P} \in \mathcal{M}^e(P)$ is called **minimal,** if

(1) $\widehat{P} = P$ on $\mathcal{A}_0$ and
(2) $L \in \mathcal{M}^2$ and $\langle L, M\rangle = 0$ P a.s. implies $L \in \mathcal{M}(\widehat{P})$.

An equivalent martingale measure $\widehat{P}$ is uniquely determined by the right-continuous martingale $\widehat{G} \in \mathcal{M}^2(P)$ defined by

$$\widehat{G}_t = E\Big(\frac{\mathrm{d}\widehat{P}}{\mathrm{d}P} \mid \mathcal{A}_t\Big). \tag{7.20}$$

With respect to P the Doob–Meyer decomposition of S equals $S = S_0 + M + A$. Therefore, the Doob–Meyer decomposition of M with respect to $\widehat{P}$ is given by

$$M = -S_0 + S + (-A). \tag{7.21}$$

Using the Girsanov transformation, it follows that the predictable process $-A \in \mathcal{V}$ can be determined with the help of $\widehat{G}$ in the form

$$-A_t = \int_0^t \frac{1}{\widehat{G}_{s-}} \mathrm{d}\langle M, \widehat{G}\rangle_s, \quad 0 \le t \le T. \tag{7.22}$$

Since $\langle M, \widehat{G}\rangle \ll \langle M\rangle = \langle S\rangle$, A is absolutely continuous with respect to $\langle S\rangle$ and, therefore, has a representation of the form

$$A_t = \int_0^t \alpha_u \mathrm{d}\langle S\rangle_u, \quad 0 \le t \le T. \tag{7.23}$$

This gives an existence and uniqueness statement for the minimal martingale measure. We assume in the following theorem that S is a continuous semimartingale.

Theorem 7.17 (Existence and Uniqueness) *Let S be a continuous semimartingale.*

(1) The minimal martingale measure $\widehat{P}$ is uniquely determined if it exists.
(2) $\widehat{P}$ exists if and only if

$$\widehat{G}_t = \exp\Big(- \int_0^t \alpha_s \, \mathrm{d}M_s - \frac{1}{2} \int_0^t \alpha_s^2 \mathrm{d}\langle S \rangle_s \Big) \tag{7.24}$$

defines an L^2-martingale under P, i.e. $\widehat{G} \in \mathcal{M}^2(P)$. In this case:

$$\frac{\mathrm{d}\widehat{P}}{\mathrm{d}P} = \widehat{G}_T. \tag{7.25}$$

(3) The minimal martingale measure $\widehat{P}$ preserves the orthogonality of L^2-martingales, i.e.: If for an $L \in \mathcal{M}^2$ it holds: $\langle L, M \rangle = 0$ with respect to P, then $\langle L, S \rangle = 0$ also with respect to $\widehat{P}$.

Proof

(1) Let $\widehat{P}$ be a minimal martingale measure and let $\widehat{G} \in \mathcal{M}^2(P)$ then according to KW

$$\widehat{G}_t = \widehat{G}_0 + \int_0^t \beta_s \, \mathrm{d}M_s + L_t, \quad 0 \le t \le T,$$

with $L \in \mathcal{M}^2(P)$, $\langle L, M \rangle = 0$, $\beta \in \mathcal{L}^2(M)$. W.r.t $\widehat{P}$ holds

$$-A_t = \int_0^t \frac{1}{\widehat{G}_{s-}} \mathrm{d}\langle M, \widehat{G} \rangle_s = \int_0^t \frac{1}{\widehat{G}_{s-}} \beta_u \mathrm{d}\langle S \rangle_u.$$

Therefore, it follows:

$$\alpha = -\frac{\beta}{\widehat{G}_-}. \tag{7.26}$$

It is $\widehat{G} > 0[P]$ because of $\widehat{P} \approx P$ and $\langle M \rangle = \langle S \rangle$ and, therefore, it holds that $\int_o^T \alpha_u^2 \mathrm{d}\langle S \rangle_u < \infty$ P a.s. Since by assumption $\widehat{P}$ is minimal, it follows $\widehat{G}_0 = E\big(\frac{\mathrm{d}\widehat{P}}{\mathrm{d}P} \mid \mathcal{A}_0\big) = 1$ and $L \in \mathcal{M}(\widehat{P})$ according to Definition 7.16. Therefore, it follows: $\langle L, \widehat{G} \rangle = 0$ and, therefore, $\langle L \rangle = \langle L, \widehat{G} \rangle = 0$, thus $L = 0$.

$\widehat{G}$, therefore, solves the equation

$$\widehat{G}_t = 1 + \int_0^t \widehat{G}_{s-}(-\alpha_s)\,\mathrm{d}M_s. \tag{7.27}$$

The solution for a continuous martingale M of (7.27) is given by the stochastic exponential $\widehat{G}$ in (7.24). From this follows the uniqueness of the minimal martingale measure

(2) Let $\widehat{G}$ defined in (7.24) be a L^2-martingale, $G \in \mathcal{M}^2(P)$. To prove that $\widehat{P}$ is minimal,it is to show :

$$L \in \mathcal{M}^2(P) \quad \text{and} \quad \langle L, M\rangle = 0 \text{ with respect to } P \Longrightarrow L \in \mathcal{M}(\widehat{P}).$$

Since $L, \widehat{G} \in \mathcal{M}^2(P)$, it follows by the maximal inequality

$$\begin{aligned} E_{\widehat{P}} \sup_{0\le t\le T} |L_t| &= E\Big(\sup_{0\le t\le T} |L_t|\Big)\widehat{G}_T \\ &\le E\Big(\sup_{0\le t\le T} L_t^2\Big)^{\frac{1}{2}} (E\widehat{G}_T^2)^{\frac{1}{2}} \le 4(EL_T^2)^{\frac{1}{2}}(E\widehat{G}_T^2)^{\frac{1}{2}} < \infty. \end{aligned}$$

From this follows that $L \in \mathcal{M}(\widehat{P})$.

(3) To prove that $\widehat{P}$ preserves the orthogonality, let $L \in \mathcal{M}^2$ and $\langle L, M\rangle = 0$ with respect to P. For semimartingales Y, Z the quadratic covariation is

$$[Y, Z] = \langle Y^c, Z^c\rangle + \sum_s \Delta Y_s \Delta Z_s.$$

Since S and A are continuous, it holds:

$$\begin{aligned} \langle L, S\rangle &= \langle L^c, S\rangle + \langle L^d, S\rangle = \langle L^c, S\rangle \\ &= \langle L, S\rangle = [L, M] + [L, A] = [L, M] \end{aligned}$$

with respect to $\widehat{P}$. Because of the continuity of M it holds $[L, M] = \langle L^c, M\rangle = \langle L, M\rangle = 0$ with respect to P and, therefore, $[L, M] = 0$ with respect to $\widehat{P}$.

□

The minimal martingale measure is described by the exponential density in (7.24). According to Corollary 6.35 is $\widehat{P}$ thus is, under the conditions there, a projection of P on the set of martingale measures with respect to the relative entropy

$$H(Q \,\|\, P) = \begin{cases} \displaystyle\int \log \frac{\mathrm{d}Q}{\mathrm{d}P}\,\mathrm{d}Q, & \text{for } Q \ll P, \\ \infty, & \text{otherwise.} \end{cases}$$

The following theorem gives an independent proof of this important result.

Theorem 7.18 (Minimal Martingale Measure and Entropy)

(a) The minimal martingale measure $\widehat{P}$ minimizes the functional

$$H(Q \parallel P) - \frac{1}{2} E_Q \int_0^T \alpha_u^2 \,\mathrm{d}\,\langle S \rangle_u \tag{7.28}$$

w.r.t. Q on the set $\mathcal{M}^e$, α the $\langle S \rangle$-density of A from (7.23).

(b) For $Q \in \mathcal{M}^e$ it holds:

$$H(Q \parallel P) \geq \frac{1}{2} E_Q \int_0^T \alpha_u^2 \,\mathrm{d}\,\langle S \rangle_u . \tag{7.29}$$

For $Q = \widehat{P}$ holds equality in (7.29).

(c) $\widehat{P}$ minimizes the relative entropy $H(\cdot \parallel P)$ on the set of martingale measures $Q \in \mathcal{M}^e$ with

$$E_Q \int_0^T \alpha_u^2 \,\mathrm{d}\,\langle S \rangle_u \geq E_{\widehat{P}} \int_0^T \alpha_u^2 \,\mathrm{d}\,\langle S \rangle_u .$$

Proof Let $Q \in \mathcal{M}^e$; then M has with respect to Q the Doob–Meyer decomposition (cf. (7.21))

$$M_t = S_t - S_0 + \Big(- \int_0^t \alpha_u \,\mathrm{d}\,\langle S \rangle_u \Big). \tag{7.30}$$

By (7.24) $\widehat{G}_T = \frac{\mathrm{d}\widehat{P}}{\mathrm{d}P} \in L^2(P)$ and, therefore, $H(\widehat{P} \parallel P) = \int \widehat{G}_T \log \widehat{G}_T \,\mathrm{d}P < \infty$.

Now let $\widehat{Q} \in \mathcal{M}^e$ be the minimal martingale measure, then it follows with $\widetilde{G}_T = \frac{\mathrm{d}\widehat{Q}}{\mathrm{d}P}$ from (7.30):

$$\begin{aligned}
H(\widehat{P} \parallel P) &= H(\widehat{P} \parallel \widehat{Q}) + \int \log \widetilde{G}_T \mathrm{d}\widehat{P} \\
&= H(\widehat{P} \parallel \widehat{Q}) + \int \Big(- \int_0^T \alpha_s \,\mathrm{d}M_s - \frac{1}{2} \int_0^T \alpha_u^2 \,\mathrm{d}\,\langle S \rangle_u \Big) \mathrm{d}\widehat{Q} \\
&= H(\widehat{P} \parallel \widehat{Q}) + \frac{1}{2} E_{\widehat{Q}} \Big(\int_0^T \alpha_u^2 \,\mathrm{d}\,\langle S_u \rangle \Big) \\
&< \infty \quad \text{by assumption.}
\end{aligned}$$

This implies

$$H(\widehat{P} \parallel P) - \frac{1}{2} E_{\widehat{Q}}\Big(\int_0^T \alpha_u^2 \, \mathrm{d} \langle S \rangle_u \Big) = H(\widehat{P} \parallel \widehat{Q}) \geq 0. \tag{7.31}$$

From this follows (7.28) and (7.29).

In Eq. (7.31) is $H(\widehat{P} \parallel \widehat{Q}) = 0$ exactly then, when $\widehat{P} = \widehat{Q}$. From this follows c). □

As a result, we now obtain that the FS decomposition can be determined via the KW decomposition with respect to the minimum martingale measure. This also gives a representation of the optimal strategy.

Theorem 7.19 (Optimal Strategy)

(a) The optimal strategy $\widehat{\Phi}$ and thus the FS decomposition in (7.15) *is uniquely determined. It can be determined with the help of the KW decomposition of H in* (7.3) *with respect to the minimal martingale measure $\widehat{P}$.*

(b) If $\widehat{V}$ is a right continuous version of the martingale $E_{\widehat{P}}(H \mid \mathcal{A}_t)$ then the optimal strategy $\widehat{\Phi}$ is given by

$$\widehat{\Phi} = (\varphi^H, \widehat{V} - \varphi^H \cdot S) \text{ with } \varphi^H = \frac{\mathrm{d}\langle \widehat{V}, S \rangle}{\mathrm{d}\langle S \rangle} \text{ w.r.t. } \widehat{P}$$

(cf. Definition 7.11).

Proof Since $H \in L^2(P)$ and $\frac{\mathrm{d}\widehat{P}}{\mathrm{d}P} \in L^2(P)$, it follows $H \in L^2(\widehat{P})$, and the martingale $E_{\widehat{P}}(H \mid \mathcal{A}_t) = \widehat{V}_t$ is well defined. Now let

$$V_t = H_0 + \int_0^t \varphi_u^H \, \mathrm{d}S_u + L_t^H \tag{7.32}$$

be an FS-decomposition of H under P, then $L^H \in \mathcal{M}(\widehat{P})$ and $\int_0^t \varphi_u^H \, \mathrm{d}S_u \in \mathcal{M}(\widehat{P})$ and so it holds $\widehat{V}_t = E_{\widehat{P}}(H \mid \mathcal{A}_t) = V_t$. With $\overline{P} = P \times \mathrm{d}\langle X \rangle$ defined on the σ-algebra $\mathcal{P}$ of the predictable sets in $\overline{\Omega} = \Omega \times [0, T]$, $\overline{P}(\mathrm{d}\omega, \mathrm{d}t) = P(\mathrm{d}\omega)\mathrm{d}\langle S \rangle_t(\omega)$ and analogously $\overline{\widehat{P}}$, it holds:

$$\langle L^H, S \rangle = 0 \, [\overline{P}]$$

and, therefore, also according to Theorem 7.17 (3): $\langle L^H, S \rangle = 0 \, [\overline{\widehat{P}}]$. Therefore, the process L^H from the FS-decomposition in (7.32) is also a $\widehat{P}$-martingale orthogonal to S. The FS decomposition in (7.32) with respect to P is, therefore, identical to the KW-decomposition of H under the minimal martingale measure $\widehat{P}$.

By Theorem 7.8, it follows that $\widehat{\Phi} = (\varphi^H, V - \varphi^H \cdot S)$ is an optimal strategy. □

Remark In Föllmer and Schweizer (1991), the present approach to optimal strategies is extended to the case with incomplete information, i.e., there is a filtration $\mathcal{A}_t \subset \widetilde{\mathcal{A}}_t$, $0 \le t \le T$ such that H is representable with respect to this larger filtration $\widetilde{\mathcal{A}}$

$$H = \widetilde{H}_0 + \int_0^T \widetilde{\varphi}_u^H \, \mathrm{d}S_u,$$

$\widetilde{\varphi}^H \in \mathcal{L}(\widetilde{\mathcal{P}})$ is predictable with respect to $(\widetilde{\mathcal{A}}_t)$. The optimal strategy is then obtained from the projection of the optimal strategy $\widetilde{\varphi}^H$ onto $\mathcal{P}$ the predictable σ-algebra with respect to $(\mathcal{A}_t)$, with respect to the minimal martingale measure

$$\varphi_u^H = E_{\overline{P}}(\widetilde{\varphi}_u^H \mid \mathcal{P}) \text{ and } \varphi_0^H = V - \varphi_u^H \cdot S, \; V_t = E_{\widehat{P}}(H \mid \mathcal{A}_t).$$

Bibliography

J.-P. Ansel and C. Stricker. Couverture des actifs contingents et prix maximum. *Ann. Inst. Henri Poincaré, Probab. Stat.*, 30(2): 303–315, 1994.

J. Azema and M. Yor. Une solution simple au problème de Skorokhod. *Lect. Notes Math.*, 721:90–115, 1979.

H. Bauer. *Wahrscheinlichkeitstheorie und Grundzüge der Maßtheorie. 4th edition* Berlin – New York: de Gruyter, 1991.

F. Bellini and M. Frittelli. On the existence of minimax martingale measures. *Math. Finance*, 12(1):1–21, 2002.

S. Biagini. Expected utility maximization: Duality methods. In R. Cont, editor, *Encyclopedia of Quantitative Finance*. Wiley & Sons, Ltd, 2010.

S. Biagini and M. Frittelli. On the super replication price of unbounded claims. *Ann. Appl. Probab.*, 14(4):1970–1991, 2004.

S. Biagini and M. Frittelli. Utility maximization in incomplete markets for unbounded processes. *Finance Stoch.*, 9(4):493–517, 2005.

S. Biagini and M. Frittelli. A unified framework for utility maximization problems: An Orlicz space approach. *Ann. Appl. Probab.*, 18(3):929–966, 2008.

S. Biagini and M. Frittelli. On the extension of the Namioka-Klee theorem and on the Fatou property for risk measures. In F. Delbaen, M. Rásonyi and C. Stricker, editors, *Optimality and Risk – Modern Trends in Mathematical Finance*, pages 1–28. Berlin – Heidelberg: Springer, 2009.

S. Biagini, M. Frittelli, and M. Grasselli. Indifference price with general semimartingales. *Math. Finance*, 21(3):423–446, 2011.

T. Björk. *Arbitrage Theory in Continuous Time*. Oxford: Oxford University Press, 4th edition, 2019.

F. Black and M. Scholes. The pricing of options and corporate liabilities. *J. Polit. Econ.*, 81(3):637–654, 1973.

J. Borwein and Q. Zhu. *Techniques of Variational Analysis*. Number 20 in CMS Books in Mathematics. Springer, 2005.

R. Carmona, editor. *Indifference Pricing: Theory and Applications*. Princeton, NJ: Princeton University Press, 2009.

C. S. Chou, P. A. Meyer, and C. Stricker. Sur les intégrales stochastiques de processus previsibles non bornes, *Lect. Notes Math.*, 784:128–139, 1980.

J. C. Cox and C. Huang. Optimal consumption and portfolio policies when asset prices follow a diffusion process. *J. Econ. Theory*, 49(1):33–83, 1989.

I. Csiszár. I-divergence geometry of probability distributions and minimization problems. *Ann. Probab.*, 3:146–158, 1975.

F. Delbaen and W. Schachermayer. A general version of the fundamental theorem of asset pricing. *Math Ann.*, 300(3):463–520, 1994.

L. Rüschendorf, *Stochastic Processes and Financial Mathematics*, Mathematics Study Resources 1, https://doi.org/10.1007/978-3-662-64711-0

F. Delbaen and W. Schachermayer. The no-arbitrage property under a change of numéraire. *Stochastics Rep.*, 53(3–4):213–226, 1995.

F. Delbaen and W. Schachermayer. The fundamental theorem of asset pricing for unbounded stochastic processes. *Math Ann.*, 312(2):215–250, 1998.

F. Delbaen and W. Schachermayer. A compactness principle for bounded sequences of martingales with applications. In *Seminar on Stochastic Analysis, Random Fields and Applications. Centro Stefano Franscini, Ascona, Italy, September 1996*, pages 137–173. Basel: Birkhäuser, 1999.

F. Delbaen and W. Schachermayer. *The Mathematics of Arbitrage.* Berlin: Springer, 2006.

R. Durrett. *Brownian Motion and Martingales in Analysis.* Belmont, California: Wadsworth Advanced Books & Software, 1984.

E. B. Dynkin. *Theory of Markov Processes.* Pergamon, 1960.

E. B. Dynkin. *Markov Processes. Volumes I, II*, volume 121/122 of *Grundlehren Math. Wiss.* Springer, Berlin, 1965.

R. J. Elliott. *Stochastic Calculus and Applications*, volume 18 of *Applications of Mathematics.* Springer, 1982.

P. Erdös and M. Kac. On certain limit theorems of probability. *Bull. Amer. Math. Soc.* 52:292–302 (1946).

H. Föllmer and Y. M. Kabanov. Optional decomposition and Lagrange multipliers. *Finance Stoch.*, 2(1):69–81, 1998.

H. Föllmer and P. Leukert. Quantile hedging. *Finance Stoch.*, 3(3):251–273, 1999.

H. Föllmer and P. Leukert. Efficient hedging: cost versus shortfall risk. *Finance Stoch.*, 4(2):117–146, 2000.

H. Föllmer and M. Schweizer. Hedging of contingent claims under incomplete information. In M. H. A. Davis and R. J. Elliott, editors, *Applied Stochastic Analysis*, pages 389–414. Gordon & Breach, 1991.

H. Föllmer and D. Sondermann. Hedging of non-redundant contingent claims. *Contributions to Mathematical Economics, Hon. G. Debreu*, pages 206–223, 1986.

M. Frittelli. The minimal entropy martingale measure and the valuation problem in incomplete markets. *Math. Finance*, 10(1):39–52, 2000.

R. Geske and H. E. Johnson. The American Put Option Valued Analytically, 1984.

P. Gänssler and W. Stute. *Wahrscheinlichkeitstheorie.* Berlin-Heidelberg-New York: Springer-Verlag, 1977.

T. Goll and J. Kallsen. Optimal portfolios for logarithmic utility. *Stochastic Processes Appl.*, 89(1):31–48, 2000.

T. Goll and L. Rüschendorf. Minimax and minimal distance martingale measures and their relationship to portfolio optimization. *Finance Stoch.*, 5(4):557–581, 2001.

S. Gümbel. Dualitätsresultate zur Nutzenoptimierung und allgemeiner Nutzenindifferenzpreis. Master thesis, Universität Freiburg, Mathematisches Institut, 2015.

J. M. Harrison and D. M. Kreps. Martingales and arbitrage in multiperiod securities markets. *J. Econ. Theory*, 20:381–408, 1979.

J. M. Harrison and S. R. Pliska. Martingales and stochastic integrals in the theory of continuous trading. *Stochastic Processes Appl.*, 11:215–260, 1981.

H. He and N. D. Pearson. Consumption and portfolio policies with incomplete markets and short-sale constraints: The finite-dimensional case. *Math. Finance*, 1(3):1–10, 1991a.

H. He and N. D. Pearson. Consumption and portfolio policies with incomplete markets and short-sale constraints: The infinite dimensional case. *J. Econ. Theory*, 54(2):259–304, 1991b.

A. Irle. *Finanzmathematik: Die Bewertung von Derivaten.* Teubner Studienbücher, 2003.

K. Itô. *Foundations of Stochastic Differential Equations in Infinite Dimensional Spaces*, volume 47. Philadelphia, PA: Society for Industrial and Applied Mathematics (SIAM), 1984.

K. Itô and H. McKean, Jr. *Diffusion Processes and their Sample Paths. 2nd printing (corrected)*, volume 125. Springer, Berlin, 1974.

S. D. Jacka. A martingale representation result and an application to incomplete financial markets. *Math. Finance*, 2(4):239–250, 1992.

J. Jacod. *Calcul Stochastique et Problèmes de Martingales*, volume 714 of *Lect. Notes Math.* Springer, Cham, 1979.

J. Jacod. Intégrales stochastiques par rapport à une semimartingale vectorielle et changements de filtration. Seminaire de Probabilitiès XIV, 1978/79: 161–172, 1980.

J. Jacod and P. Protter. *Probability Essentials*. Berlin – Heidelberg: Springer, 2004.

J. Jacod and A. N. Shiryaev. *Limit Theorems for Stochastic Processes*, volume 288. Springer, Berlin, 1987.

O. Kallenberg. *Foundations of Modern Probability*. New York, NY: Springer, 2nd edition, 2002.

J. Kallsen. Optimal portfolios for exponential Lévy processes. *Mathematical Methods of Operations Research*, 51(3):357–374, 2000.

I. Karatzas and S. E. Shreve. *Brownian Motion and Stochastic Calculus*. Springer, 1991.

I. Karatzas and S. E. Shreve. *Methods of Mathematical Finance*. Springer, 1998.

I. Karatzas, J. P. Lehoczky, S. E. Shreve, and G.-L. Xu. Martingale and duality methods for utility maximization in an incomplete market. *SIAM J. Control Optim.*, 29(3):702–730, 1991.

A. Klenke. *Wahrscheinlichkeitstheorie*. Berlin: Springer, 2006.

D. O. Kramkov. Optional decomposition of supermartingales and hedging contingent claims in incomplete security markets. *Probab. Theory Relat. Fields*, 105(4):459–479, 1996.

D. O. Kramkov and W. Schachermayer. The asymptotic elasticity of utility functions and optimal investment in incomplete markets. *Ann. Appl. Probab.*, 9(3):904–950, 1999.

H. Kunita and S. Watanabe. On square integrable martingales. *Nagoya Math J.*, 30:209–245, 1967.

H.-H. Kuo. *Introduction to Stochastic Integration*. Springer, 2006.

F. Liese and I. Vajda. *Convex Statistical Distances*, volume 95 of *Teubner-Texte zur Mathematik*. Leipzig: BSB B. G. Teubner Verlagsgesellschaft, 1987.

J. Memin. Espaces de semi martingales et changement de probabilité. *Z. Wahrscheinlichkeitstheorie Verw. Geb.*, 52:9–39, 1980.

R. C. Merton. Optimum consumption and portfolio rules in a continuous-time model. *J. Econ. Theory*, 3:373–413, 1971.

R. C. Merton. Theory of rational option pricing. *Bell J. Econ. Manage. Sci.*, 4(1):141–183, 1973.

M. P. Owen and G. Žitković. Optimal investment with an unbounded random endowment and utility-based pricing. *Math. Finance*, 19(1):129–159, 2009.

P. Protter. *Stochastic Integration and Differential Equations. A New Approach*, volume 21 of *Applications of Mathematics*. Berlin etc.: Springer-Verlag, 1990.

S. T. Rachev and L. Rüschendorf. Models for option prices. *Theory Probab. Appl.*, 39(1):120–152, 1994.

S. T. Rachev and L. Rüschendorf. *Mass Transportation Problems. Vol. 1: Theory. Vol. 2: Applications*. New York, NY: Springer, 1998.

D. Revuz and M. Yor. *Continuous Martingales and Brownian Motion*, volume 293. Berlin: Springer, 3rd (3rd corrected printing) edition, 2005.

T. Rheinländer and J. Sexton. *Hedging Derivatives*, volume 15 of *Adv. Ser. Stat. Sci. Appl. Probab.* Hackensack, NJ: World Scientific, 2011.

R. Rockafellar. *Convex Analysis*. Princeton University Press, 1970.

L. C. G. Rogers and D. Williams. *Diffusions, Markov Processes and Martingales. Vol. 2: Itô Calculus*. Cambridge: Cambridge University Press, 2nd edition, 2000.

L. Rüschendorf. On the minimum discrimination information theorem. *Suppl. Issue, Stat. Decis.*, 1:263–283, 1984.

L. Rüschendorf. *Wahrscheinlichkeitstheorie*. Heidelberg: Springer Spektrum, 2016.

W. Schachermayer. Optimal investment in incomplete markets when wealth may become negative. *Ann. Appl. Probab.*, 11(3):694–734, 2001.

W. Schachermayer. A super-martingale property of the optimal portfolio process. *Finance Stoch.*, 7(4):433–456, 2003.

W. Schachermayer. Aspects of mathematical finance. In M. Yor, editor, *Aspects of Mathematical Finance. Académie des Sciences, Paris, France, February 1, 2005*, pages 15–22. Berlin: Springer, 2008.

H. H. Schaefer. *Topological Vector Spaces*, volume 3. Springer, New York, NY, 1971.

M. Schweizer. Risk-minimality and orthogonality of martingales. *Stochastics Stochastics Rep.*, 30(2):123–131, 1990.

M. Schweizer. Option hedging for semimartingales. *Stochastic Processes Appl.*, 37(2):339–363, 1991.

A. N. Shiryaev. *Probability*, volume 95 of *Graduate Texts in Mathematics*. New York, NY: Springer-Verlag, 2nd edition, 1995.

D. W. Stroock and S. R. S. Varadhan. *Multidimensional Diffusion Processes*, volume 233 of. *Grundlehren Math. Wiss*. Springer, Berlin, 1979.

M. Yor. Sous-espaces denses dans L^1 ou H^1 et représentation des martingales (avec J. de Sam Lazaro pour l'appendice). *Lect. Notes Math.*, 649:265–309, 1978.

K. Yosida. *Functional Analysis*, volume 123 of *Grundlehren Math. Wiss*. Springer, Berlin, 4th edition, 1974.

Index

L. Rüschendorf, *Stochastic Processes and Financial Mathematics*, Mathematics Study Resources 1, https://doi.org/10.1007/978-3-662-64711-0

Printed in the United States
by Baker & Taylor Publisher Services